Frontiers of Chemical Dynamics

NATO ASI Series

Advanced Science Institutes Series

A Series presenting the results of activities sponsored by the NATO Science Committee, which aims at the dissemination of advanced scientific and technological knowledge, with a view to strengthening links between scientific communities.

The Series is published by an international board of publishers in conjunction with the NATO Scientific Affairs Division

A Life Sciences **B Physics**	Plenum Publishing Corporation London and New York
C Mathematical and Physical Sciences **D Behavioural and Social Sciences** **E Applied Sciences**	Kluwer Academic Publishers Dordrecht, Boston and London
F Computer and Systems Sciences **G Ecological Sciences** **H Cell Biology** **I Global Environmental Change**	Springer-Verlag Berlin, Heidelberg, New York, London, Paris and Tokyo

PARTNERSHIP SUB-SERIES

1. Disarmament Technologies	Kluwer Academic Publishers
2. Environment	Springer-Verlag / Kluwer Academic Publishers
3. High Technology	Kluwer Academic Publishers
4. Science and Technology Policy	Kluwer Academic Publishers
5. Computer Networking	Kluwer Academic Publishers

The Partnership Sub-Series incorporates activities undertaken in collaboration with NATO's Cooperation Partners, the countries of the CIS and Central and Eastern Europe, in Priority Areas of concern to those countries.

NATO-PCO-DATA BASE

The electronic index to the NATO ASI Series provides full bibliographical references (with keywords and/or abstracts) to more than 50000 contributions from international scientists published in all sections of the NATO ASI Series.
Access to the NATO-PCO-DATA BASE is possible in two ways:

– via online FILE 128 (NATO-PCO-DATA BASE) hosted by ESRIN, Via Galileo Galilei, I-00044 Frascati, Italy.

– via CD-ROM "NATO-PCO-DATA BASE" with user-friendly retrieval software in English, French and German (© WTV GmbH and DATAWARE Technologies Inc. 1989).

The CD-ROM can be ordered through any member of the Board of Publishers or through NATO-PCO, Overijse, Belgium.

Series C: Mathematical and Physical Sciences – Vol. 470

Frontiers of Chemical Dynamics

edited by

Ersin Yurtsever
Middle East Technical University,
Chemistry Department,
Ankara, Turkey

Springer Science+Business Media, B.V.

Proceedings of the NATO Advanced Study Institute on
Frontiers of Chemical Dynamics
Kemer, Antalya, Turkey
5–16 September 1994

A C.I.P. Catalogue record for this book is available from the Library of Congress.

ISBN 978-94-010-4153-9 ISBN 978-94-011-0345-9 (eBook)
DOI 10.1007/978-94-011-0345-9

Printed on acid-free paper

CONTENTS

Preface

This volume contains the lectures presented at the NATO Advanced Study Institute (ASI) on " Frontiers of Chemical Dynamics ", held in the Club Alda, Kemer, Turkey, from 5th September to 16th September 1994.

The Kemer area, famous for its pristine beaches and craggy mountains provided an excellent atmosphere for an intellectually and socially active meeting. The first class facilities of Club Alda allowed the participants to concentrate on the scientific activities without any outside interferences and disturbances.

The main objective of the meeting was to bring experts of chemical dynamics to discuss problems from both experimental and theoretical points of view. The organizing committee has helped a great deal to collect an impressive list of lecturers, although there were quite a number of other scientists whom we would have liked to invite. Unfortunately, the number of lecturers is limited and we had to leave out some of them.The selection of the lecturers from a very long list was a difficult process and those who are approached in our very first attempt were all known for giving very good lectures. The purpose of the ASI's are mainly educational even though they may be at a very high level and it is essential to keep in mind the pedagogical aspects of the meeting without sacrificing the scientific quality. This point was underlined several times in our communications with lecturers. The lectures covered a variety of important problems of dynamical nature and they are arranged to follow a certain line of thought with the idea that this compilation can be used as a supplementary textbook for a graduate course.

The Advanced Study Institute was financed by a major grant from NATO and we would like to express our sincere gratitude to the NATO Scientific Committee. We also would like to thank NATO representatives in Greece, Portugal and Turkey whose grants made the participation of several students from these countries possible. The Oyak Tourism organization contributed to conference by an additional grant to cover the expenses of a number of participants. Finally Middle East Technical University provided its facilities for communication and printing purposes.

The local organization was covered by Oyak Tourism who handled the job very professionally and allowed us concentrate on the scientific aspects of the conference. Together with the staff of Club Alda, they provided a very warm social environment for which I am very grateful.

Finally I would like to express my thanks to my colleague Turgay Uzer who has spent a tremendous amount of time on all the organizational details and suffered with me when we thought we had problems. The success of the organization is mostly due to his efforts.

Ersin Yurtsever, Editor and Director

List of Participants

Director:

Prof.Dr.Ersin Yurtsever
Middle East Technical University
Chemistry Department
06531 Ankara
TURKEY

Lecturers:

Prof.Lucas Lathouwers
Dept. of Mathematics
Univ. of Antwerp-Ruca
Groenenborgerhan
171-2020 Antwerp
BELGIUM

Prof.Andre D.Bandrauk
Univ. de Sherbrooke, Que, JIK 2RI,
Dep. de Chimie
CANADA

Prof.Osman Atabek
Centre National de la Recherche
Scientifique, Laboratorie de Photo
Physique Moleculaire Bol. 213
Universite De Paris-Sud, Campus
d'Orsay, 91405
Orsay FRANCE

Prof.Jürgen Brickmann
Institut für Physikalische Chemie I
Technische Hochschule Darmstadt
Petersen str. 20, 64287
Darmstadt GERMANY

Prof.Lorenz Cederbaum
Theoretiche Chemie, Universitaet
Heidelberg, Im Neunheimer
Felt 253, 69121 Heidelberg
GERMANY

Prof.Franco Gianturco
Universita Degli Studi Di Roma,
Dipartimento di Chimica,
Citta Universitaria, 00185 Roma
ITALY

Prof. R.D.Levine
The Fritz Haber Research Center,
Jerusalem 91904
ISRAEL

Prof. Moshe Shapiro
Dept. of Chemical Physics, Weizmann
Institute of Science,
Rehovot 76100
ISRAEL

Prof. F. Borondo
Dep. Quimica, CXIV, Universidad
Autonoma de Madrid, Cantoblanco
28049 Madrid
SPAİN

Prof.Dr.Atilla Aşkar
Koç University
Fen ve İnsani Bilimleri Fak.
Çayır Cad . No:5
80860 İstinye
İstanbul,
TURKEY

Prof. A.J. McCaffery
The Univ. Sussex, The School of
and Mol. Sciences, Brighton
BN 1 9 QJ
UK

Prof. Roger E.Miller
Dep. Chem. Univ. N.Carolina, Chapel
Hill, North Carolina 27599,
USA

Prof. W.H. Miller
Department of Chemistry,
University of California, Berkeley,
Ca 94720
USA

Prof. Herschel Rabitz
Princeton University,
Department of Chemistry, Princeton,
New Jersey 08544-1009
USA

Prof.Michael White
Chemistry Dep.
Brookhaven National Lab.
Upton NY11973
USA

Prof. Turgay Uzer
Georgia Institute of Technology
School of Physics
Atlanta, Georgia
30332-0430
USA

Participants:

G.N.Vayssilov
Uni.Sofia, Faculty of Chemistry
1126 Sofia,
BULGARIA

Claude Dion
Laboratoire de Chimie Theorique
Deptement de Chimie Universite de
Sherbrooke, J1K 2 RI
CANADA

Tucker Carrington
Dept de Chimie, University of
Montreal CP6128 Succ Centre ville
Montreal QC H3CJ7
CANADA

Andre Conjusteau
Laboratoire de Chimie Theorique
Deptement de Chimie Universite de
Sherbrooke, J1K 2 RI
CANADA

Roberto Numico
Laboratoire de Photophysique
Moleculaire de CNRS
Bat 213, Universite Paris Sdu.
Campus d'Orsay 91 405 Orsay
FRANCE

Tassaing Thiery
University Bordeauxi
Laboratoire de Spectroscopie
Moleculare at Cristalline
351 cours de la liberation
33405 Talence
FRANCE

Thomas Gunkel
Institut für Physikalische Chemie I
Technische Hochshule
Petersen Str. 20
64287 Darmstadt
GERMANY

J.Hinze
Uni.Bielefeld, Fak.für Chemie
Bielefeld, 33615
GERMANY

Hilmar Hamann
Institut Phys. Chemie
Tammann Strage 6
37077 Göttingen
GERMANY

Andreas Jackle,
Universitat Heidelberg Im
Neuenheimer, Theoretische Chemie
Feld 253 D-69120 Heidelberg
GERMANY

Andreas Kumpf
Theoretical Chemistry
Universitat Gesamthochschule Siegen
D-57068 Siegen
GERMANY

Stefan Linkersdoerfer
Institut Fuer Physikalische Chemie
Tammannstr. 6
37077 Goettingen
GERMANY

Christina Jansch
Max-Plack-Institut
für Biopysikalische- Chemie
Am Fassberg D-37077
Göttingen
GERMANY

Roger G. Schmidt
Institut für Physikalische Chemie I
Technische Hochschule Darmstadt
Petersen Str. 20,
D-64287 Darmstadt
GERMANY

Udo Schmitt
Institut für Physikalische Chemie I
Technische Hochschule Darmstadt
Petersen Str. 20,
D-64287 Darmstadt
GERMANY

C.A. Nicolaides
Theoretical and Physical Chemistry
Institute
46. Vassileos Constantinou Avenue
Athens 116 35
GREECE

Manolis Founargiotakis
Univ of Crete, Deg. of Chemistry
PO. Box 1527, 71110
GREECE

Savas Georgiou
Institute of Electronic Structure
and Laser Foundation for
Research and Technology
PO. Box 1527 Heraklion, Crete 71110
GREECE

Aliki Vegiri
Teoretical and Physical Chemistry
Ins., NHRF
48 Constantinov Avenie,
Athens, 11635
GREECE

Michal Ben-Nun
The Fritz Haber Research Center
for MolecularDynamics,
The Hebrew University of Jerusalem,
Jerusalem 91904
ISRAEL

Alexander G. Abrashkevich
Department of Chemical Physics
The Weizmann Institute of science
Rekorot 76100
ISRAEL

Cecilia Coletti
Universita di Perugia
Dept. of Chemistry
Via Eice di Solto, 8 I-06123
Perugia,
ITALY

Daniela Ascenzi
Universita di Perugia
Dept. of Chemistry
Via Eice di Solto, 8 I-06123 Perugia
ITALY

Marcello Coreno
Universita la Sapienza
Dip. Chimica NEC
P.le A. Moro, 5
I-00185 Rome
ITALY

S.Nakabayashi
Department of Chemistry Faculty of
Science Hokkaido Uni., Kita 10 Nishi
8, Kita-ku Sapporo 060
JAPAN

A.J.H.M. Meijer
Institute of Theoretical Chemistry
University of Nijmegen
Toernooiveld
6525 ED Nijmegen
THE NETHERLAND

Dorota Bielinska
Institute of Pysics, N. Copernicus
University Torun
Instytut Fizgki UMK Grudzizdzka
5; PL-87-100 Torun
POLAND

Paulo Vieira
Centro Fisica Molecular
Av. Rovisco Pais, Complexo I-IST
1000-lisboa
PORTUGAL

Jorgo Marquez Gonçalves
Departmento de Quimica Faculdade
de Ciencias da Universidade do Porto
Praça Gomez Teixeira 4000 Porto
PORTUGAL

Adelina Sporea
Research Institute for Rare and
Radioactive Metals,
Bd. Carol I, nr. 78 Sector2
PO.Box 31-53 Bucharest
ROUMANIA

Gennadii Milnikov
The Institute of Structural
Macrokinetics, Chernogolovka,
Moscow Region 142432
RUSSIA

Raul Guatens
Universidad Autonam De Madrid,
Facultad De Ciencias
C-IX 611
28049 Cantoblanco, Madrid
SPAIN

L.S. Jaoa Damas
Universidad Autonama De Madrid
Fac. Ciencia
28049 CANTOBLANCO, Madrid
SPAIN

Gonzalo Garcia de Polavieja
Dep Quimica, C-IX, Universida
Autonama de Madrid
CANTOBLANCO 28049, Madrid
SPAIN

Rosa M. Benito
Universidad Politecnica De Madrid
pto. De fisica aplica T.S.T. De
Telecomuncacion 28040, Madrid
SPAIN

Nihat Baysal
Boğaziçi University
Chemical Eng. Dept.
80815 Bebek İSTANBUL
TURKEY

Safiye Sağ Erdem
Bogaziçi Univ. Dept. of Chemistry
80 815 Bebek İSTANBUL
TURKEY

Canan Baysal
Boğaziçi University
Polymer Research Center
80 815 Bebek İSTANBUL
TURKEY

Pemra Doruker
Boğaziçi University
Polymer Research Center
80 815 Bebek İSTANBUL
TURKEY

Türkan Haliloğlu
Polymer Research Center
Boğaziçi University
80 815 Bebek İSTANBUL
TURKEY

İlker Özkan
Middle East tech. Univ.
Sci. Ed. Dep.
Ankara,
TURKEY

Nuran Elmacı
Middle East Tech. Univ.
Chemistry Dep.
Ankara,
TURKEY

Mine Yurtsever
Middle East Tech. Univ.
Chemistry Dep.
Ankara,
TURKEY

Hasan KARAASLAN
Middle East Tech. Univ.
Chemistry Dep.
Ankara,
TURKEY

Cengiz S. AŞKUN
Middle East Tech. Univ.
Chemistry Dep.
Ankara,
TURKEY

Pınar SARGIN
Middle East Tech. Univ.
Chemistry Dep.
Ankara,
TURKEY

Z.C.Kuruoğlu
Bilkent Uni.,Chemistry Dep.
Ankara,
TURKEY

Sinan Hüsrevoğlu
Bilkent Uni.,Chemistry Dep.
Ankara,
TURKEY

Erhan Demirbaş
University of Sheffield,
Departement of Chemistry
Brookhill D-9 Sheffield
UK

Ruth Wilson
University of Sussex
Pig Pigeon Holes, Dept Chem
University of Sussex, Falmer Brighton
Sussex,
UK

Zaid Rawi
University of Sussex
School of molecular Sciences
Falmer , Brighton
E. Sussex BNI 2QJ
UK

Keith Atkins
Department of Chemistry
Durham UNiversity
South road
Durham DH13LE
UK

Ernestine Lee
Utah State University,
Department of Chemistry and
Biochemistry
Logan Utah 84322
USA

Kevin Lee
Utah State University,
Department of Mathematics
Logan Utah 84322
USA

Daniele Romanini
Princeton University
Chemistry Department
Washington Rd. Princeton,
NJ, 08544
USA

Andrea Callegari
Princeton Uni.,Dep.of Chem.
Washington Rd. Princeton NJ
08544
USA

M.Iken
School of Physics
Georgia Ins.Technology
Atlanta,
USA

Ray Bemish
Dept. of Chemistry
University of North Carolina
B-5 Venable Hall
Chapel Hill, NC 27599
USA

David Farrely
Utah State University
Department of Chemistry
Logan UT 84322-0300
USA

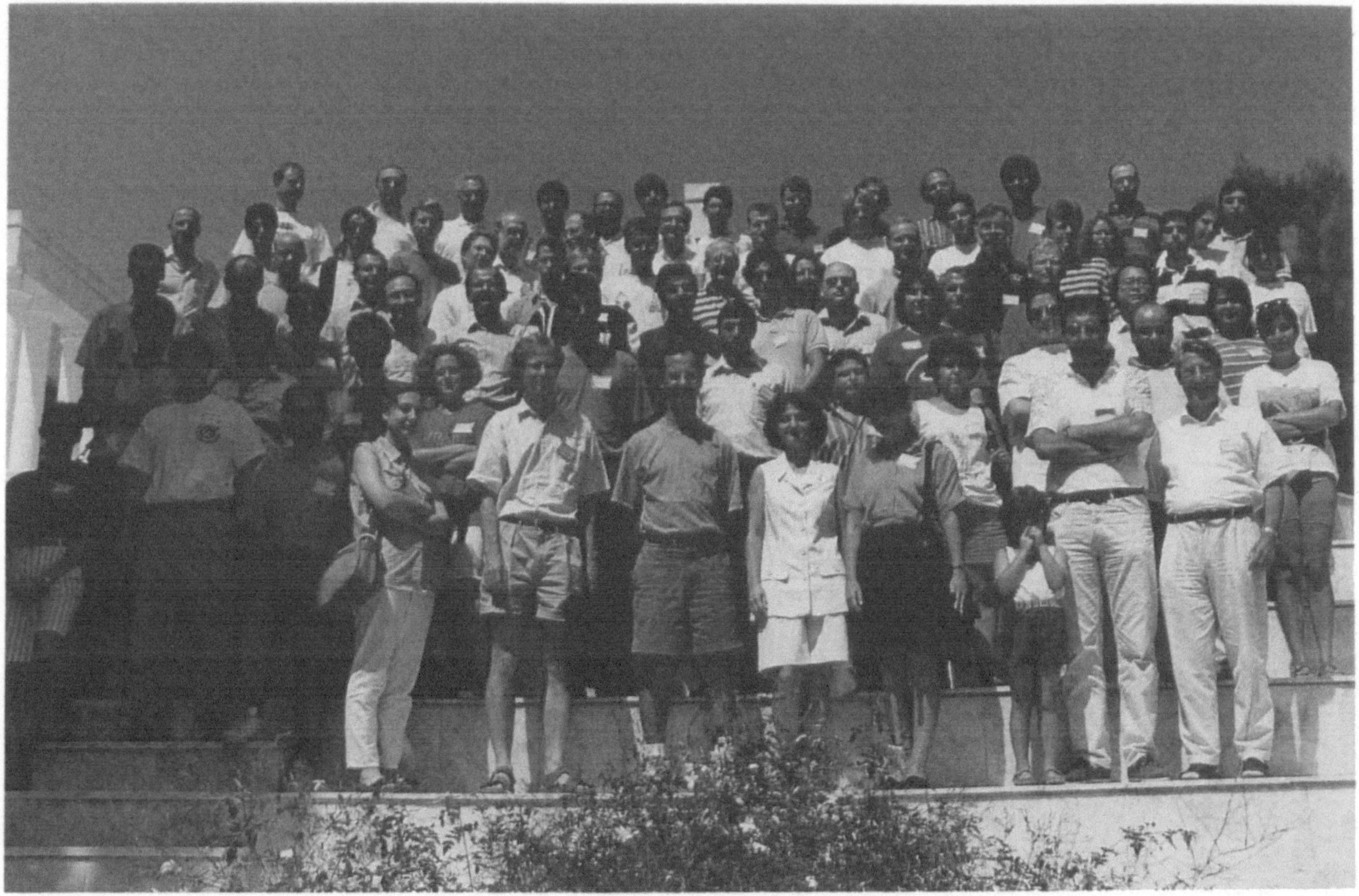

Group Photograph

THEORETICAL OVERVIEW OF CHEMICAL DYNAMICS

W. H. MILLER
Department of Chemistry, University of California, and
Chemical Sciences Division, Lawrence Berkeley Laboratory
Berkeley, California 94720 USA

1. Introduction

Chemical dynamics is the link between the potential energy surface (or surfaces) and physically observable chemical phenomena. The potential surface comes in principle from an *ab initio* quantum chemistry calculation (within the Born-Oppenheimer approximation) though in practice it is often constructed by some more approximate model, e.g., semiempirical quantum chemistry or totally empirical "force field" models. The purpose of this Overview is to give a brief snapshot of the present state of the methodology and scope of applications in this area. I will concentrate on chemical dynamics in the gas phase, though much of the methodology (and mentality!) of this field has carried over to the study of dynamical processes in condensed phases, gas-surface collision processes, and also dynamics in biomolecular systems. In these latter fields of application there is obviously strong input from and overlap with statistical mechanics.

Even the field of gas phase theoretical chemical dynamics is too large, however, to give anything but a cursory treatment in a brief overview. I will thus attempt to point out what I think are presently some of the major currents in theoretical chemical dynamics, with references to useful reviews, and I apologize beforehand that the overview will necessarily be biased to areas with which I am most familiar. Sections 2, 3 and 4 discuss molecular collision processes (i.e., scattering), perhaps the most well-defined and rigorous approach to studying chemical dynamics, and Sections 5 and 6 discuss intramolecular dynamics and laser-induced processes.

The four volume set edited by Baer[1] is a recent collection of more detailed reviews of the field, and the two volume set edited by Miller[2] an earlier one. Other more recent collections of reviews are the volume by Bernstein[3] and that by Bowman.[4]

E. Yurtsever (ed.), Frontiers of Chemical Dynamics, 1–20.

2. Elastic Scattering

The study of elastic scattering is today quite *passe*, but in the early days of chemical dynamics (the 1960's) it served in important "test bed" role, both experimentally and theoretically, for developing the tools to be used to study more interesting processes. E.g., elastic scattering of the rare gas atoms[5] with each other in the early 1970's was the first test of Y. T. Lee's new "universal" detector in a crossed molecular beam apparatus, applications of which to this and many other processes led to his sharing the 1986 Nobel Prize in Chemistry. Lee's measurements of the differential scattering cross sections allowed the definitive determination of the intermolecular potential energy function V(r) of essentially all the rare gas atoms with each other.

The fully rigorous quantum mechanical elastic differential cross section is given by[6]

$$\sigma_E(\theta) = \left| \frac{1}{2ik} \sum_{\ell=0}^{\infty} (2\ell+1) P_\ell(\cos\theta)(e^{2i\eta_\ell(E)}-1) \right|^2 , \tag{2.1}$$

where the phase shift $\eta_\ell(E)$ for orbital angular momentum ℓ and energy E is determined from the asymptotic form of the regular solution of the radial Schrödinger equation,

$$\left(-\frac{\hbar^2}{2\mu}\frac{d^2}{dr^2} + V(r) + \frac{\hbar^2 \ell(\ell+1)}{2\mu r^2} - E \right) f_{\ell E}(r) = 0 , \tag{2.2}$$

with asymptotic boundary condition

$$\lim_{r\to\infty} f_{\ell E}(r) \propto \sin\left(kr - \frac{\pi\ell}{2} + \eta_\ell(E) \right) . \tag{2.3}$$

The procedure is, given the potential function V(r) and a value of the energy E, to solve the Schrödinger equation, Eq. (2.2), for all values of ℓ to obtain the phase shifts, and then to compute the cross section from Eq. (2.1). Even though hundreds (or thousands) of values of ℓ's may be needed for the sum in Eq. (2.1) to converge, this is nevertheless a trivial calculation nowadays. The potential function V(r) is typically determined by assuming a functional form involving a set of parameters, carrying out the calculation of $\sigma_E(\theta)$ as described, and adjusting the parameters in the potential in a least squares procedure to fit the experimental cross section. Though this least squares fitting procedure is not as elegant as formal "inversion" approaches, it is in practice the most general and straight-forward approach.

Elastic scattering was also the theoretical workhorse in earlier years for developing many *approximate* theoretical methods, and though these approximations are no longer needed for elastic scattering calculations they still serve as useful guides for dealing with more complex processes.

One of the most important examples of this is the *semiclassical* approximation to quantum mechanics: in 1959 Ford and Wheeler[7] described the explicit sequence of approximations by which the rigorous quantum cross section of Eqs. (2.1)-(2.3) degenerates to the completely classical cross section,

$$\sigma_{CL}(\theta) = \sum_k \frac{b_k}{\sin\theta|\Theta'(b_k)|} , \tag{2.4}$$

where $b_k \equiv b_k(\theta)$ are the values of impact parameters b which classically scatter at angle θ, i.e., the roots of the equation

$$|\Theta(b)| = \theta , \tag{2.5}$$

where $\Theta(b)$ is the classical deflection angle as a function of b,

$$\Theta(b) = \pi - \int_{r_0}^{\infty} dr \frac{2b}{r^2}\left(1 - \frac{V(r)}{E} - \frac{b^2}{r^2}\right)^{-1/2} , \tag{2.6}$$

and r_0 is the classical turning point (the largest value r for which the integrand of Eq. (2.6) is zero.) The essential approximations are to use the WKB approximation for the phase shift and the Legendre polynomial, to replace the sum over ℓ by an integral, and then evaluate the integral by the stationary phase approximation. The result is

$$\sigma_{SC}(\theta) = \left|\sum_k \left(\frac{b_k}{\sin\theta|\Theta'(b_k)|}\right)^{1/2} e^{i\phi_k/\hbar}\right|^2 , \tag{2.7}$$

where $\{b_k\}$ are the same as in the classical cross section; they emerge semiclassically as the values of ℓ ($\ell \equiv kb$, $k = \sqrt{2mE/\hbar^2}$) for which the phase in the integral over ℓ is stationary and thus make the dominate contribution to the sum/integral. The phases $\{\phi_k\}$ are the classical action for the kth trajectory that classically scatters at angle θ.

Eqs. (2.1), (2.4), and (2.7) show the generic structure relating quantum, classical, and the semiclassical correction in scattering cross sections. Eq. (2.7) shows clearly that

$$\sigma_{SC}(\theta) = \sigma_{CL}(\theta) + \text{interference terms}\,, \tag{2.8}$$

so that it is interference between the different classical contributions to the cross section that is the most fundamental effect of quantum mechanics. Bernstein's 1966 review[6] of the semiclassical description of quantum effects in elastic scattering is still one of the best and most comprehensive, and that by Berry and Mount[8] is also insightful. Recently Child[9] has reviewed semiclassical theory more generally.

3. Inelastic Scattering

Inelastic scattering of atoms and molecules is clearly of more physical interest than elastic scattering. This includes excitation or relaxation of rotational and vibrational degrees of freedom and also of electronic states. Rotational/vibrational inelasticity is important in understanding classical transport phenomena in gases, and the vibrational relaxation of highly excited molecules is of crucial importance for describing "unimolecular" reaction in the Lindeman mechanism,[10]

$$A + B \rightarrow A^* + B \tag{3.1a}$$

$$A^* + B \rightarrow A + B \tag{3.1b}$$

$$A^* \rightarrow \text{products}\quad . \tag{3.1c}$$

I.e., after collisional excitation of A to A* by the bath gas B — or A* may be produced "hot" as the product of some preceding chemical reaction — the important consideration is the competition between collisional stabilization of A* by the bath gas, Eq. (3.1b), and the unimolecular decomposition of A*, Eq. (3.1c). Currently, therefore, there is much effort devoted to learning more quantitatively about vibrational de-activation of highly excited vibrational states.[11] Electronically inelastic collisions are important in many gas laser systems, in the upper atmosphere, and in plasmas. Many of these applications involve one of the collision partners being an ion.

The quantum mechanical description of an inelastic scattering process is straightforward. Leaving aside details involving angular momentum, the wavefunction for the generic situation is expanded as

$$\psi_i(r,\xi) = \sum_j \phi_j(\xi) f_{j\leftarrow i}(r) , \tag{3.2}$$

where $\{\phi_j(\xi)\}$ denotes the bound-state eigenfunctions for the internal (rotational, vibrational, and electronic) degrees of freedom (with ξ collectively denoting the appropriate coordinates) of the colliding molecules, and r is the radial (translational) coordinate, i.e., the distance between the centers of mass of the two molecules. The radial function (matrix) $\{f_{j\leftarrow i}(r)\}$ is determined by the coupled-channel (channel being the historical term for the various internal states) Schrödinger equation,

$$\left(-\frac{\hbar^2}{2\mu}\frac{d^2}{dr^2} - E_j\right) f_{j\leftarrow i}(r) + \sum_{j'} V_{j,j'}(r) f_{j'\leftarrow i}(r) = 0 , \tag{3.3}$$

where $V_{j,j'}(r)$ is the matrix of the interaction potential $V(\xi,r)$ with respect to the basis of internal states,

$$V_{j,j'}(r) = \int d\xi\, \phi_j(\xi)^* V(\xi,r) \phi_{j'}(\xi) , \tag{3.4}$$

and $E_j = E - \varepsilon_j$, where E is the total energy and ε_j the energy eigenvalue corresponding to ϕ_j (i.e., E_j is the available translational energy for channel j). The boundary conditions for the radial functions are that they be regular for $r \rightarrow 0$, and for large r have an incoming radial wave in initial channel i and outgoing waves in all open channels,

$$\lim_{r\to\infty} f_{j\leftarrow i}(r) \sim -\frac{e^{-ik_j r}}{v_j^{1/2}} \delta_{j,i} + \frac{e^{ik_j r}}{v_j^{1/2}} S_{j,i}(E) , \tag{3.5}$$

where $v_j = \sqrt{2E_j/\mu}$ is the translational velocity for channels j, and $S_{j,i}$ is the S-matrix, in terms of which all scattering cross sections can be expressed. Using Eq. (3.5) in Eq. (3.2) shows that the full wavefunction has the following form are large r,

$$\psi_i(r,\xi) \sim -\phi_i(\xi) \frac{e^{-ik_i r}}{v_i^{1/2}} \quad \sum_j \phi_j(\xi) \frac{e^{ik_j r}}{v_j^{1/2}} S_{j,i} , \tag{3.6}$$

from which one more clearly sees that the initial channel i has an incoming radial wave, and all channels have outgoing (scattered) waves. The S-matrix is a unitary matrix and has the interpretation as the probability amplitude for the i→j transition; the transition probability is thus

$$P_{j\leftarrow i} = |S_{j,i}|^2 \ , \tag{3.7}$$

When angular momentum — the orbital angular momentum of relative motion (i.e., the angular degrees of freedom of the translational coordinate *vector* $\vec{r}$) plus any angular momentum from the internal degrees of freedom — is taken account of, one finds that the total angular momentum J is conserved (because of the isotropy of space) and is thus a diagonal label of the S-matrix, $S^J_{j,i}(E)$. The state-to-state differential cross section — the most detailed possible scattering observable — is then given by

$$\sigma_{j\leftarrow i}(\theta) = \left| \frac{1}{2ik_i} \sum_J (2J+1)\, d^J_{m_j,m_i}(\theta) \left(S^J_{j,i}(E) - \delta_{ji} \right) \right|^2 \tag{3.8}$$

where $d^J_{mm'}(\theta)$ is the Wigner rotation function, and m_i, m_j are the projections of the total angular momentum onto the initial and final relative velocity vectors. One can easily recognize the sense in which Eq. (3.8) is the generalization of Eq. (2.1) for the elastic scattering of two structureless particles; in the latter case $J \rightarrow \ell$ (i.e., the orbital angular momentum ℓ is the total angular momentum), m_j and $m_i \rightarrow 0$, $d^J_{00}00(\theta) = P_\ell(\cos\theta)$, and

$$S^J_{j,i}(E) \rightarrow e^{2i\eta_\ell(E)} \ . \tag{3.9}$$

The theoretical task, therefore, is to solve Eq. (3.3) with the boundary conditions of Eq. (3.5) to obtain the S-matrix, in terms of which the inelastic cross sections are given by Eq. (3.8). Such calculations are relatively straight-forward nowadays,[12] although they can be time consuming if the number of channels is large; the largest calculations which are reasonable at present involve up to ~1000 coupled channels. This may seem like a large number, but consider the number of rotational/vibrational states of a diatomic molecule that have an energy below an energy E,

$$N(E) = \sum_{v=0} \sum_{j=0} (2j+1)\, h(E - \varepsilon_{vj}) \ . \tag{3.10}$$

Using a simple rigid rotor-harmonic oscillator approximation for the energy levels,

$$\varepsilon_{vj} \cong \hbar\omega(v+\tfrac{1}{2}) + Bj(j+1) ,$$

where ω is the vibrational frequency and B the rotation constant, gives

$$N(E) \cong \frac{E^2}{2\hbar\omega B} . \tag{3.11a}$$

For E = 0.1 eV, $\hbar\omega = 2000\ cm^{-1}$, and $B = 1\ cm^{-1}$, which are typical of a relatively small diatomic molecule (O_2, N_2, CO), Eq. (3.11a) gives N ~ 160, not too large a number of channels. But for I_2 ($\hbar\omega = 215\ cm^{-1}$, $B = 0.037\ cm^{-1}$) at this energy one has $N \cong 40{,}000$! And this is the number of channels for only *one* diatomic molecule; for the collision of two diatomics at energy E one has

$$N(E) \cong \frac{E^4}{4!\hbar\omega_1\hbar\omega_2 B_1 B_2} , \tag{3.11b}$$

$$\cong \frac{1}{6} N_1 N_2 , \tag{3.11c}$$

where N_1 and N_2 are the number of channels for molecules 1 and 2, respectively. Thus the number of channels involved can become unmanageably large very quickly!

Typically, however, the more channels there are that are strongly coupled in an inelastic collision the better it is to approximate the dynamics by classical mechanics; i.e., there are more channels the heavier the particles, but this is also the limit in which classical mechanics is a better approximation. Thus there have been many *classical trajectory simulations* of inelastic collision processes.[13] These have the advantage that no approximations other than the use of classical mechanics need be made, and the number of classical equations of motion to be solved (Hamiltonian's equations) grows *linearly* with the number of particles, while the numbers of coupled channels in the coupled-channel Schrödinger equation grows *exponentially* with this number.

There also exist a wide variety of approximate quantum mechanical and semiclassical theories.[14] In various limits some of the degrees of freedom can be treated as slow or fast compared to others, leading to sudden or adiabatic approximations, and in some cases the coupling between translational and internal motion can make perturbation theory a useful approximation.

Classical S-matrix theory[15,9] is a "rigorous" semiclassical theory, rigorous in that it incorporates the full classical mechanics for all degrees of freedom without approximation; it may be viewed as the generalization of the Ford and Wheeler semiclassical description of elastic scattering discussed in Section 2. Thus the inelastic transition probability in this theory is of the form

$$P^{SC}_{j\leftarrow i} = \left| \sum_k P^{(k)}_{j,i}{}^{1/2}\, e^{i\phi^{(k)}_{j,i}/\hbar} \right|^2 , \tag{3.12}$$

where $P^{(k)}_{j,i}$ is the purely classical contribution to the i→j transition from the kth trajectory which leads to it — i.e., the *completely* classical transition probability is

$$P^{CL}_{j\leftarrow i} = \sum_k P^{(k)}_{j,i} \tag{3.13}$$

— and $\phi^{(k)}_{i,j}$ is a classical action integral along the corresponding trajectory. The semiclassical transition probability thus has the same structure as in Eq. (2.8),

$$P^{SC}_{j\leftarrow i} = P^{CL}_{j\leftarrow i} + \text{interference.} \tag{3.14}$$

Interference between the different classical trajectories which contribute to the i→j transition will thus cause interference effects (e.g., "rainbows") in the product distribution of internal states analogous to such effects in the angular distribution (i.e., differential cross section) for elastic scattering.

4. Reactive Scattering

Quantum mechanical reactive scattering[16,17,3,4] provides the fundamental and rigorous description of chemical reactions and is thus the type of collision process of most interest to us. Unfortunately it is also the most complicated to deal with because of the lack of one physically appropriate set of coordinates for "translation" and "internal" degrees of freedom. I.e., the natural coordinates for describing translational and internal degrees of freedom for the reactant molecules are not the natural ones for describing those of the products. See reference 16 for a fairly detailed discussion of this "coordinate problem" for reactive scattering.

Because of this problem with coordinates, most of the modern ways of carrying out quantum reactive scattering calculations do not use the straight-forward coupled-channel expansion of Section 3 (though the use of hyperspherical coordinates[18] does allow this approach). Rather a *variational* method[19-21] is used to calculate the S-matrix, and this allows one to use basis functions expressed in terms of different coordinates in a straight-forward manner.

The variational approach received a major boost also when it was realized[19a] that the simplest variational method — the Kohn variational principle, which is essentially the Rayleigh-Ritz variational principle for eigenvalues modified to incorporate scattering boundary conditions — is free of anomalous (i.e., spurious, unphysical) singularities if it is formulated with S-matrix type boundary conditions rather than standing wave boundary conditions as had been typically used previously. It is useful first to state the Kohn variational approach for the general inelastic scattering case of Section 3. Thus the variational expression for the S-matrix is

$$S[\psi_f,\psi_i] = S_{fi}^{(0)} + \frac{i}{\hbar}\langle\psi_f|H\text{-}E|\psi_i\rangle \, , \tag{4.1}$$

where ψ_f and ψ_i are variational ("trial") wavefunctions of the form that satisfy the correct boundary conditions [cf. Eq. (3.6)]

$$\psi_\ell(\xi,r) \sim -\frac{e^{-ik_\ell r}}{v_\ell^{1/2}}\phi_\ell(\xi) + \sum_j \frac{e^{ik_j r}}{v_j^{1/2}}\phi_j(\xi)\, S_{j,\ell}^{(0)} \, , \tag{4.2}$$

for ℓ = i and f. In practice we have taken the trial function to be of the form

$$\psi_\ell(\xi,r) = \Phi_\ell(\xi,r) + \sum_j \Phi_j(\xi,r)^*\, C_{j,\ell} + \sum_k \chi_k(\xi,r)\, C_{k,\ell} \, , \tag{4.3}$$

where $\Phi_\ell(\xi,r)$ is an asymptotically incoming radial wave in channel ℓ,

$$\Phi_\ell(\xi,r) \sim \frac{e^{-ik_\ell r}}{v_\ell^{1/2}}\phi_\ell(\xi) \, , \tag{4.4}$$

and Φ_ℓ^* is a corresponding outgoing wave. The functions $\{\chi_k\}$ in Eq. (4.3) are an L^2 basis that describes the interaction (small r) region of the composite molecular system. The coefficients $\{C_{j,\ell}\}$ and $\{C_{k,\ell}\}$ in Eq. (4.3) are the variational parameters with respect to which Eq. (4.1) for the S-matrix is extremized; i.e., substituting the functions ψ_f and ψ_i of the form of Eq. (4.3) into Eq. (4.1) gives the S-matrix as a quadratic function of these coefficients, and the variational condition

$$0 = \frac{\partial}{\partial C_\lambda} S(\{C_{j,f}\},\{C_{\ell,f}\},\{C_{j,i}\},\{C_{\ell,i}\}) , \tag{4.5}$$

for each of the coefficients $C_\lambda = C_{j,f}$, $C_{\ell,f}$, etc., leads to linear equations for the coefficients which are solved by matrix inversion. Using these variationally optimum coefficients in the expression for the S-matrix then gives the following variationally optimum result (within the given basis set) for the S-matrix,

$$S_{f,i} = \frac{i}{\hbar} (M_{f,i} - \mathbf{M}_f^T \bullet \mathbf{M}^{-1} \bullet \mathbf{M}_i) , \tag{4.6}$$

where

$$M_{f,i} = \langle\Phi_f|H-E|\Phi_i\rangle , \tag{4.7a}$$

$$\mathbf{M}_\ell = \begin{Bmatrix} \langle\Phi_j^*|H-E|\Phi_\ell\rangle \\ \langle\chi_k|H-E|\Phi_\ell\rangle \end{Bmatrix} , \tag{4.7b}$$

for ℓ = i and f, and

$$\mathbf{M} = \begin{Bmatrix} \langle\Phi_{j'}^*|H-E|\Phi_j\rangle , \langle\Phi_{j'}^*|H-E|\chi_k\rangle \\ \langle\chi_{k'}|H-E|\Phi_j\rangle , \langle\chi_{k'}|H-E|\chi_k\rangle \end{Bmatrix} . \tag{4.7c}$$

Thus the scattering problem has been cast in the form of a standard quantum mechanical calculation, i.e., computing matrix elements of the Hamiltonian with respect to a set of basic functions and then performing a linear algebra calculation (i.e., the matrix inverse in Eq. (4.6)).

The power of this variational result is that it applies as written also for reactive scattering provided one expands the definition of the channel label also to include an arrangement label. E.g., for an atom-diatom reaction, A+BC(vjm) → AB(v′j′m′)+C, the initial channel index i denotes the collection ≡ (a,v,j,m), labeling the arrangement a and quantum state (v,j,m) of that arrangement; the final channel label in this case is f = (c,v′,j′,m′). The only modification of the above formulae in Eqs. (4.1)-(4.7) is in Eq. (4.4), noting that the coordinates appropriate for asymptotic channel ℓ may be different for different channels, i.e.,

$$\Phi_{\ell} \sim -\frac{e^{-ik_{\ell}r_{\ell}}}{v_{\ell}^{1/2}} \phi_{\ell}(\xi_{\ell}) . \tag{4.4'}$$

Reference 16 discusses how this strategy of using different coordinates for different basis functions is the same as that in the LCAO approach for constructing molecular orbitals in electronic structure theory by using atomic orbitals which are expressed in terms of spherical coordinates referenced to (i.e., "centered on") various nuclei.

Many of the recent accurate quantum reactive scattering calculations have utilized the above approach;[19-22] methods using hyperspherical coordinates have been the primary alternative.[23,24,18] Complete state-to-state differential cross section calculations have been carried out for the $H+H_2$ reaction and its isotopic variants,[19c,22] i.e.,

$$H+H_2(\text{ortho}) \rightarrow H_2(\text{para})+H$$

$$D+H_2 \rightarrow HD+H$$

$$H+D_2 \rightarrow HD+D ,$$

and also for the $F+H_2 \rightarrow HF+H$ reaction.[23] Calculations for many other reaction have been carried out for J=0 only.

It is also useful to note that methods have been developed which allow one to calculate the *cumulative* reaction probability (CRP) for a reaction directly, without having to solve for the individual state-to-state S-matrix elements.[25] This is important because the thermally averaged rate constant for a reaction can be expressed as the Boltzmann average of the CRP N(E),

$$k(T) = [2\pi\hbar Q_r(T)]^{-1} \int_{-\infty}^{\infty} dE\, e^{-E/kT} N(E) . \tag{4.8}$$

The CRP is a sum of reactive probabilities, i.e., squares of S-matrix elements, over all open reactant and product channels,

$$N(E) = \sum_{n_r,n_p} |S_{n_p,n_r}(E)|^2 , \tag{4.9}$$

where $n_r(n_p)$ labels the asymptotic channel states of the reactant (product). Miller *et al.*[26] showed that the CRP can also be expressed as

$$N(E) = \frac{1}{2}\,(2\pi\hbar)^2\ \mathrm{tr}[\hat{F}\delta(E-\hat{H})\hat{F}\delta(E-\hat{H})]\ , \qquad (4.10)$$

where $\hat{H}$ is the Hamiltonian operator and $\hat{F}$ a flux operator. Eq. (4.10) forms the basis for a "direct" calculation since it is explicitly independent of individual reactant and product states. The heart of such calculations is finding a useful way to represent the microcanonical density operator, $\delta(E-\hat{H})$, and several efficient ways have been developed.[26-29]

Finally, any of the inelastic and reactive collision processes may also involve changes in the electronic state.[30] Formally, this requires only the addition of the electronic state index to the channel label in the coupled-channel equations in Eq. (3.3) or basis functions in Eqs. (4.7). In practice, however, it requires that one know the non-adiabatic coupling (the Hamiltonian matrix elements non-diagonal in the electronic state index) as well as the different potential energy surfaces (the diagonal matrix elements).

5. Intramolecular Dynamics and Unimolecular Reactions

The resurgence of modern research in high resolution spectroscopy has stimulated theoretical studies of intramolecular dynamics. New experimental techniques make it possible to excite molecules to much higher energy states than ordinary one-photon absorption spectroscopy. The underlying vibrational motion is therefore much more complex and not usefully described by the standard normal mode picture.

The theoretical approach for characterizing intramolecular dynamics is in principle quite straight-forward:[31] since the molecular system is bounded, its quantum description is that of stationary states, i.e., eigenfunctions and eigenvalues of the Hamiltonian. One thus just (!) needs to choose an appropriate set of basis functions, form the Hamiltonian matrix, and diagonalize it, and this has indeed been carried out for a variety of small molecules, H_2O, HCN, ... Such calculations allow one to analyze the intramolecular motion in complete detail.

Because the ability to carry out these rigorous quantum calculations is limited to small molecular systems — very much the same kind of limitations as noted above for reactive scattering — there are a variety of approximate methods which are used. Some of these are still within a quantum framework — e.g., various perturbative approximations, sudden/adiabatic separations — and others use classical mechanics,[13] either completely classically or within a semiclassical framework.

The important questions in intramolecular dynamics have to do with how the energy moves between different degrees of freedom. E.g., if a laser excites motion that is predominantly CH stretch motion in benzene, how quickly does this energy move into other degrees of freedom in the molecule, and which of the other degrees of freedom are most effective in accepting this energy?[32] This question becomes even more interesting if some degrees of freedom lead to bond-breaking, in which case one is talking about ***unimolecular*** reactions.[10]

The rigorous quantum mechanical description of unimolecular reactions is that of Siegert eigenstates.[33] These are eigenfunctions of the Schrödinger equation with *outgoing wave boundary conditions*, and because of the complex boundary condition the eigenvalues are complex, $\{En\text{-}i\Gamma_n/2\}$. The real part of the eigenvalue is the energy of the metastable state of the molecule, and its unimolecular decay rate is given in terms of the imaginary part, $\Gamma_n/\hbar$. (This latter relation is only true if these resonance states are non-overlapping, i.e., $\Gamma_n<\Delta E$ on the average, which means that the molecule on the average lives for at least a few bound state motions before it decays.) The unimolecular decay of the formaldehyde molecule,

$$H_2CO^{\ddagger} \rightarrow H_2 + CO\ , \tag{5.1}$$

is one of the best characterized unimolecular reactions, a large number of its complex eigenvalues $\{En\text{-}i\Gamma_n/2\}$ having been determined experimentally.[34]

Another major question regarding intramolecular dynamics and unimolecular reactions is the extent to which the underlying dynamics can be described as *chaotic* or not.[35] Chaotic dynamics is a concept from classical mechanics, and the quantum analog is that the set of eigenstates (bound or Siegert) in some energy interval are "strongly mixed", i.e., representable only as a long expansion in any separable basis. The opposite limit of chaotic dynamics is completely integrable (or separable) motion, which in quantum mechanics means that in some set of coordinates the wavefunction would be a product of factors, one for each degree of freedom. Reality, of course, typically lies somewhere between these extremes, with the molecular dynamics appearing to behave chaotically or not depending on the level of excitation and the particular degrees of freedom which are excited. Most of the discussions of chaotic dynamics have been theoretical, for it is very difficult to find a well-defined experimental signature of chaos. It is often stated that such a signature is that a spectrum is "intrinsically unassignable", but this is of course very hard to distinguish from a spectrum that is simply very difficult to assign.

For molecules undergoing unimolecular decomposition, one measure of chaotic dynamics is the *distribution* of the unimolecular decay rates for individual molecular eigenstates that all have (approximately) the same energy (and total angular momentum). The more chaotic the underlying dynamics the narrower will be this distribution about the average, but there will nevertheless be some fluctuations about the average even if the dynamics is completely chaotic.[34,36] This type of analysis has recently been carried out for the unimolecular decay of formaldehyde noted above, and the experimental distributions are consistent with the interpretation that the dynamics is indeed chaotic.

Finally, it should be noted that the enormous amount of experimental work on van der Waals clusters has stimulated much theoretical effort in describing these systems.[37] Especially interesting has been the vibrational predissociation[38] caused when a laser is used to excite one of the "tight" molecular vibrations, e.g.,

$$\mathrm{Ar} \cdot\cdot\, \mathrm{HC\ell}(v=0) + h\nu \rightarrow \mathrm{Ar} \cdot\cdot\, \mathrm{HC\ell}(v=1) \tag{5.2a}$$

$$\mathrm{Ar} \cdot\cdot\, \mathrm{HC\ell}(v=1) \rightarrow \mathrm{Ar} + \mathrm{HC\ell}(v=0)\ . \tag{5.2b}$$

6. Photon-Induced Processes

A variety of photon-induced processes other than simple spectroscopy are also under study nowadays. *Photodissociation*[39] with infrared lasers, which cause vibrational excitation, and also with visible and UV lasers, which cause electronic excitation, are important processes in their own right as well as being useful ways of probing reaction dynamics.

Although photodissociation is often referred to as a "half collision", its quantum mechanical description requires the full scattering wavefunction for the dissociative products. Thus, if $|\psi_{i,1}\rangle$ is the wavefunction for the initial state, in electronic state 1, then the state-to-state photodissociation cross section is given by

$$\sigma_{f\leftarrow i}(\omega) = \frac{4\pi^2\omega}{c}\, |\langle\psi_{E,f,2}|\vec{\mu}\cdot\hat{\varepsilon}|\psi_{i,1}\rangle|^2\ , \tag{6.1}$$

where $\psi_{E,f,2}$ is the scattering wavefunction for total energy $E = E_i + \hbar\omega$ in electronic state 2, and where i and f denote the quantum numbers of the nuclear degrees of freedom in the initial and final states. ($\vec{\mu}$ is the dipole operator and $\hat{\varepsilon}$ the polarization direction of the electronic field vector.) The primary task in computing such cross sections is carrying out the scattering calculation on the final potential energy surface (i.e., electronic state 2) to obtain the scattering wavefunction that appears in Eq. (6.1).

Photodissociation, as described above, corresponds to a continuous wave (CW) laser with frequency ω and polarization $\hat{\varepsilon}$. Eq. (6.1) is also a perturbative result, valid for one-photon dipole allowed transitions. (It is derived via the long time limit of first order time-dependent perturbative theory, leading to the usual Golden Rule expression.) Existence of high power lasers, and more importantly pulsed lasers with controlled wave forms, however, has created interest recently in exploring more general ways that electromagnetic radiation can be used to study and to influence molecular behavior. Within the standard semiclassical approximation for treating the interaction of the molecular system and the laser, one considers the following time-dependent Hamiltonian,

$$\hat{H}(t) = \hat{H}_{mol} - \hat{\mu} \cdot \vec{E}(t) \,. \tag{6.2}$$

This is an operator within the space of the molecular degrees of freedom, $\hat{H}_{mol}$ being the field-free Hamiltonian of the molecular system and $\hat{\mu}$ its dipole operator. (Eq. (6.2) obviously also involves the dipole approximation for the coupling between the laser and the molecular degrees of freedom.) For a laser *pulse*, one has $\vec{E}(t) \to 0$ for $t \to -\infty$ and $+\infty$, so that one can consider transition probabilities $P_{f\leftarrow i}$ induced between initial and final states $|i\rangle$ and $|f\rangle$ that are eigenstates of $\hat{H}_{mol}$. $P_{f\leftarrow i}$ is a functional of the electric field vector,

$$P_{f\leftarrow i}[\vec{E}(t)] \,, \tag{6.3}$$

and one can consider the possibility of optimizing the field to enhance specific transitions. Several different strategies have been pursued for choosing the electric field $\vec{E}(t)$, i.e., the laser pulse (or pulses), to affect transitions of interest. Some methods are based on ultrashort (femtosecond, 10^{-15} sec) pulses,[40] and others rely on the coherence of two (or more) CW lasers.[41] A sequence of pulses[42] (stimulated Raman) has also been discussed as a means of enhancing specific dynamical processes. The optimization problem has been considered quite generally as a variational problem,[43,44] i.e., maximizing the functional in Eq. (6.3) subject to some constraints (e.g., a given total pulse power). These approaches all suggest interesting future directions for theoretical studies as the experimental technologies progress to make "laser control" a practical reality.

7. The Future

Predictions of future scientific developments are often quite worthless, but I believe that one can at least envision further progress in theoretical chemical dynamics along two fronts: (1) extension of rigorous methodology to provide benchmark results for larger, though still small, molecular systems, and (2) application of approximate dynamical treatments to evermore complex chemical systems. Prediction (1) will happen without question if for no other reason than the fact that computer power will continue to grow. I was amazed to see a recent advertisement for a new portable 'notebook' type computer that performs at 8 to 10 megaflops (floating point operations per second) with 16 megabytes of internal memory. I saw in some old notes of mine that the CDC 7600, the top supercomputer all through the 1970's, performed at 3 megaflops! The latest IBM RS 6000 workstation performs at more than 100 megaflops. In addition to increased computer power, though, one should also see advances in methodology that will make the application of rigorous treatments possible for more interesting chemical systems.

There will nevertheless be interest in modeling chemical dynamics of molecular systems much beyond the capacity of rigorous quantum treatments. For some years already, many groups model biomolecular systems using classical trajectory simulation methods. The ideal situation would be to *combine* the accurate quantum treatment of a small sub-system, which involves the primary chemical process of interest, with a more approximate treatment of the much larger remainder of the system. Because classical mechanics is feasible for treating many degrees of freedom, this suggests using it as the "more approximate" method.

One thus seeks ways to combine the accurate quantum treatment of a few degrees of freedom with a classical or semiclassical treatment of the many remaining degrees of freedom. There are many such approaches that already exist, but no single one has emerged as the most generally accurate and useful model. Maybe there is no one universal such approach, but this is an area of intense interest and effort, and one expects to see progress.

8. Acknowledgment

This work was supported by the Director, Office of Energy Research, Office of Basic Energy Sciences, Chemical Sciences Division, of the U.S. Department of Energy under Contract No. DE-AC03-76SF00098 and also by the National Science Foundation, Grant CHE-9422559.

9. References

1. M. Baer, ed., *Theory of Chemical Reaction Dynamics*, Vols. I-IV, CRC Press, Boca Raton, FL, 1985.
2. W. H. Miller, ed., *Dynamics of Molecular Collisions*, Parts A and B (Vols. 1 and 2 of *Modern Theoretical Chemistry*), Plenum, NY, 1976.

3. R. B. Bernstein, ed., *Atom-Molecule Collision Theory: a Guide for the Experimentalist*, Plenum, NY, 1979.
4. J. M. Bowman, ed., *Advances in Molecular Vibrations and Collision Dynamics*, Vol. 2, JAI Press, Greenwich, CT, 1994.
5. Y. T. Lee, *et al.*, J. Chem. Phys. **53**, 2123 (1970); **53**, 3755 (1970); **55**, 5762 (1971); **56**, 1511 (1972); **56**, 5801 (1972); **59**, 601 (1973).
6. R. B. Bernstein, Adv. Chem. Phys. **10**, 75 (1966).
7. K. W. Ford and J. A. Wheeler, Ann. Phys. **7**, 259, 287 (1959).
8. M. V. Berry and K. E. Mount, Rept. Prog. Phys. **35**, 315 (1972).
9. (a) M. S. Child, *Semiclassical Mechanics with Molecular Applications*, Oxford U.P., 1991.
   (b) Also, see W. H. Miller, Science **233**, 171 (1986).
10. (a) P. A. Robinson and R. A. Holbrook, *Unimolecular Reactions*, Wiley, NY, 1972.
    (b) W. Forst, *Theory of Unimolecular Reactions*, Academic Press, NY, 1973.
    (c) R. G. Gilbert and S. C. Smith, *Theory of Unimolecular and Recombination Reactions*, Blackwell, Oxford, 1990.
11. E.g., see J. Troe, J. Chem. Phys. **97**, 288 (1992), and references therein.
12. (a) W. A. Lester, ref. 2, p. 1.
    (b) J. C. Light, ref. 3, p. 239.
    (c) See, e.g., D. Secrest, in ref. 3, pp. 265, 377.
13. (a) R. N. Porter and L. M. Raff, ref. 2, pg. 1.
    (b) W. H. Hase, ref. 2, p. 121.
    (c) M. D. Pattengill, ref. 3, p. 359.
    (d) L. M. Raff and D. L. Thompson, ref. 1, vol. III, p. 2.
14. E.g., (a) D. A. Micha, ref. 2, p. 81.
    (b) H. K. Shin, ref. 2, p. 131.
    (c) R. E. Wyatt, ref. 3, p. 477.
    (d) J. Jellinek and D. J. Kouri, ref. 1, Vol. II, p. 1.
15. W. H. Miller, Adv. Chem. Phys. **25**, 69 (1974); **30**, 77 (1975).
16. W. H. Miller, Ann. Rev. Phys. Chem. **41**, 245 (1990).
17. D. E. Manolopoulos and D. C. Clary, Ann. Rep. C Roy. Soc. of Chem. **80**, 95 (1989).
18. A. Kuppermann, in ref. 4, and earlier references therein.
19. (a) W. H. Miller and B. M. D. D. Jansen op de Haar, J. Chem. Phys. **86**, 6213 (1987).
    (b) J. Z. H. Zhang, S. I. Chu, and W. H. Miller, J. Chem. Phys. **88**, 6233 (1988).

(c) J. Z. H. Zhang and W. H. Miller, J. Chem. Phys. **91**, 1528 (1989).

20. (a) D. G. Truhlar and D. J. Kouri, *et al.*, J. Chem. Phys. **91**, 1643 (1989).
    (b) D. G. Truhlar and D. J. Kouri, *et al.*, J. Phys. Chem. **92**, 3202 (1988).
21. D. E. Manolopoulas and R. E. Wyatt, Chem. Phys. Lett. **152**, 23 (1988); **159**, 123 (1989); J. Chem. Phys. **91**, 6096 (1989); **93**, 403 (1990).
22. S. L. Mielke, D. G. Truhlar, and D. W. Schwenke, J. Phys. Chem. **98**, 1053 (1994).
23. J. M. Launay and M. le Dournef, Chem. Phys. Lett. **169**, 473 (1990).
24. R. T Pack, E. A. Butcher, and G. A. Parker, J. Chem. Phys. **99**, 9310 (1993).
25. W. H. Miller, Accts. Chem. Res. **26**, 174 (1993).
26. (a) W. H. Miller, S. D. Schwartz, and J. W. Tromp, J. Chem. Phys. **79**, 4889 (1983).
    (b) T. Seideman and W. H. Miller, J. Chem. Phys. **96**, 4412 (1992); **97**, 2499 (1992).
27. U. Manthe and W. H. Miller, J. Chem. Phys. **99**, 3411 (1993).
28. D. Thirumalai, B. C. Garrett, and B. J. Berne, J. Chem. Phys. **83**, 2972 (1985).
29. V. A. Mandelshtrom and H. S. Taylor, J. Chem. Phys. **99**, 222 (1993).
30. (a) M. Baer and C. Y. Ng, eds., Advances in Chemical Physics, **82**, Part 2, 1992.
    (b) M. S. Child, in ref. 3, p. 427.
31. (a) J. C. Light and S. E. Choi, J. Chem. Phys. **97**, 7031 (1992).
    (b) M. Quack, Ann. Rev. Phys. Chem. **41**, 839 (1990).
    (c) J. M. Bowman and B. Gazdy, J. Chem. Phys. **94**, 454 (1991).
32. E. L. Sibert, W. P. Reinhardt, and J. T. Hynes, J. Chem. Phys. **81**, 1115 (1984).
33. See, e.g., W. H. Miller, Chem. Rev. **87**, 19 (1987).
34. (a) W. F. Polik, D. R. Guyer, W. H. Miller, and C. B. Moore, J. Chem. Phys. **92**, 3471 (1990).
    (b) W. H. Miller, R. Hernandez, C. B. Moore, and W. F. Polik, J. Chem. Phys. **93**, 5657 (1990).
35. D. W. Noid, M. L. Koszykowski, and R. A. Marcus, Ann. Rev. Phys. Chem. **32**, 267 (1981).
36. R. D. Levine, Adv. Chem. Phys. **70**, 53 (1988).

37. (a) R. C. Cohen and R. J. Saykally, Ann. Rev. Phys. Chem. **42**, 369 (1991).
(b) J. M. Hutson, Ann. Rev. Phys. Chem. **41**, 123 (1990).
38. (a) J. A. Beswick *et al.*, J. Chem. Phys. **72**, 3653 (1980); **73**, 3018 (1980).
(b) D. H. Zhang and J. Z. H. Zhang, Chem. Phys. Lett. **199**, 187 (1992).
39. (a) M. N. R. Ashfold and J. E. Baggott, eds., *Molecular Photodissociation Dynamics*, Roy. Soc. Chem., Letchworth, 1987.
(b) R. Schinke, *Photodissociation Dynamics*, Cambridge U. P., 1993.
40. S. Mukamel, Ann. Rev. Phys. Chem. **41**, 647 (1990).
41. P. Brumer and M. Shapiro, Ann. Rev. Phys. Chem. **43**, 257 (1992).
42. D. J. Tannor and S. A. Rice, Adv, Chem. Phys. **70**, 180 (1988).
43. (a) L. Shen, S. Shi, and H. Rabitz, J. Phys. Chem. **97**, 12114 (1993).
(b) D. Neuhauser and H. Rabitz, Accts. Chem. Res. **26**, 496 (1993).
44. K. R. Wilson, S. Mukamel, *et al.*, J. Phys. Chem. **97**, 2320, 12602 (1993).

Photodissociation of Weak Bonds: The Spectroscopy and Vibrational Dynamics of Molecular Complexes

R.E. Miller
Department of Chemistry
University of North Carolina
Chapel Hill, N.C. 27599.

Abstract

We discuss the application of high resolution infrared spectroscopy to the study of weakly bound molecular complexes formed in a molecular beam. The focus is on the use of the optothermal method to study the spectroscopy of the parent complex and to characterize the fragments resulting from their photodissociation. In the former case we consider several examples were the spectroscopy of the complex is directly sensitive to the associated wide amplitude motion. We consider the influence of large external electric fields on the eigenstates of these complexes. In the photofragmentation experiments we make use of the orientation imposed by this electric field to aid in the characterization of the final state distributions of the fragments. We also discuss the possibility of modifying the structure and dynamics of these complexes using an external field.

(1) Introduction

An ongoing pursuit in the field of experimental gas phase chemical reaction dynamics has been the development of better methods for characterizing the initial and final states of systems undergoing either unimolecular or bimolecular reactions. In parallel with these efforts has been the development of femtosecond technology that now allows us to follow the dynamics in real time [1,2]. There is ongoing discussion concerning the relative merits of these two approaches, commonly referred to as frequency and time domain experiments, respectively. Intuition might first suggest that the time domain experiments would always provide more information than those designed to characterize only the initial and final states of the system, much as watching an entire movie is always more informative than seeing only the credits at either end. This analogy is a poor one, however, since the uncertainty principle imposes limitations on both frequency and time domain experiments. In practice neither approach gives complete information and both occupy an important place in our attempts at understanding chemical dynamics.

In the present paper we discuss a class of experiments, based upon high resolution infrared spectroscopy in molecular beams, designed to study unimolecular reactions associated with weakly bound complexes at the state-to-state level. This pursuit is complicated by the fact that many of the properties of the system are vectorial and some are correlated. For example, conservation of energy requires that the all of the degrees of

E. Yurtsever (ed.), Frontiers of Chemical Dynamics, 21–41.

freedom of the fragments be correlated such that the overall energy sums to that of the parent molecule. On the other hand, it is all of this rich detail that helps to better define the nature of the dynamical processes. Therefore, once we achieve "complete" characterization of the system, meaningful comparisons can be made with theory.

An issue of central importance in chemical reaction dynamics is the nature of the energy transfer processes within and between reactants and products. At a fundamental level, bond rupture and formation can be understood in terms of the transfer of energy into the reaction coordinate, causing a bond to break, and then relaxation of the energy away from the newly formed bonds in the product molecules to the other degrees of freedom of the system. By their very nature, these processes are highly anharmonic and their detailed characterization remains a formidable challenge. In recent years spectroscopists have taken on the challenge of trying to characterize the quantum states of a molecule at the high vibrational energies corresponding to the chemically interesting regime. At these energies the density of states becomes extremely high and the coupling between the states very strong, the result being that the vibrations can no longer be characterized in terms of simple isolated local or normal modes. In the extreme limit, where RRKM theory [3] applies, there is rapid energy redistribution that tends to statistically sample the available states. Although we are far from having a complete understanding of the quantum state dynamics of such systems in this regime, the recent progress that has been made in both experiment [4-8] and theory [9,10] is helping to better define the important processes. Ultimately, the detailed characterization of the intramolecular couplings in a molecule would provide us with a fundamental understanding of the chemistry, in both the statistical and nonstatistical regimes. After all, energy transfer from one vibrational mode of a molecule to another is determined by the intermode couplings, which, in the ground electronic states of molecules, are predominantly due to anharmonic and/or coriolis effects. Of course the problem becomes even more challenging when one moves from the realm of isolated molecules to solvated systems. Finite sized molecular clusters can provide us with new ways of approaching many of these complex problems.

The often overwhelming complexity [11] that is associated with the spectroscopy of molecules at high energies is in large part simply due to the high density of states in this regime, corresponding to the large number of ways the energy can be rearranged among the available vibrational degrees of freedom. On the other hand, it is the coupling of the energy between the vibrational modes of the molecule which is of fundamental interest, and ultimately can provide us with an understanding of molecular energy transfer. It therefore seems appropriate to look for systems where the strong anharmonic and coriolis coupling is present but the density of states is low. The hope is that by isolating just a few interacting states, a more detailed characterization of the important coupling terms and dynamical mechanisms can be obtained.

Weakly bound molecular complexes represent such a class of photochemical systems. The specific dynamics of interest here is vibrational predissociation, which involves excitation of a high frequency vibration associated with one of the monomer units within the complex and subsequent energy transfer into the weak bond. Due to the presence of this fragile intermolecular bond, the energy threshold for dissociation of these systems is often very low. This fact, when combined with the relatively weak coupling between the initially excited intramolecular vibration and the intermolecular modes, often gives rise to dynamics which is in the highly non-statistical regime [12-22]. Studies of these systems can thus provide important insights into the nature of the associated dynamics since the deviations from statistical behavior can reveal the "preferences" the system has for particular photodissociation channels. The weakness and floppiness of the bond between the constituent molecules (< 5 kcal/mol.) provides the anharmonic and coriolis coupling that gives rise to the vibrational dynamics of interest. Single quantum vibrational excitation of one of the constituent monomer units is usually sufficient for dissociation to occur.

In the present work we examine a number of infrared laser based experiments that address the two important aspects of this type of study, namely the spectroscopic characterization of the parent complex and the determination of the rate of photodissociation of the complex and the final state distribution of the resulting fragments. Each of these are illustrated by several examples from the author's laboratory.

(2) The Parent Molecule - Characterization and Control

Recent developments in the field of infrared laser-molecular beam spectroscopy have lead to a wealth of information on both the structure [23-26] and potential energy surfaces [27-31] associated with these weakly bound molecular complexes in the ground electronic state[22,25,32-38]. For semi-rigid complexes, the ro-vibrational structure observed in an infrared spectrum yields accurate rotational constants, which can often be used to unambiguously determine the structure of the complex. For more floppy systems, the rotational/vibrational/tunneling spectroscopy can provide direct information on the associated potential energy surface. The basic approach is to identify as many bound states of the van der Waals complex as possible [39-42] and to compare these with the results of multidimensional quantum calculations based upon an assumed potential energy surface, with the eventual goal of fitting the parameterized potential to the experimental data [43,44]. In many cases, *ab initio* calculations can be useful in constraining the potential surface in regions that are not sampled by the available spectroscopy.

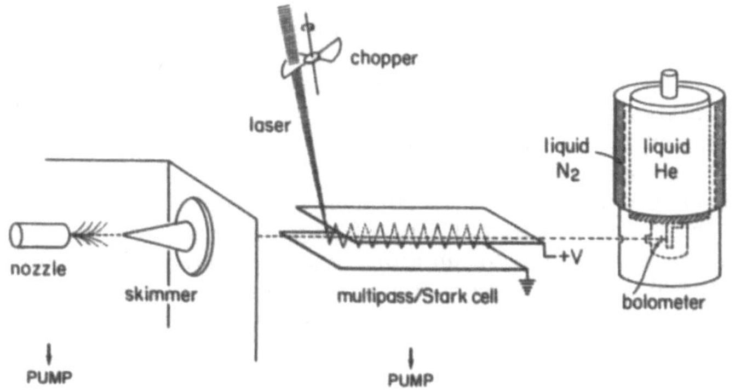

Figure 1: A schematic diagram of the optothermal spectroscopy apparatus. An F-center laser is used to pump transitions in the parent complex which leads to dissociation. The multipass cell, which is used to increase the efficiency of laser excitation, also serves as a set of electrodes for the pendular spectroscopy experiments. Alternatively, a second set of electrodes can be installed to apply the DC field parallel to the laser polarization direction.

The infrared laser - optothermal method has been discussed in detail in a number of previous publications [37,38]. Figure 1 shows the configuration used in our laboratory for obtaining high resolution infrared spectra, the basic idea being to use a tunable infrared laser (in the present case an F-center laser tunable from 2900 - 4500 cm^{-1}) to excite molecules in a well collimated cw molecular beam. Detection is based upon monitoring the associated change in the molecular beam energy using a liquid helium cooled bolometer. The multipass cell shown in the figure is used to increase the interaction volume between the laser and the molecular beam.

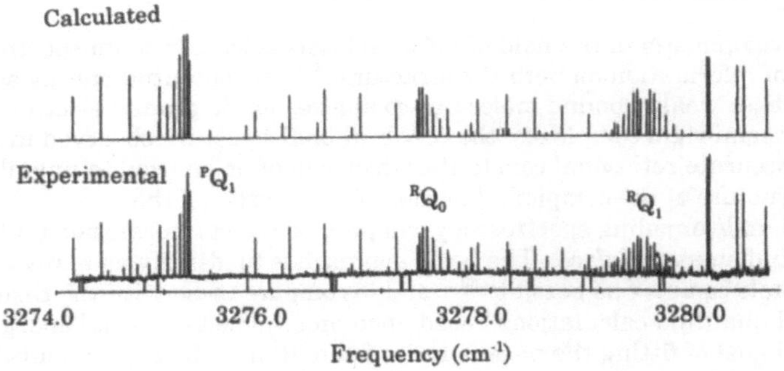

Figure 2: The infrared spectrum of the asymmetric C-H stretching vibration in C_2H_2-HF. The calculated spectrum was obtained using a rigid rotor Hamiltonian, indicating that the spectroscopy of this complex is not sensitive to wide amplitude motions.

Although the complexes of interest are loosely bound and inherently floppy, in many cases the associated rotational spectra are rather well described by a semi-rigid rotor Hamiltonian. This "apparent rigidity" has been discussed previously in the literature [45]. Consider for example, the spectroscopy of the "T-shaped" acetylene-HF complex, which has an equilibrium structure corresponding to the HF hydrogen bonded to the π electrons of the acetylene. Figure 2 shows the optothermal near infrared spectrum associated with the asymmetric C-H stretch on the acetylene sub-unit. Since the transition moment associated with this vibrational band is along the acetylene axis, essentially perpendicular to the A-axis of the complex (which passes through the F atom and the center-of-mass of the acetylene), the resulting spectrum is characterized as a perpendicular-type band ($\Delta K_a=\pm1$) [46]. Three of the corresponding sub-bands are indicated in the figure, along with those calculated from a rigid rotor Hamiltonian. The fact that the agreement is very good shows that the spectroscopy is rather insensitive to the wide amplitude intermolecular motions in this complex.

Before going on to consider cases where the spectroscopy is sensitive to the large amplitude motions, we can use the spectrum shown in Figure 2 to illustrate several important experimental features. First, the numerous transitions of opposite sign to those of the complex are due to fundamental and hot band transitions associated with the acetylene monomer. The opposite sign is indicative of the fact that the complexes dissociate upon vibrational excitation, resulting in a decrease in the molecular beam flux reaching the bolometer, while the monomer does not dissociate in the excited vibrational state and thus delivers extra internal energy to the detector. The resolution of the method is clearly very high, due to the nearly orthogonal crossing of the laser and the collimated molecular beam. In fact, the instrumental linewidth is small enough to enable us to measure the homogeneous contribution to the lineshape due to dissociation of the complex. In this case the lifetime is found to be 3.6 ns [47]. As it turns out, the lifetimes of these complexes fall in a very favorable range, namely short enough to be measured with the available resolution and not so short that the rotational fine structure is lost. As we will see in the following section, this enables us to study the dissociation dynamics of these complexes at the rotational quantum state resolved level. In addition, the lifetime is sufficiently short so that the molecules do not move appreciably from the point of excitation to dissociation, which is also important in the state-to-state experiments discussed in the following section.

The rare gas - diatomic complexes have become model systems in the study of wide amplitude motions owing, in part, to the large spectroscopic data base that is now available for such systems [39,48]. Assuming the diatom is rigid, the vibrational motions of the complex occur on a two dimensional intermolecular surface corresponding to the intermolecular distance and the bend angle. The fact that the theoretical methods are well developed for solving such 2D problems for the bound state energies has also contributed to the advances in this area [28,49,50]. This approach for determining intermolecular potential energy surfaces is particularly powerful in view of the fact that excitation of various intermolecular bending and stretching states of the complexes provides sensitivity to different parts of the surface. This is a result of the fact that these systems are indeed floppy, so that the vibrational wavefunctions are quite different from one intermolecular vibrational state to another and thus sample different regions of the surface.

Since the potential surfaces remain 2-dimensional for any linear molecule - rare gas complex, the same theoretical framework can be applied to the study of interactions between rare gas atoms and polyatomic linear molecules. Our interest in this regime comes from the need to know how to deal with interactions between extended molecular systems. In the limit of interactions between large molecules, it is obviously important to move beyond single center interaction potentials, more in the direction of atom-atom interactions, as is typically done in molecular modeling. Studies of the type considered here can therefore be extremely useful in testing the various approaches for distributing interactions using the detailed spectroscopic data.

An example of such a system is Ar-acetylene, which provides an interesting comparison with the acetylene-HF case discussed above. Figure 3 shows the experimental and calculated spectra associated with the asymmetric C-H stretching vibration of this "T-shaped" complex, which again gives rise to a perpendicular-type band. Although much of the spectrum resembles that of a rigid "T-shaped" molecule [51-53], the weak transitions that appear at the high frequency end of the spectrum cannot be reproduced by a rigid rotor calculation. These transitions correspond to excitation of a combination band involving the low frequency bending mode of the acetylene against the argon atom. In this case, the calculated spectrum is based upon a full two dimensional calculation using a potential energy surface that allows the acetylene sub-unit to execute this wide amplitude bending motion. In this case, the collocation approach [50,54] was used to solve for the energy levels and wavefunctions of a two dimensional non-rigid rotor and the transition intensities were determined by calculating the appropriate electric dipole matrix elements. The parameters defining the potential energy surface were then varied to reproduce the spectrum [53]. As expected, the intensity of this combination band, which correlates in the free rotor limit with a rotational excited state of acetylene sub-unit, increases with the amplitude of the bend. The stiffer bond in acetylene-HF results in a smaller bending amplitude and thus a combination band that is too weak to be observed in our experiment.

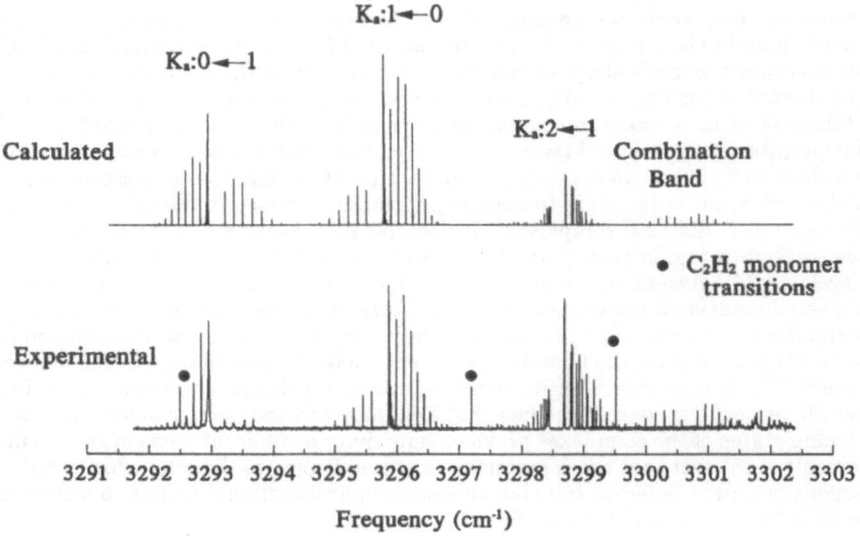

Figure 3: The infrared spectrum of the asymmetric C-H stretching vibration in C_2H_2-Ar. The calculated spectrum was obtained from a full 2-D dynamical calculation and clearly shows features that are sensitive to the wide amplitude bending motion.

For the 2-dimensional systems discussed above, the methods for calculating the bound state energies and wavefunctions are now efficient enough to permit least squares fitting of the potential to the data. However, the extension of this approach to larger systems is by no means straightforward. For example, the intermolecular potential between two rigid linear molecules is four dimensional, while the general case of two rigid non-linear molecules is six dimensional. These multidimensional surfaces present new challenges for both experiment and theory. On the theoretical side, there is the difficulty associated with solving the multidimensional quantum mechanical problem in a timely fashion, such that the potential can be adjusted to fit the data. The experimental issue is one of uniqueness. As the dimensionality of the potential surface increases, the number of features that need to be defined increases dramatically and the question becomes, "Is there enough data to uniquely define the potential in so many dimensions?". To a certain extent, this problem can be addressed by exploring many different intermolecular vibrational states of the complex, each one providing a somewhat different view of the potential energy surface, as discussed above. In fact, far infrared spectroscopy is now being used in this way with considerable success [55,56]. Unfortunately, the experimentalist does not have complete freedom in exploring the potential in this way due to the fact that many of the vibrational states of interest have very small transition moments connecting them to the populated states. As a result, the practical situation is that for many systems the data base is insufficient to constrain the entire potential energy surface. In such cases, it is often necessary to focus on only a few internal coordinates, namely those that correspond to the most floppy motions. Fortunately, the spectroscopy cooperates in this regard, owing to the fact that it is often this motion that results in deviations from rigid rotor like behavior.

Consider, for example, the case of the water-carbon dioxide complex [57], which has a planar, symmetric equilibrium structure with the oxygen of the water attracted to the carbon atom. *Ab initio* calculations [57] show that the only feasible wide amplitude tunneling motion corresponds to internal rotation of the water sub-unit (C=14.512 cm^{-1}) about the symmetry axis of the complex. The carbon dioxide is much to heavy (B=0.395 cm^{-1}) to tunnel through this barrier and all other internal rotational motions have high barriers. As a result, the problem can be easily separated into one dealing with the overall rotation of the complex, treated as a pseudo-rigid rotor, and one dealing with the tunneling of the water rotor through the barrier. The latter motion shifts of the rigid rotor subbands to higher and lower frequencies for even and odd K_a'', respectively. The magnitude of this shift provides information on the height of the barrier along this coordinate. Once again, if the spectroscopy of the corresponding bending excited states can also be obtained, then more detailed information on the shape of the barrier can be obtained.

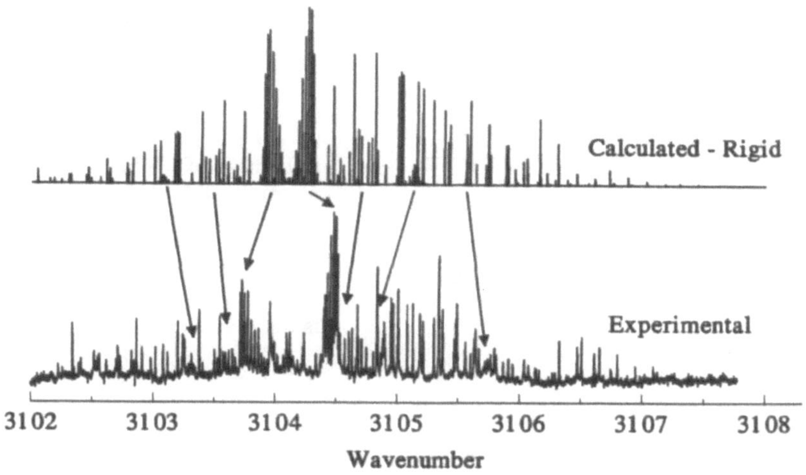

Figure 4: An opto-thermal spectrum of the CO_2-C_2H_4 complex showing the effect of internal rotation. For comparison, the calculated spectrum of the corresponding rigid complex is also shown.

In view of the fact that all of the intermolecular vibrational degrees of freedom are rather low in frequency, there is no reason to think that this separation will always be valid. Indeed, the full range of cases must exist, namely from that considered above where only one tunneling coordinate exists and is separable from overall rotation, to the limit where all of the intermolecular vibrational motions are strongly coupled to overall rotation of the complex. Consider for a moment an intermediate case where two internal degrees of freedom are strongly mixed in a way that makes the separation of overall rotation and internal motion invalid. In a recent study we have identified such a system, namely the ethylene-carbon dioxide complex. In this case the carbon dioxide lies above the plane of the ethylene in a symmetric fashion. The complication arises from the fact that the B rotational constant of the carbon dioxide (B=0.395 cm^{-1}) is of the same order of magnitude as the C rotational constant of ethylene (C=0.828 cm^{-1}). As a result, if the barrier to internal rotation about this axis of the complex is low, both monomer units will be free to internally rotate. The problem can then no longer be viewed as a rotating frame upon which the lighter monomer unit is tunneling, as was the case for CO_2-H_2O. The result is an unusual pattern of sub-bands for this system, as illustrated in Figure 4, which compares the experimental perpendicular-type spectrum, correlating with the ν_9 band of the ethylene monomer, with a rigid rotor spectrum based upon the parallel geometry. The seemingly random pattern of subband shifts is quite different from that of the water-carbon dioxide system discussed above.

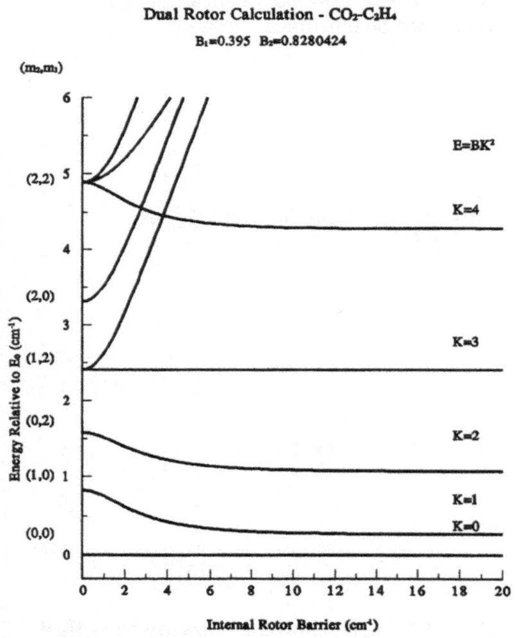

Figure 5: A double rotor correlation diagram for the CO_2-C_2H_4 complex showing the lowest few internal rotor states as a function of the barrier to internal rotation. The states that rapidly move to high energy correlate with the excited bending states of the complex. For a barrier greater than 20 cm^{-1} the internal rotation is quenched.

This pattern can be explained, however, by considering the correlation diagram shown in Figure 5. This variational calculation was carried out for the case of two rotors sharing a common axis. The left extreme of the graph shows the free rotor limit, where the barrier to internal rotation is zero. Note that only even carbon dioxide states exist, owing to the associated nuclear spin statistics. As the barrier increases, the levels shift and eventually converge on a simple rigid rotor set of "K type" levels. It is now clear that the unusual pattern of levels arises from the fact that the two sets of rotor levels are intertwined. The pattern observed in the spectrum can then be reproduced by choosing the correct value of the barrier height, in this case approximately 8 cm^{-1}. Such a low barrier is necessary to explain the fact that the internal monomer rotations, which involve the motion of heavy atoms, are feasible.

Although the ultimate goal of these spectroscopic studies of weakly bound complexes is to obtain the full multidimensional potential surface, there are clearly many cases where the data is insufficient to uniquely define such. It is clear from the present discussion, however, that some progress can still be made by making appropriate separations of the motions. As the next section illustrates, all of this spectroscopic data on the associated intermolecular dynamics on these multidimensional surfaces provides a firm starting point for studying the predissociation dynamics of these complex.

So far in this section we have discussed the use of infrared spectroscopy in the characterization of the rotational and vibrational states of these weakly bound complexes. We now turn our attention from characterization to control. By control we mean the ability to modify the states of a molecule using an external influence in such a way as to provide new information on either the potential energy surfaces involved or the dynamics on these surfaces. The fact that the rigid rotor levels are not always the optimum states of the system to study can be illustrated by considering that molecular rotation tends to average over orientation. As a result, properties of the system that are dependent upon its orientation in space can only be measured in an average way by studying the system in its free rotor states. A wide variety of approaches have been developed for imposing either alignment [58-64] or orientation [65-72] on a molecular ensemble. The approach considered here is straightforward in principle, namely that of orienting molecules through the interaction of their electric dipole with a large external electric field. This "pendular state" method has been discussed in detail previously [70]. Its viability results from the fact that molecules can be strongly cooled in a molecular beam so that their rotational energy is small with respect to the interaction between the dipole moment of the molecule and the electric field. In this limit, overall rotation of the molecule is quenched and the molecules undergo pendular-type motion about the field direction. The situation is particularly favorable for the case of many weakly bound complexes which have large dipole moments and small rotational constants [72,73]. In the next section we discuss the usefulness of this orientational control in the study of molecular photodissociation.

Before leaving this section we need to address one final issue dealing with the external control of molecular states. In the above discussion we implicitly made the assumption that the external field only effected the orientation of the molecule in space. In effect, the field served to transform the molecule fixed z axis into the space fixed frame. However, since many of these complexes are extremely floppy, the possibility also exists for making the interaction between the dipole moments of the monomer units in the complex and the field larger than the anisotropy associated with the motion of one monomer relative to the other. In such cases, the external field will also result in a deformation of the parent molecular complex. This opens up the interesting possibility of exploring the intermolecular potential by distorting the complex from its "equilibrium" geometry. More precisely, we would be changing the vibrational averaging over the potential by application of the field. This may also provide us with a method for studying the influence of structure on the vibrational predissociation dynamics. We must emphasize that these influences will only be significant for very floppy systems, since the interaction energy between typical dipoles (1-5 D) and reasonably achieved electric fields (150 kV/cm) is only of the order of 10 cm^{-1}. The first successful demonstration of this effect is for the HF dimer. Figure 6 shows a series of infrared spectra recorded as a function of applied electric field. At zero field the spectrum has two P and two R branches, this doubling arising from the interchange tunneling motion of HF dimer [74]. As the electric field is increased, however, the two minina are no longer equivalent, since one corresponds to the dipole being oriented parallel to the electric field, while interchange results in the dipole pointing against the field. As a result, the tunneling splitting is observed to decrease with increasing electric field, until at the highest fields the tunneling is quenched. Work is presently underway to solve for the multidimensional problem in an external field to ascertain the sensitivity this type of experiment has to various features of the potential energy surface. These results do show, however, that experiments of this type can be used to obtain some control over the intermolecular dynamics.

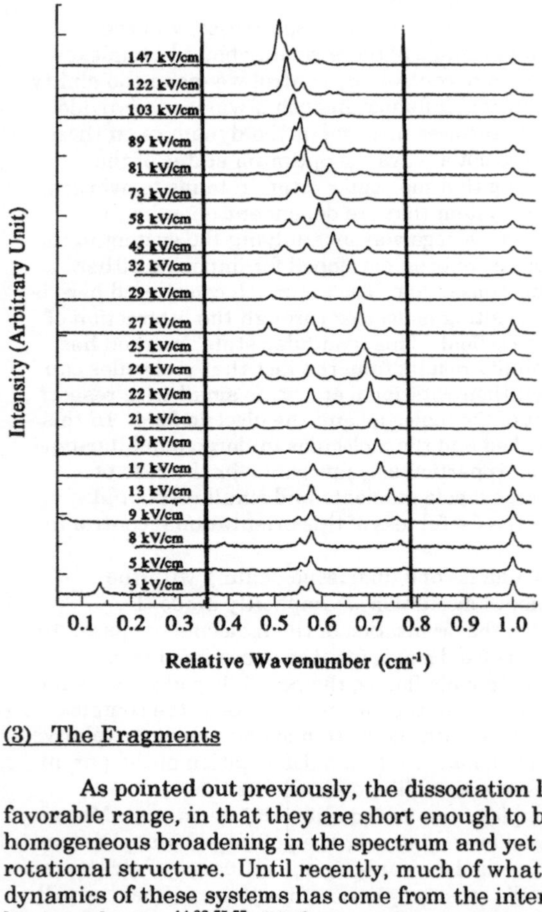

Figure 6: Pendular state spectra of the HF dimer as a function of electric field showing the quenching of the tunneling motion.

<u>(3) The Fragments</u>

As pointed out previously, the dissociation lifetimes of these complexes fall in a favorable range, in that they are short enough to be determined by measurement of the homogeneous broadening in the spectrum and yet are not so short as to blur the detailed rotational structure. Until recently, much of what we have learned about the dissociation dynamics of these systems has come from the interpretation of the homogeneous linebroadening [14,22,75-77]. Unfortunately, experiments of this type always leave one to speculate concerning (1) the final outcome of the photochemical event and (2) the detailed dynamical events leading up to dissociation. To address these questions we clearly need more sophisticated experiments that allow us to determine the condition of the photofragments. The "ideal" for such an experiment is one where the initial parent state is fully defined in terms of its velocity, orientation, vibrational and rotational state and where the internal and translational states of all the products are also measured. This includes not only the scalar properties, such as the product internal state distributions, but also the various scalar and vector correlations which may exist [59,61,78].

A number of methods have recently been developed and applied to the study of the final state distributions resulting from infrared photodissociation of complexes [16,79-84]. Since theoretical calculations of these dynamical quantities are extremely computationally intensive [85-88] our best opportunity for making comparisons between experiment and theory will come as we develop experimental methods that can be applied to relatively simple systems. To date these include a number of rare gas - diatom systems studied in electronically excited states [89-92] and the ground state studies of HF dimer [93-95] and H_2,D_2-HF [86-88,96]. The latter four atom systems already push the limits of the present theoretical methods.

The approach we have taken to study the state-to-state dissociation of these complexes on the ground electronic state potential is to measure the photofragment angular distributions using the optothermal technique. The experimental apparatus used for this purpose is shown in Figure 7. The molecular beam source is rotated about the photolysis point so that the photofragment angular distribution can be measured using the bolometer detector. Given that the spectroscopy of the complex is already assigned, the pump laser (a cw color center laser) can be tuned into resonance with a single ro-vibrational transition. In this way only complexes in a well defined ro-vibrational state are dissociated, ensuring the selection of the initial state of the parent complex.

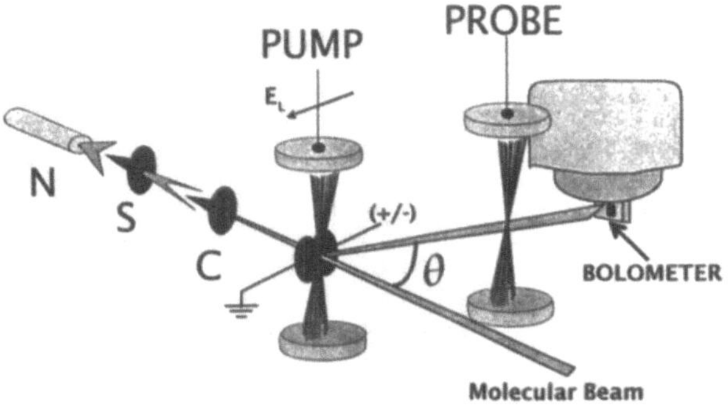

Figure 7: A schematic diagram of the apparatus used to measure photofragment angular distributions. Two F-center lasers are used, one to pump the complex and the second to probe the resulting fragments.

The detailed shape of the resulting angular distributions is determined by a large number of factors, including many dealing with the geometry of the apparatus [93]. The important dynamical information is contained in the recoil energy of the photofragments, which can be determined if the individual channels can be resolved. In general, the photofragments are ejected over a range of angles in the center-of-mass frame so that, even for a single recoil energy, the laboratory frame angular distribution is somewhat broad. Nevertheless, each photofragment channel gives rise to a "peak" in the laboratory frame angular distribution owing to the fact that there is a wide range of center-of-mass angles that contribute to the signal at laboratory angles corresponding to the recoil velocity vector being approximately orthogonal to the stream velocity of the parent molecule [81]. When the photofragment state density is low, which is to say that there are very few states of the monomer fragments energetically (or dynamically) accessible, the resulting peaks in the laboratory frame angular distribution can be resolved. In such cases, their relative intensities provide direct information on the probabilities of populating these states, while their angular position determines the corresponding recoil kinetic energy.

The difficulty with the above approach is that as soon as the density of states becomes moderately high the assignment of the peaks in the angular distribution becomes ambiguous and eventually the structure disappears entirely. Under these conditions, individual final states cannot be assigned and only the average kinetic energy release can be determined. The probe laser shown in Figure 7 has been added to help overcome this difficulty. By tuning the probe laser through the various monomer transitions, the relative populations in the corresponding lower states can be determined. Here again, detection is done by the bolometer, which in this case detects the increase in energy of the fragments due to their vibrational excitation by the probe laser. In this way the relative state-to-state probabilities can be determined without the need for resolving the peaks in the angular distribution.

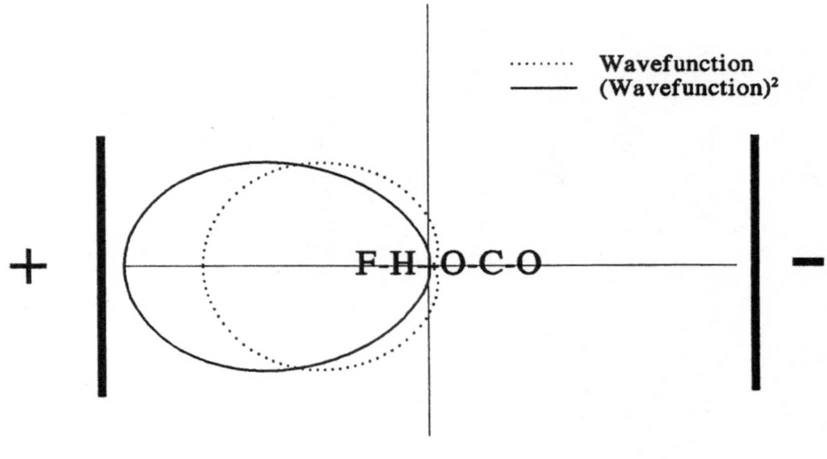

Figure 8: The orientation imposed on CO_2-HF by an external electric field.

For binary complexes formed from monomers of unequal mass there is the added complication that, due to conservation of momentum, the two fragments scatter to different angles in the laboratory frame. As a result, each photodissociation channel gives rise to two peaks in the angular distribution. This added complexity makes it even more difficult to obtain a unique assignment of the angular distribution for such cases. It is here that we can make good use of the pendular orientation method discussed above. This is done by applying an orienting field to the photolysis region, as shown in Figure 7, so that the molecules can be oriented prior to dissociation. Consider, for example, the case shown in Figure 8. The solid curve is a cut through the square of the lowest energy (m=0) pendular state wavefunction (proportional to the probability distribution) for the case of a linear molecule with rotational constant B and $\omega = \mu(D)\times\mathbf{E}(kV)\times 0.0167916/\mathbf{B}(cm^{-1})=5$. For the case of CO_2-HF shown in the figure, this corresponds to an electric field of 8.6 kV/cm, which is easily achieved. It is clear from this figure that under these conditions the probability of a molecule being antioriented with the electric field at this value of ω is very low. It is important to note that pendular states with $m\neq 0$ are less well oriented. As a result, for the best possible orientation some form of state selection is necessary. In the present case this is easily achieved by making use of the spectroscopy to select only m=0. For a linear molecule, that in zero field would not show a Q branch in the infrared spectrum, the selection rules appropriate for the laser polarization direction parallel to the electric field ($\Delta m=0$) are such that the transition associated with the pendular states appear near the vibrational origin. As a result, they are easily separated from all of the other transitions in the spectrum. This is illustrated in Figure 9 which shows such a "Q branch-like" feature for the case of N_2-HF at an electric field of 16 kV/cm.

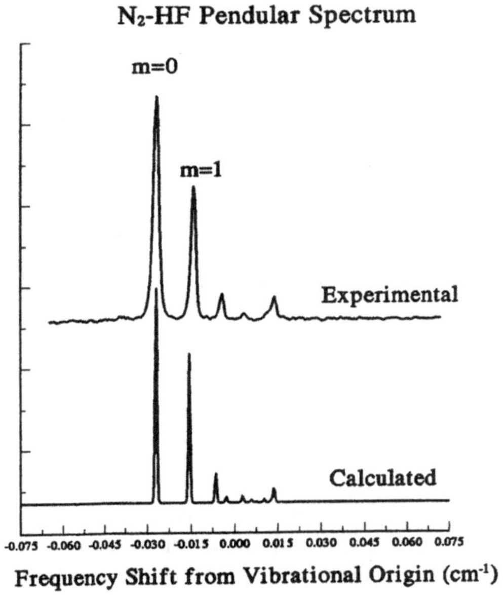

Figure 9: Pendular state spectra of the linear N_2-HF complex.

Returning to the schematic in Figure 8, if we now assume that the complex dissociates in such a way that the fragments essentially recoil along the axis of the parent molecule, it is obvious that the HF fragment will be ejected towards the positive electrode, while the CO_2 fragment will recoil towards the negative electrode. Thus two separate angular distributions can be measured, one for each fragment, by rotating to either side of the molecular beam. This can be very helpful in uncovering new features in the angular distribution that would normally be lost due to the overlapping of the two fragment distributions. This is illustrated for the case of CO_2-HF in Figure 10. This zero field angular distribution shows two prominent features that can be explained in terms of the mass difference between the two fragments, the lighter HF fragment giving rise to the large angle peak and the CO_2 producing the peak at small angles. In an earlier study [97] we were able to confirm this assignment by positioning the bolometer to detect the HF fragment and using the second F-center laser to probe the various rotational states of this fragment. We found that a single HF rotational state was populated in this case, namely j_{HF}=6. A reasonable estimate of the dissociation energy of the complex suggested, by conservation of energy, that the CO_2 fragment was produced in the (00^01) vibrational state. Since both the ν_3 mode of CO_2 and the HF stretching vibrations are strongly dipole allowed, we were led by this result to the conclusion that the dissociation of this complex was dominated by a dipole-dipole coupled intermolecular V-V energy transfer process that preferentially produced the CO_2 fragment in only (00^01) and the HF in j=6.

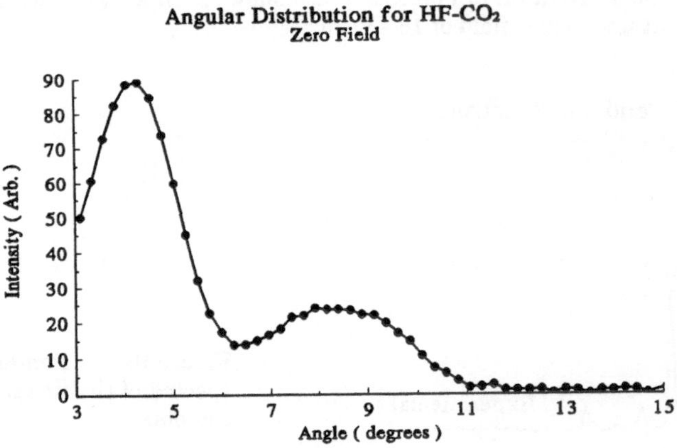

Figure 10: A zero field angular distribution for the CO_2-HF complex showing the angular separation of the HF and CO_2 fragments.

More recent experiments using the pendular state approach reveals that the situation is more complex than the above picture implies. Figure 11 shows an angular distribution corresponding to detection of only the CO_2 fragment. In agreement with the zero field result, there is a large peak at small angles which results from the j_{HF}=6 channel

However, it is now clear that the HF fragment peak that is present in the zero field experiment obscured other structure in the angular distribution which can only come from other channels. By eliminating the overlap between the two fragments it is now clear that there are at least three different channels contributing to the dynamics. Preliminary fits to this data suggest that at least two CO_2 vibrational states and three HF rotational states are involved. As our attention turns more towards complexes with higher densities of states, this use of the pendular state method will clearly be very important.

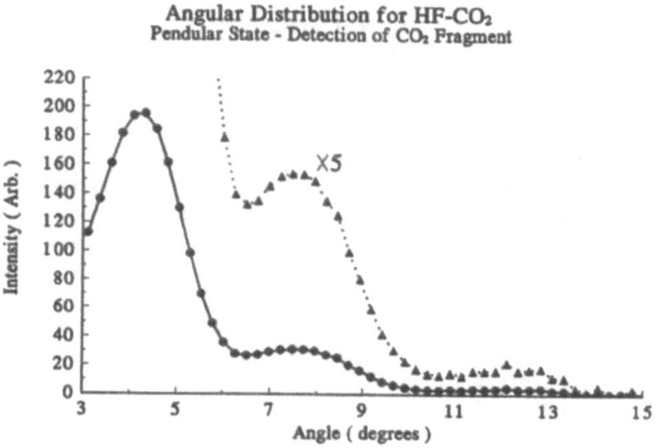

Figure 11: A pendular state angular distribution for CO_2-HF corresponding to detection of only the CO_2 fragment. From this it is clear that there are at least three open channels that contribution to the dissociation process.

A more fundamentally important application of this pendular state method is in the determination of the nature of the photofragment recoil distribution. In the above discussion we assumed that the molecules dissociated along the axis of the parent molecule. This axial recoil approximation is used routinely in the description of photodissociation, despite the fact that very little is actually known about the molecular frame angular probability distributions ($f(\theta_m,\phi_m)$ [98-100]). The problem arises from the fact that the nature of the alignment produced by single photon excitation with a linearly polarized laser is such that the laboratory frame angular distributions are only sensitive to the P_0 and P_2 terms in the Legendre expansion of the true molecule fixed frame angular distribution [98-100]. For such cases, the best that can be done is to measure a single anisotropy parameter, normally designated β in the equation for the laboratory angular distribution [98-100]:

$$I(\theta_l, \phi_l) = \frac{1+\beta P_2(\cos\theta_l)}{4\pi} \tag{1}$$

If the excitation process is not rotationally state selective, even this single anisotropy parameter β can be diminished by overall rotation of the excited molecules. Although this limitation can be partially overcome by making use of multiphoton methods, as best illustrated in the photoelectron literature [101-104], there have been no applications of this latter approach to photodissociation.

The situation is quite different for the case of an m=0 selected pendular state. Since the pendular orientation distribution can be made arbitrarily sharp by going to larger fields, the laboratory frame measurements can become sensitive to the higher and odd order moments of the $f(\theta_m,\phi_m)$ function. A formal treatment of the problem, which involves transforming the pendular state wavefunctions and molecule fixed frame distributions into the laboratory fixed frame [105], yields a modified form of Eq. 1, namely:

$$I(\theta_l, \phi_l) = \sum_{k=0}^{\infty} b_k P_k(\cos\theta_l) A_k \qquad \text{Eq. 2}$$

where the values of A_k are determined by the strength of the electric field and the properties of the given system. As the electric field is increased the higher order A_k terms become significant, providing sensitivity to the corresponding Legendre terms in the expansion of the $f(\theta_m,\phi_m)$ function. The b_k's are the expansion coefficients that can be thought of as generalized β parameters, one for each moment of the distribution.

These effects can be seen more clearly in graphical form. Figure 12 shows a series of laboratory frame (translating with the center-of-mass of the complex) angular distributions $(I(\theta_l,\phi_l))$ as a function of the electric field. The bold line shows the $f(\theta_m,\phi_m)$ function that was assumed in these calculations. At low fields, the m=0 state is approximately spherically symmetric so that, even though the molecule fixed frame distribution has angular structure, the resulting laboratory frame angular distribution is spherically symmetric. As the electric field is increased, the laboratory frame distribution begins to show the features that are present in the $f(\theta_m,\phi_m)$ function.

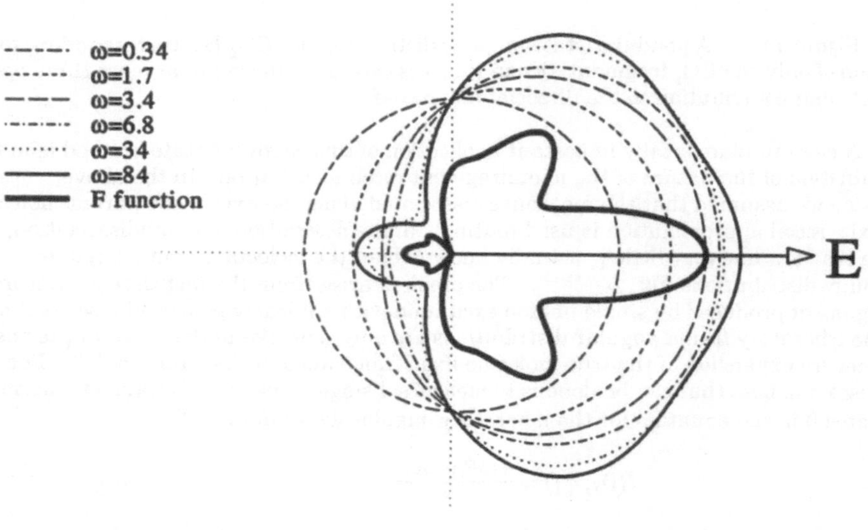

Figure 12: Calculated angular distributions as a function of electric field for f(θ,φ) (the F function) shown in the figure.

The calculations shown in Figure 12 clearly show that the pendular state method can provide sensitivity to features in the photofragment angular distribution that are not available from the more conventional laser polarization alignment experiments. The question we now wish to address is, "How will these effects be manifested in various types of experiments?". Let us begin with the angular distribution experiments considered above. The first experimental evidence for this comes from our recent study of the N_2-HF complex [106] where the state-to-state rotational/vibrational probability distributions have been determined by fitting the angular distribution obtained at zero field. These probabilities were then used to calculate the pendular state angular distributions at a given electric field, which were then compared with the experimental measurements. Although the results were in qualitative agreement, the experimental distributions were not reproduced in this way to within the experimental uncertainty. These differences arise from the fact that the pendular state experiment is indeed sensitive to the detailed shape of the $f(\theta_m,\phi_m)$ function, while the zero field results are not. For a detailed discussion of the implication of these differences the reader is referred to the original work [106].

Doppler spectroscopy has been used extensively in the determination of anisotropy parameters in molecular photodissociation [60,78,84,107] so that it is worth considering the implication of the present results in this type of experiment. Figure 13 shows the Doppler profiles calculated assuming a particular laser geometry for two different $f(\theta_m,\phi_m)$ functions, namely a truncated Legendre expansion of a delta function (F1) and the function (F2) shown in Figure 12. In these calculations we assumed that the velocity of recoil (**v**) is perpendicular to the rotational angular momentum vector of the probed fragment (**J**) and **J** is parallel to the transition moment of the parent molecule (**μ**). In addition the propagation direction of the probe laser is chosen to be parallel to the electric field direction of the pump and the electric field direction of the pump laser is assumed to be parallel to the DC electric field. The differences between the zero field and pendular results are clearly evident in these Doppler profiles, reflecting the shape of the differential photofragment distribution defined by $f(\theta_m,\phi_m)$. To date, however, there have been no experimental verifications of these effects in Doppler spectroscopy experiments.

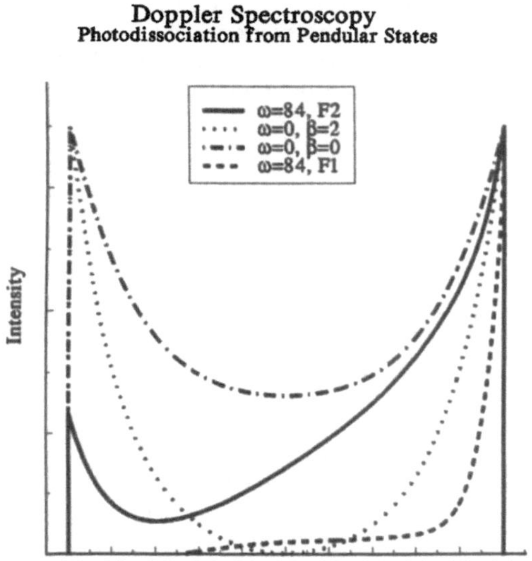

Figure 13: Calculated Doppler profiles corresponding to dissociation from pendular states. These are compared with those calculated at zero electric field.

The importance of this type of measurement can be appreciated if one considers that the electric field effectively changes the relative contributions to the angular distribution from the various partial waves contributing to the dissociation process. As higher j states are mixed together to form the pendular state, they contribute with the appropriate phases, to define the photofragment angular distribution function. Experiments that provide sensitivity to such phase information, revealed by the associated quantum mechanical interferences, although rare, are needed to provide the most complete comparisons with theory [108].

(4) Summary

In this overview we have discussed some of the contributions that have been made by optothermal infrared spectroscopy to our understanding of the structure, and more generally the intermolecular potential energy functions, of weakly bound complexes. When used in conjunction with a rotatable molecular beam apparatus, these spectroscopic methods can also provide detailed information on the final state distributions of the fragments, from which it is possible to obtained new insights into the nature of the dissociation process. We have also considered the control of the orientation, structure and dynamics of these complexes using externally applied electric fields. These method can be used to obtained new information on the photofragment angular distributions in the molecule fixed frame. At large fields the molecules are essentially held fixed in the laboratory frame, eliminating much of the angular averaging that is usually associated with molecular rotation. These methods should provide us with the experimental tools needed to measure differential photodissociation cross sections analogous to the differential scattering cross sections determined from cross beam experiments. The possibility of modifying the vibrational dynamics of these systems in a controlled manner has been demonstrated, but it remains to be seen just how useful an approach this will become.

Acknowledgments

The author is grateful to the students, postdoctoral associates and visitors with whom he has had the pleasure of working and who have made this work possible. Support for this research is gratefully acknowledged from the National Science Foundation (Grant No. CHE-93-18936) and the Donors of the Petroleum Research Fund (administered by the ACS).

References

(1.) Zewail, A.H.; *Science* **1988,** *242,* 1645.
(2.) Rosker, M.J.; Dantus, M.; Zewail, A.H.; *Science* **1988,** *241,* 1200.
(3.) Wardlaw, D.M.; Marcus, R.A.; *Adv. Chem. Phys.* **1987,** *70,* 231.
(4.) Parmenter, C.S.; *Faraday Discuss. Chem. Soc. ,(75)* **1983,** *7-22,*
(5.) Parmenter, C.S.; *J. Phys. Chem.* **1982,** *86,* 1735.
(6.) Smalley, R.E.; *J. Phys. Chem.* **1982,** *86,* 3504.
(7.) Felker, P.M.; Zewail, A.H.; *J. Chem. Phys.* **1985,** *82,* 2961.
(8.) Go, J.; Bethardy, G.A.; Perry, D.S.; *J. Phys. Chem.* **1990,** *94,* 6153.
(9.) Uzer, T.; *Physics Reports* **1991,** *2,* 73.

(10.) Stuchebrukhov, A.A.; Marcus, R.A.; *J. Chem. Phys.* **1993,** *98,* 6044.
(11.) Pique, J.; Chen, Y.; Jonas, D.M.; Lundberg, J.K.; Hamilton, C.E.; Adamson, G.W.; Silbey, R.J.; Field, R.W.; *AIP Conf. Proc. ,Volume Date 1988* **1989,** *191,* 673.
(12.) Jucks, K.W.; Miller, R.E.; *J. Chem. Phys.* **1988,** *88,* 6059.
(13.) Dayton, D.C.; Miller, R.E.; *Chem. Phys. Lett.* **1989,** *156,* 578.
(14.) Dayton, D.C.; Miller, R.E.; *Chem. Phys. Lett.* **1988,** *143,* 181.
(15.) Butz, K.W.; Catlett, D.L., Jr.; Ewing, G.E.; Krajnovich, D.; Parmenter, C.S.; *J. Phys. Chem.* **1986,** *90,* 3533.
(16.) Dayton, D.C.; Jucks, K.W.; Miller, R.E.; *J. Chem. Phys.* **1989,** *90,* 2631.
(17.) Lovejoy, C.M.; Nesbitt, D.J.; *J. Chem. Phys.* **1990,** *93,* 5387.
(18.) Nesbitt, D.J.; Lovejoy, C.M.; Lindeman, T.G.; ONeil, S.V.; Clary, D.C.; *J. Chem. Phys.* **1989,** *91,* 722.
(19.) Lovejoy, C.M.; Nelson, D.D., Jr.; Nesbitt, D.J.; *J. Chem. Phys.* **1987,** *87,* 5621.
(20.) Lovejoy, C.M.; Nelson, D.D., Jr.; Nesbitt, D.J.; *J. Chem. Phys.* **1988,** *89,* 7180.
(21.) Fraser, G.T.; Pine, A.S.; *J. Chem. Phys.* **1989,** *91,* 633.
(22.) Pine, A.S.; Fraser, G.T.; *J. Chem. Phys.* **1988,** *89,* 6636.
(23.) Klemperer, W.; *Springer Ser. Chem. Phys.* **1978,** *3,* 398.
(24.) Klemperer, W.; *NATO ASI Ser. ,Ser. C* **1987,** *212,* 455.
(25.) Jucks, K.W.; Huang, Z.S.; Miller, R.E.; Fraser, G.T.; Pine, A.S.; Lafferty, W.J.; *J. Chem. Phys.* **1988,** *88,* 2185.
(26.) Lovejoy, C.M.; Nesbitt, D.J.; *J. Chem. Phys.* **1987,** *87,* 1450.
(27.) Hutson, J.M.; *Annu. Rev. Phys. Chem.* **1990,** *41,* 123.
(28.) Hutson, J.M.; *J. Chem. Phys.* **1988,** *89,* 4550.
(29.) Cohen, R.C.; Saykally, R.J.; *J. Phys. Chem.* **1992,** *96,* 1024.
(30.) Cohen, R.C.; Saykally, R.J.; *J. Chem. Phys.* **1991,** *95,* 7891.
(31.) Cohen, R.C.; Saykally, R.J.; *Annu. Rev. Phys. Chem.* **1991,** *42,* 369.
(32.) Kleiner, I.; Fraser, G.T.; Hougen, J.T.; Pine, A.S.; *J. Mol. Spectrosc.* **1991,** *147,* 155.
(33.) Fraser, G.T.; Pine, A.S.; *J. Chem. Phys.* **1989,** *91,* 3319.
(34.) Nesbitt, D.J.; Lovejoy, C.M.; *J. Chem. Phys.* **1990,** *93,* 7716.
(35.) McIlroy, A.; Lascola, R.; Lovejoy, C.M.; Nesbitt, D.J.; *J. Phys. Chem.* **1991,** *95,* 2636.
(36.) Nesbitt, D.J.; Lovejoy, C.M.; *Faraday Discuss. Chem. Soc.* **1988,** *86,* 13.
(37.) Gough, T.E.; Miller, R.E.; Scoles, G.; *Appl. Phys. Lett.* **1977,** *30,* 338.
(38.) Miller, R.E.; *Science* **1988,** *240,* 447.
(39.) Lovejoy, C.M.; Nesbitt, D.J.; *J. Chem. Phys.* **1989,** *91,* 2790.
(40.) Farrell, J.T.; Sneh, O.; Knight, A.E.W.; Nesbitt, D.J.; unpublished.
(41.) Dvorak, M.A.; Reeve, S.W.; Burns, W.A.; Grushow, A.; Leopold, K.R.; *Chem. Phys. Lett.* **1991,** *185,* 399.
(42.) Elrod, M.J.; Host, B.C.; Steyert, D.W.; Saykally, R.J.; Mol. Phys. **1973,** 79,245.
(43.) Hutson, J.M.; *J. Chem. Phys.* **1992,** *96,* 6752.
(44.) Nesbitt, D.J.; Child, M.S.; Clary, D.C.; *J. Chem. Phys.* **1989,** *90,* 4855.
(45.) Nesbitt, D.J.; Naaman, R.; *J. Chem. Phys.* **1989,** *91,* 3801.
(46.) Herzberg G. *Molecular spectra and molecular structure. II. Infrared and raman spectra of polyatomic molecules;* Van Nostrand Reinhold: New York, 1945.
(47.) Huang, Z.S.; Miller, R.E.; *J. Chem. Phys.* **1989,** *90,* 1478.
(48.) Saykally, R.J.; *Acc. Chem. Res.* **1989,** *22,* 295.
(49.) Hutson, J.M.; *J. Chem. Phys.* **1989,** *91,* 4455.
(50.) Peet, A.C.; Yang, W.; *J. Chem. Phys.* **1989,** *90,* 1746.
(51.) Hu, T.A.; Prichard, D.G.; Sun, L.H.; Muenter, J.S.; Howard, B.J.; *J. Mol. Spectrosc.* **1992,** *153,* 486.

(52.) DeLeon, R.L.; Muenter, J.S.; *J. Chem. Phys.* **1980,** *72,* 6020.
(53.) Bemish, R.J.; Block, P.A.; Pedersen, L.G.; Yang, W.T.; Miller, R.E.; *J. Chem. Phys.* **1993,** *99,* 8585.
(54.) Peet, A.C.; Yang, W.; Miller, W.H.; *J. Chem. Phys.* **1989,** *91(12),* 7537.
(55.) Cohen, R.C.; Saykally, R.J.; *J. Phys. Chem.* **1990,** *94,* 7991.
(56.) Cohen, R.C.; Saykally, R.J.; *J. Chem. Phys.* **1993,** *98,* 6007.
(57.) Block, P.A.; Marshall, M.D.; Pedersen, L.G.; Miller, R.E.; *J. Chem. Phys.* **1993,** *98,* 10107.
(58.) Gericke, K.H.; Klee, S.; Comes, F.J.; Dixon, R.N.; *J. Chem. Phys.* **1986,** *85,* 4463.
(59.) Vasudev, R.; Zare, R.N.; Dixon, R.N.; *J. Chem. Phys.* **1984,** *80,* 4863.
(60.) Loo, R.O.; Hall, G.E.; Haerri, H.P.; Houston, P.L.; *J. Phys. Chem.* **1988,** *92,* 5.
(61.) Houston, P.L.; *J. Phys. Chem.* **1987,** *91,* 5388.
(62.) Hall, G.E.; Sivakumar, N.; Ogorzalek, R.; Chawla, G.; Haerri, H.P.; Houston, P.L.; Burak, I.; Hepburn, J.W.; *Faraday Discuss. Chem. Soc.* **1986,** *82,* 13.
(63.) Thoman, J.W.; Chandler, D.W.; Parker, D.H.; Janssen, M.H.M.; *Laser Chem.* **1988,** *9,* 27.
(64.) Chandler, D.W.; Houston, P.L.; *J. Chem. Phys.* **1987,** *87,* 1445.
(65.) Kramer, K.H.; Bernstein, R.B.; *J. Chem. Phys.* **1965,** *42,* 767.
(66.) Gandhi, S.R.; Xu, Q.; Curtiss, R.J.; Bernstein, R.B.; *J. Phys. Chem.* **1987,** *91,* 5437.
(67.) Beuhler, R.J.; Bernstein, R.B.; Kramer, K.H.; *J. Amer. Chem. Soc.* **1966,** *88:22,* 5331.
(68.) Choi, S.; Bernstein, R.; *J. Chem. Phys.* **1986,** *85,* 150.
(69.) Friedrich, B.; Herschbach, D.R.; *Nature* **1991,** *353,* 412.
(70.) Friedrich, B.; Pullman, D.P.; Herschbach, D.R.; *J. Phys. Chem.* **1991,** *95,* 8118.
(71.) Loesch, H.J.; Remscheid, A.; *J. Chem. Phys.* **1990,** *93,* 4779.
(72.) Rost, J.M.; Griffin, J.C.; Friedrich, B.; Herschbach, D.R.; *Phys. Rev. Lett.* **1992,** *68,* 1299.
(73.) Block, P.A.; Bohac, E.J.; Miller, R.E.; *Phys. Rev. Lett.* **1992,** *68,* 1303.
(74.) Pine, A.S.; Lafferty, W.J.; Howard, B.J.; *J. Chem. Phys.* **1984,** *81,* 2939.
(75.) Dayton, D.C.; Block, P.A.; Miller, R.E.; *J. Phys. Chem.* **1991,** *95,* 2881.
(76.) Jucks, K.W.; Miller, R.E.; *J. Chem. Phys.* **1987,** *86,* 6637.
(77.) Huang, Z.S.; Jucks, K.W.; Miller, R.E.; *J. Chem. Phys.* **1986,** *85,* 6905.
(78.) Hall, G.E.; Sivakumar, N.; Chawla, D.; Houston, P.L.; Burak, I.; *J. Chem. Phys.* **1988,** *88,* 3682.
(79.) Bohac, E.J.; Marshall, M.D.; Miller, R.E.; *J. Chem. Phys.* **1992,** *97,* 4890.
(80.) Casassa, M.P.; Stephenson, J.C.; King, D.S.; *NATO ASI Ser. ,Ser. B* **1988,** *171,* 367.
(81.) Marshall, M.D.; Bohac, E.J.; Miller, R.E.; *J. Chem. Phys.* **1992,** *97,* 3307.
(82.) Bohac, E.J.; Marshall, M.D.; Miller, R.E.; *J. Chem. Phys.* **1992,** *97,* 4901.
(83.) Hetzler, J.R.; Casassa, M.P.; King, D.S.; *J. Phys. Chem.* **1991,** *95,* 8086.
(84.) Shorter, J.H.; Casassa, M.P.; King, D.S.; *J. Chem. Phys.* **1992,** *97,* 1824.
(85.) Clary, D.C.; *J. Chem. Phys.* **1992,** *96,* 90.
(86.) Zhang, D.H.; Zhang, J.Z.H.; Bacic, Z.; *J. Chem. Phys.* **1992,** *97,* 927.
(87.) Zhang, D.H.; Zhang, J.Z.H.; Bacic, Z.; *J. Chem. Phys.* **1992,** *97,* 3149.
(88.) Zhang DH and Zhang JZH.; Chem. Phys. Lett. **1992**, 199, 187.
(89.) Waterland, R.L.; Lester, M.I.; Halberstadt, N.; *J. Chem. Phys.* **1990,** *92,* 4261.
(90.) Drobits, J.C.; Skene, J.M.; Lester, M.I.; *AIP Conf. Proc. ,Volume Date 1987* **1988,** *172,* 567.
(91.) Halberstadt, N.; Beswick, J.A.; Roncero, O.; Janda, K.C.; *J. Chem. Phys.* **1992,** *96,* 2404.

(92.) Halberstadt, N.; Serna, S.; Roncero, O.; Janda, K.C.; *J. Chem. Phys.* **1992,** *97,* 341.
(93.) Bohac, E.J.; Marshall, M.D.; Miller, R.E.; *J. Chem. Phys.* **1992,** *96,* 6681.
(94.) Zhang, D.H.; Zhang, J.Z.H.; *J. Chem. Phys.* **1993,** *99,* 6624.
(95.) Bemish, R.J.; Wu, M.; Miller, R.E.; *Faraday Discuss. Chem. Soc.* **1994,** *96,*
(96.) Bohac, E.J.; Miller, R.E.; *J. Chem. Phys.* **1993,** *98,* 2604.
(97.) Bohac, E.J.; Miller, R.E.; *Phys. Rev. Lett.* **1993,** *71,* 54.
(98.) Yang, S.; Bersohn, R.; *J. Chem. Phys.* **1974,** *61,* 4400.
(99.) Zare, R.N.; *Ber. Bunsenges. Phys. Chem.* **1982,** *86,* 422.
(100.) Greene, C.H.; Zare, R.N.; *Annu. Rev. Phys. Chem.* **1982,** *33,* 119.
(101.) Rudalph, H.; Mckoy, V.; *J. Chem. Phys.* **1989,** *91,* 2235.
(102.) Wiedmann, R.T.; Tonkyn, R.G.; White, M.G.; Wang, K.; Mckoy, V.; *J. Chem. Phys.* **1992,** *97,* 768.
(103.) Braunstein, M.; Mckoy, V.; Dixit, S.N.; *J. Chem. Phys.* **1992,** *96,* 5726.
(105.) Wu, M.; Bemish, R.J.; Miller, R.E.; *J. Chem. Phys.*, in press.
(106.) Bemish, R.J.; Bohac, E.J.; Wu, M.; Miller, R.E.; *J. Chem. Phys.*, in press.
(107.) Foy, B.R.; Casassa, M.P.; Stephenson, J.C.; King, D.S.; *J. Chem. Phys.* **1990,** *92,* 2782.
(108.) Chen, Z.; Brumer, P.; Shapiro, M.; *Chem. Phys. Lett.* **1992,** *198,* 498.

MOLECULAR PHOTOIONIZATION DYNAMICS AT THRESHOLD ENERGIES

Michael G. White

Chemistry Department, Brookhaven National Laboratory, Upton, NY 11973-5000

1 Introduction

Photoinduced molecular fragmentation processes offer a unique opportunity for studying collision dynamics in which the angular momenta, relative energies and spatial orientations of the interacting particles are constrained. Examples include neutral photodissociation which leads to chemical bond cleavage, and photoionization which results in the formation of a free electron and molecular cation. The half-collision view of photofragmentation processes provides the link between the highly detailed state-to-state information derived from such studies and the less tractable full collision problems such as reactive scattering and inelastic electron scattering. Central to these collision processes is the nature of the close range complex or "transition state" whose time-evolution determines the fragment yields and their state distributions. Except for a few experiments involving simple diatomic molecules, however, time-resolved measurements of the photoexcited collision complex are not usually feasible.[1] Instead, the dynamics of the fragmentation process are inferred from measurements of the asymptotic product state distributions. To obtain an adequate description of the fragmentation dynamics, it becomes necessary to completely characterize the quantum state distributions of the fragments as well as their relative translational energies and spatial correlations. Much recent effort in photodissociation dynamics involves the development of probe schemes, e.g. Doppler profile analyses, which can provide a maximum amount of product information (scalar and vector) with the minimum number of measurements and experimental configurations.[2]

Molecular photoionization, resulting in only two fragments (photoelectron and cation), should in principle be easier to characterize than most

E. Yurtsever (ed.), Frontiers of Chemical Dynamics, 43–77.

neutral photodissociation problems since all the linear recoil momentum is carried away by the photoelectron (a very good approximation for the low kinetic energies considered in this work). Conservation of energy then requires that the internal quantum state distribution of the molecular cation be given by the kinetic energy distribution of the photoelectron. This relationship is formally given by the familiar expression

$$KE = h\nu - (E_i^+ - E_g'')$$

where $h\nu$ is the ionizing photon energy and where i, g refer to specific rovibronic states of the molecular cation and neutral ground state. Photoelectron spectroscopy (PES) is simply based on the measurement of the photoelectron kinetic energy spectrum (KE) to derive the cation energy levels ($E_i^+ - E_g''$) and their population distributions. In the conventional application of PES, shown diagrammatically in Fig. (1a), a molecule is ionized with a fixed frequency VUV or x-ray photon. For these measurements, an energy dispersive electron analyzer using electrostatic or magnetic deflection, or time-of-flight (TOF) is used to measure the photoelectron kinetic energy spectrum. The remarkable contributions that "conventional" PES has made to our current understanding of photoionization dynamics, and molecular electronic structure in general is well documented elsewhere.[3–5] Just as remarkable, however, is that in spite of technological improvements in spectrometer systems over the last two decades, only the H_2^+ cation, which has a very large rotational constant ($B_e^+ = 66.2$ cm^{-1}), has been rotationally resolved by single-photon PES.[6–8] As will be seen later, the ability to obtain rotationally resolved photoionization spectra can provide a great deal of new information on molecular photoionization dynamics.

An alternative approach to PES which avoids many of the drawbacks of dispersive analyzers was developed by several groups in the early 1970's and is based on the detection of photoelectrons with near zero kinetic energy.[9] These researchers realized that photoelectrons with $KE \sim 0$ can be made to travel along nearly straight paths in a weak electric field (1–5 V/cm) and therefore can be collected with high efficiency (especially important for coincidence measurements) and with little additional energy spread. Furthermore, as such electrons are produced just above an ionization threshold, the energy resolution will ultimately be given by the bandwidth of the VUV radiation source. A photoelectron spectrum is obtained by scanning the VUV radiation and measuring the threshold photoelectron yield (see Fig. (1b)).

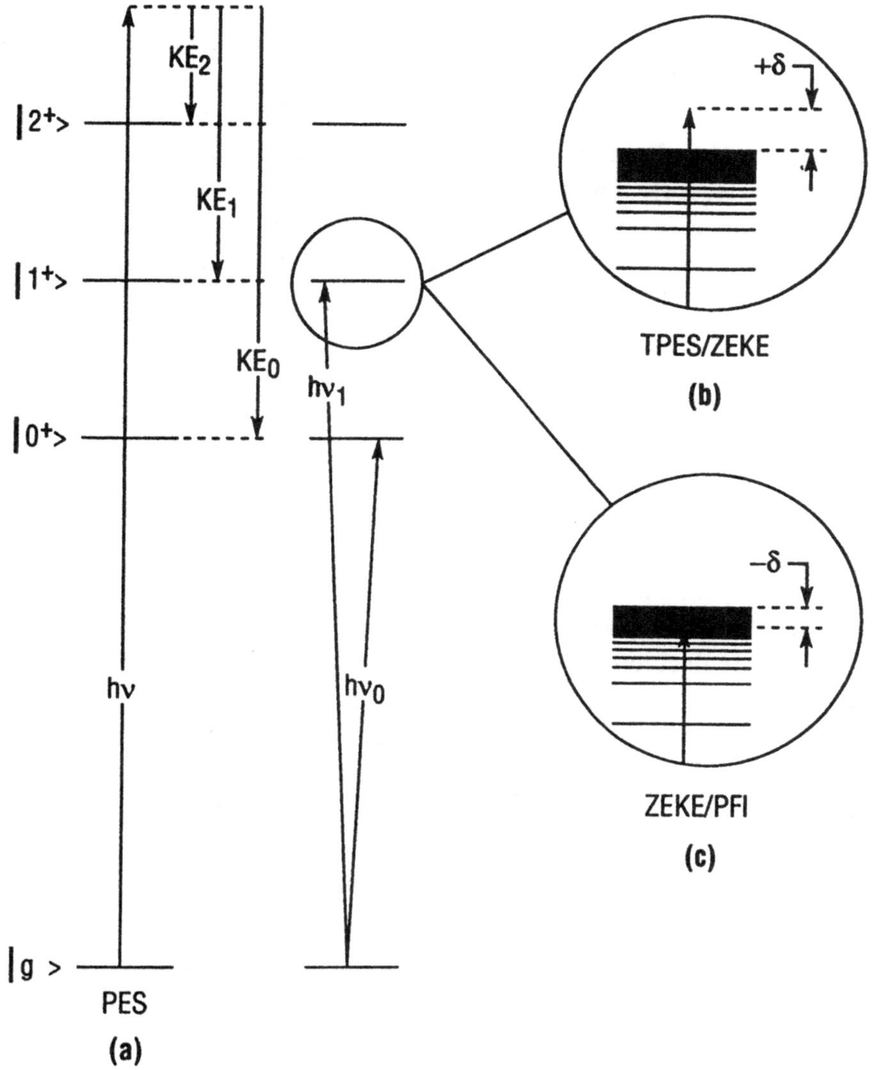

Figure 1: Schematic diagram illustrating various approaches to photoelectron spectroscopy: (a) conventional photoelectron spectroscopy (PES) where it is necessary to resolve differences in kinetic energy among the photoelectrons; (b) threshold photoelectron spectroscopy (TPES) where photoelectrons with kinetic energies less than $+\delta$ are detected and spectra are obtained by scanning the VUV radiation (hν). ZEKE (zero kinetic energy) refers to delayed pulsed extraction of above threshold electrons in the limit of $+\delta \rightarrow 0$; (c) pulsed field ionization (PFI) approach to threshold photoelectron spectroscopy. A delayed, pulsed electric field ionizes metastable high-n Rydberg states lying just below ($-\delta$) the cation threshold.

The energy window which defines the energy resolution in threshold photoelectron spectroscopy (TPES) is shown as $+\delta$ in Fig. (1b). In practice, the resolution of TPES is often limited by energetic electrons whose momentum transverse to the extraction field is small and cannot be discriminated from true threshold photoelectrons by the angular acceptance alone (steradiancy analyzers). The most effective threshold spectrometer designs have been developed for use with pulsed synchrotron radiation sources and use a combination of angular and TOF energy analysis to minimize the contribution of the "hot electron tail." In this way, groups working at the LURE synchrotron radiation facility are able to obtain TPES spectra with an overall resolution of $\sim$ 5 meV (40 cm^{-1}) and a collection efficiency approaching 50%.[10]

In an important series of papers, Müller-Dethlefs, Schlag and co-workers extended the TPES approach to laser-based, resonant multiphoton ionization (REMPI) measurements.[11–15] Early experiments on NO demonstrated that delayed, pulsed extraction of zero kinetic energy (ZEKE) electrons following threshold ionization of an intermediate state rotational line resulted in photoelectron spectra with linewidths approaching the laser bandwidth (< 1 cm^{-1}). ZEKE photoelectron techniques differ from conventional TPES in that the weak extraction field is applied as a voltage pulse after a fixed time delay following the ionization laser pulse (typically 500 nsec to 2 μsec). In actuality, ZEKE measurements are most sensitive to field ionization of very high-n Rydberg states which lie just below the ionization threshold and are field ionized by the extraction pulse ($n \geq 150$ for a 1 V/cm extraction field).[15] These Rydberg states are very long lived as a consequence of their very small interaction with the molecular ion core (n^{-3} dependence) and field induced Stark mixing.[16] Consequently, these highly excited Rydberg molecules survive the long delay time between excitation and pulsed extraction. In practice, prompt energetic electrons along with the true ZEKE electrons with small but positive kinetic energy ($+\delta$) are swept from the ionization volume by small external DC fields and the field ionized electrons lying below the threshold ($-\delta$) are detected by time gated TOF analysis. The relationship between field ionization and ZEKE spectroscopy was first made by Reiser *et al.*[15] and has led to the more appropriate description, pulsed field ionization (PFI) or ZEKE-PFI. The ZEKE-PFI approach to high resolution threshold photoelectron spectroscopy is illustrated in Fig. (1c).

With the development of ZEKE-PFI threshold spectroscopy, the ability to rotationally resolve photoionization transitions from *ground* electronic states requires pulsed VUV radiation with a bandwidth of 1 cm^{-1} or better. Although synchrotron radiation is ideal in terms of tunability, a dispersive monochromator with a resolving power of $\geq 200,000$ is very large (5 meter or better), very sensitive to alignment and very expensive when one includes the input optics required for efficient coupling to the synchrotron radiation source. A more robust high resolution VUV source, albeit with far more tuning difficulty, is based on VUV laser harmonic generation in free jet expansions. This technique, pioneered by Kung and coworkers,[17-20] utilizes the high pressure region of a pulsed gas beam as the medium to generate third-order harmonics of input visible and/or UV laser beams. Through a combination of non-resonant third harmonic generation ($3\omega_1$, $2\omega_1+\omega_2$) and resonantly-enhanced four wave sum/difference mixing techniques ($2\omega_R\pm\omega_1$) it is possible to cover the region from 150 nm to 75 nm using the rare gases Ar, Kr, and Xe as well as molecular nitrogen. With commercial dye lasers, a bandwith of 0.5–1 cm^{-1} is readily attained although a resolution as high as 0.007 cm^{-1} has been demonstrated.[20] The latter corresponds to a resolving power of $\lambda/\Delta\lambda \sim 10^7$ which is well beyond the capabilites of any present scanning monochromator designs. As shown in the following sections, the combination of ZEKE-PFI and coherent VUV radiation leads to a threshold spectroscopy with sufficient resolution ($\lesssim 1$ cm^{-1}) to reach the goal of measuring single quantum state distributions of molecular cations following photoionization.

2 Photoionization Dynamics

2.1 General Aspects

As noted in the Introduction, photofragmentation dynamics refers primarily to the nature and fate of the collision complex that is produced by photoexcitation. Because photoabsorption takes place at short range, the "photoionization collision complex" consists of the molecular ion core and a strongly coupled photoexcited electron. The short range final states are governed by the usual bound-bound selection rules governing dipole transitions, e.g. $\Delta J = 0, \pm 1$, as well as other propensity rules which depend on specific angular momentum coupling cases. As the electron escapes the molecular ion potential, the electron-core interaction weakens and the angular momentum coupling becomes more appropriate to a free photoelectron of well defined orbital angular momentum and a specific rovibronic level of the ion core. This uncoupling can be viewed as a scattering process in which the photoexcited electron exchanges energy and angular momentum with the ion core. The net result is that changes in core angular momentum well beyond $\Delta J = 0, \pm 1$ are possible. The conservation of angular momentum requires only that

$$\Delta J = J^{+} - J'' = l + 3/2, l + 1/2, \ldots, -l - 3/2, \qquad (1)$$

where l is the orbital angular momentum associated with $ks, kp, kd, \ldots$ outgoing partial waves and where the photoelectron spin ($s = \pm 1/2$) and photon angular momentum ($\vec{1}$) result in the $\pm 3/2$ term. For example, an outgoing kd wave could result in eight rotational branches, with changes in core angular momentum as large as $\pm 7/2$. Consequently, measurements of the relative rotational branch intensities reflect the probability for the various partial waves of the outgoing electron. ***This is the key result and is the motivation for pursuing rotationally resolved measurements.*** Such information is a sensitive probe of the ionization dynamics as the partial wave distribution can be strongly influenced by electron correlations and interactions of the photoexcited electron with the nuclear degrees of freedom, e.g. shape resonances and vibronic coupling.

The terminology used above to describe the photoionization process owes its origin in large part to quantum defect theory (QDT) which is based on partitioning a scattering problem into inner- and outer-core regions.[21] At long range, QDT treats the escaping photoelectron as a modified Coulomb wave of specific orbital angular momentum and treats its interaction with the core by a single quantity, the quantum defect (μ). The latter is simply related to the scattering phase shift, $\pi\mu$, for positive energies (photoionization) and gives the energy positions of Rydberg states below the ionization threshold through the expression $E_n = E_\infty - Ry/(n-\mu)^2$. The magnitude of the quantum defect is determined by the close range interaction of the excited electron with the motions of the core electrons and nuclei. A complete long range fragmentation channel, $|i\rangle$, is represented by an antisymmetrized combination of an ion core wavefunction in a specific quantum state and an outgoing electron wave of specific l. The short range QDT wavefunctions are most naturally described in terms of eigenchannels, $|\alpha\rangle$, which are Born-Oppenheimer states of the neutral molecule appropriate for specific coupling cases. QDT provides the mathematical framework by which the correct boundary conditions for the total wavefunction can be applied in moving from the short-range eigenchannel ($|\alpha\rangle$) description to the long-range fragmentation channels ($|i\rangle$). It is this change in basis or *frame transformation* which can be identified with the scattering of the photoelectron from the ion core. From a spectroscopic standpoint, the frame transformation is analogous to l-uncoupling involved in going from a Hund's case (a) or case (b) coupled basis appropriate for low-n Rydberg states, to case (d) Rydberg levels at high-n (or non-penetrating series with $\mu_l \sim 0$). This approach has been used quite succesfully for analyzing rotational photodetachment transitions of diatomic anions (OH^-, OD^-) where only ks ($l = 0$) photodetachment channels need to be considered.[22] Likewise, the frame transformation concept provides a means of interpreting ZEKE-PES by connecting rotationally resolved photoionization spectra ($|i\rangle$) to bound-state spectroscopy ($|\alpha\rangle$) for which the selection rules and perturbations are well known. This is especially useful for polyatomic systems for which the increased number of internal degrees of freedom allow for additional angular momentum couplings between the photoexcited electron and ion core. Child and Jungen[23] have used this approach to establish state-to-state correlation diagrams to follow the case (b) to case (d) frame transformation for photoionization of non-linear polyatomic molecules such as H_2O.

2.2 Rotational Photoionization Transitions: Line Strengths and Selection Rules

In practice, QDT calculates the final state wavefunctions by imposing boundary conditions on expansions of well defined photoelectron radial wavefunctions. An alternative theoretical approach to photoionization is based on Hartree-Fock (HF) methodology which allows for the direct calculation of continuum wavefunctions. The latter are most conveniently described by single-center expansions about the center-of-mass and these models readily yield expressions for rotational photoionization line strengths. The first such treatment was published in 1970 by Buckingham, Orr and

Sichel (BOS) who derived rotational line strength formulae for one-photon photoionization of diatomic molecules.[24] Due to experimental limitations discussed in the Introduction, the validity of these expressions was not to be tested against experimental data until almost 20 years later in connection with the VUV/ZEKE-PFI spectrum of O_2.[25]

The BOS rotational partial photoionization cross sections are expressed as a sum of terms in the orbital angular momentum (l) of the outgoing electron

$$\sigma_{J^+\leftarrow J''} \propto \sum_l \sum_{j_t=|l-1|}^{l+1} (2j_t+1)Q(j_t; J'', J^+)|A(j_t, l; \Lambda'', \Lambda^+)|^2 \quad , \tag{2}$$

where the coefficients $|A|^2$ are electronic bound-to-continuum dipole transition amplitudes. Electronic states are described by the projection onto the molecular axis of the electronic orbital angular momentum (Λ'', Λ^+), the total angular momentum (J'', J^+) and the total spin (S'', S^+), where the double primes refer to the initial neutral ground state. The variable j_t can be identified with the orbital angular momentum transferred to the ion core, $\vec{j_t} = \vec{l} - \vec{1}$ (Ref. [26]). In general, the BOS expression must be modified to ensure that only parity allowed transitions contribute to the partial cross sections.[26–28] Following Lefebvre-Brion,[26] Eqn. (1) must be multiplied by the factor $\frac{1}{2}[1-(-1)^P]$, where P is a parity index which can be evaluated for all possible combinations of Hund's cases from expressions given by Dixit and McKoy[27] and Xie and Zare.[28]

The general form for the parity selection rule connecting specific rovibronic states of the neutral and ion is given by[27–29]

$$\Delta J + \Delta S + \Delta p + l = even \tag{3}$$

where $\Delta S = S^+ - S''$ and where $\Delta p = p^+ - p$ represents the change in Kronig parity of the initial and final electronic states ($p = 0$ for Λ^+ or 1 for Λ^-). The utility of this expression is that it relates different photoelectron partial waves ($l = even$ or odd) to different branch transitions corresponding to a specific ΔJ and parity change. Specifically, if one spectrally resolves the parity components of the upper and lower rotational levels, it is possible to separate the l = *even* and l = *odd* contributions to the photoelectron continuum. Examples of molecules where rotational levels of different parity can be resolved include H_2 and CO, both of which involve $^2\Sigma^+ \leftarrow {}^1\Sigma^+$ photoionization transitions. For upper and lower Hund's case (b) Σ^+ states, Eqn. (3) reduces to $\Delta N + l = odd.$[27] Here, N refers to the total core angular momentum excluding spin and $\Delta N = N^+ - N''$. This selection rule dictates that the Q branch ($\Delta N = 0$) be associated with *odd* partial waves while the P and R branches ($\Delta N = \pm 1$) reflect contributions of *even* partial waves in the continuum. Extensions of parity selection rules to non-linear molecules will be discussed in detail in Sec. 4.

Returning to the BOS line strength expression (Eqn. (2)), we note that all of the dependence on rotational angular momentum is contained in the $Q(j_t; J'', J^+)$ terms. Explicit expressions for the later are given in Ref.[24] for various Hund's coupling cases describing the neutral and cation states. Unlike bound-bound electronic transitions, these rotational factors also depend on the electronic transition strengths through the angular momentum transfer, j_t. Consequently, calculations of rotational photoionization intensities require prior knowledge of the $|kl\rangle$ distribution of the photoelectron. In the absence of calculated dipole transition strengths, the BOS expressions are still very useful for extracting dynamical information through empirical "fits." For this purpose, we use the fact that the rotational factors do not explicitly depend on l and condense Eqn. (2) into a single sum, i.e.

$$\sigma_{J^+ \leftarrow J''} \propto \sum_{j_t} Q(j_t; J'', J^+) C_{j_t} \quad , \tag{4}$$

where the coefficients C_{j_t} are linear combinations of the transition amplitudes, $|A|^2$, and represent the relative probabilities of transferring $j_t \pm \frac{1}{2}$ units of angular momentum to the ion core. Although this expression has lost the explicit dependence on l, one can evaluate the $Q(j_t; J'', J^+)$ rotational factors (weighted by a ground state Boltzmann distribution of rotational levels, J'' or N'') and simulate an experimental spectrum to extract relative values of the dynamical coefficients, C_{j_t}. This procedure has been used to infer the nature of the partial wave distribution through simulations of the VUV ZEKE-PFI spectra of O_2, N_2O and CO_2. Ironically, the inability to obtain satisfactory BOS simulations for systems such as N_2O is also valuable for exposing the importance of final state interactions, e.g. autoionization, which are not included in such one-electron models.[30]

Fully *ab initio* methods for calculating rotational state distributions of cations following one-photon or (n+1) REMPI photoionization have been extensively developed by the McKoy group at Cal Tech.[27,29,31,32] Their approach uses single-center expansions to describe the photoelectron wavefunction which are obtained from solutions of a one-electron Schrödinger equation employing a Hartree-Fock molecular ion potential and numerical Schwinger variational methods. Shape resonance phenomena are accounted for by allowing for R-dependence of the molecular ion potential, $V_{ion}(r, R)$. A detailed description of the methodology and applications can be found in Ref.[32]. Overall, this approach has met with considerable success in explaining the underlying dynamics in the photoionzation of a number of diatomic molecules and has recently been generalized to include non-linear polyatomics as well. The latter will be demonstrated in Sec. 4 in regards to the ground state photoionization of the non-linear hydrides, e.g. H_2O and CH_3.

3 Experiment

The VUV photoionization apparatus used by the authors for ZEKE-PFI measurements[33] is shown schematically in Fig. 2. This apparatus has elements common to VUV spectrometers of other groups,[19,34,35] all of which include: (1) a section for VUV harmonic generation in which free jet expansions of various gases act as the non-linear conversion media; (2) collection, separation and VUV beam transport optics; (3) a TOF electron spectrometer with electrostatic field plates (or grids); (4) a molecular beam source for sample preparation. Four separate, differentially pumped vacuum chambers house the VUV source, the interaction-scattering region, the TOF electron/ion spectrometer and molecular beam source in the apparatus shown in Fig. 2. This arrangement provides maximum flexiblity for sample preparation, low background signals, and high VUV conversion efficiencies.

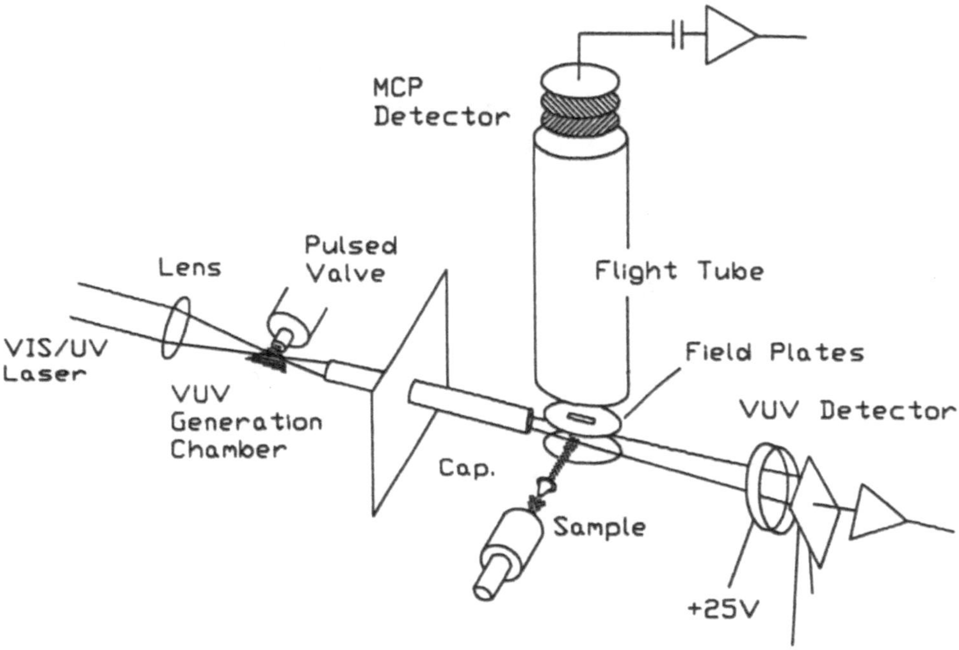

Figure 2: Schematic diagram of the experimental apparatus used by the author to obtain photoelectron, ZEKE-PFI or photoion spectra.

Tunable, narrow-band VUV radiation suitable for high resolution photoionization can be produced by various 3-photon harmonic generation schemes using atomic and molecular gases as the conversion medium.[36,37] Through a combination of non-resonant third harmonic generation ($\omega_3 = 3\omega_1$) as well as resonant and non-resonant four-wave sum and difference frequency mixing ($\omega_3 = 2\omega_1 \pm \omega_2$) it is possible to cover nearly the entire VUV region from 160 nm to 73 nm using the rare gases (Ne, Ar, Kr, Xe) and molecular nitrogen as the conversion media. This wavelength range covers the ionization energies (8.0 eV–17.7 eV) of the outer valence orbitals of nearly all molecules. Although a static gas cell could be used for generation of wavelengths longer than the LiF window cut-off (105 nm), the arrangement typified in Fig. 2 provides full spectral coverage without change of configuration. Fig. 3 shows the approximate tuning ranges for non-resonant and resonant 3-photon harmonic generation schemes using the rare gases Ar, Kr and Xe. This figure is based on practical experi-

ence in this laboratory as well the work of Kung and coworkers (Stanford, Berkeley)[17–20] and Wallenstein and co-workers (Bielefeld).[38] The tuning ranges are based on input wavelengths available from commercial Nd:YAG-pumped dye lasers using 532 nm and 355 nm pumping as well as doubling and mixing techniques to produce input radiation below 400 nm.

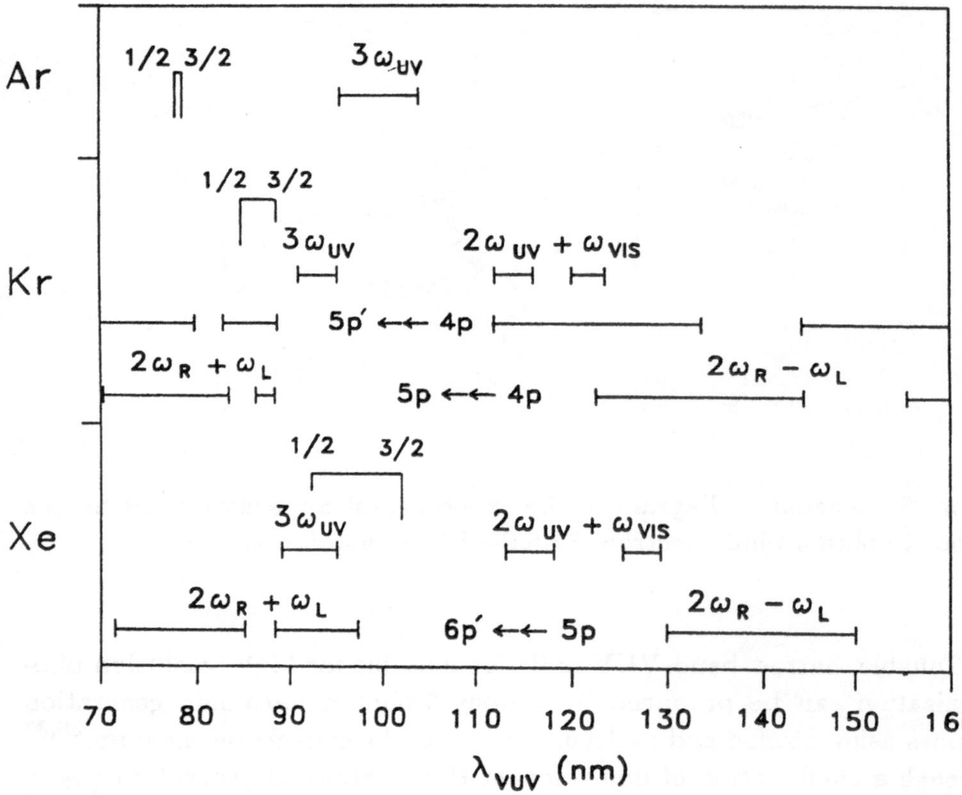

Figure 3: Approximate tuning ranges for VUV or XUV generation in the rare gases argon, krypton, and xenon by various 3-photon harmonic generation techniques. One laser methods include non-resonant third harmonic generation ($3\omega_{UV}$) and non-resonant four-wave sum generation ($2\omega_{UV} + \omega_{VIS}$). More efficient VUV generation is accomplished by two-laser, resonant four-wave sum/difference mixing ($2\omega_R \pm \omega_L$) where one laser is scanned (ω_L) and the other is tuned to a two-photon atomic transition ($2\omega_R$). The resonant wavelengths are 204.194 nm and 216.666 nm for the 5p′ ←←

4p and 5p←← 4p transitions in Kr, respectively, and 222.567 nm for the 6p′ ←← 5p transition in Xe. Also shown are the rare gas $^2P_{1/2}$ and $^2P_{3/2}$ ionization thresholds.

A unique feature of our apparatus is the use of a capillary light guide which transports the laser generated VUV radiation to the interaction region of the spectrometer. This technique was adapted from gas-phase photoelectron spectrometers used at synchrotron radiation facilites where capillary light guides were found to be the most efficient way to separate the relatively high pressure of the spectrometer (typically $1 - 10 \times 10^{-5}$ Torr) from the ultra-high vacuum environment ($\sim 10^{-9}$ Torr) of the upstream VUV monochromator.[39] Experience at synchrotron facilities showed that standard Pyrex capillaries could transport VUV radiation via internal grazing reflections with tolerable losses (~ 50%) while retaining the full polarization characteristics of the input source. Similar performance is obtained with coherently generated VUV and the capillary provides a source size adequate for one-photon ionization (~ 1 mm dia.) without intervening refocussing and harmonic rejection optics.[33] Because the radiation exiting the capillary is unfocussed, multiphoton processes (ionization, dissociation) induced by the very intense, unconverted ω_1 and/or ω_2 input beams do not occur. The unconverted input beams, however, are very useful for detecting long lived excited states produced by VUV absorption via (1+1′) REMPI. This (VUV+UV/VIS) REMPI process is especially usefull for VUV wavelength calibration using H_2 and the rare gases (Ar, Kr, Xe) for which many excited state energies are known to high precision.[40,41] The capillary is also effective as a gas filter which separates the sum and difference VUV frequencies obtained in four-wave mixing schemes. The long path length of the capillary (35.5 cm) results in total absorption of the sum frequency radiation when a suitable rare gas is directed into its center at low pressure ($\leq$ 400 mTorr). Diffraction grating monochromators such as that employed by Softley and coworkers[35] serve much the same function as the capillary but are capable of complete rejection of the unconverted input UV/VIS beams and also provide a refocussed VUV beam at the sample. Refocussing compensates in large part for the relative inefficiency of the diffraction grating ($\leq$ 5%).

3.1 ZEKE-PFI Techniques

As noted in the previous Sections and illustrated in Fig. (1c), very high resolution threshold photoelectron spectra can be obtained by the delayed, pulsed field ionization method referred to as ZEKE-PFI. Implementation of this technique is detailed in a recent review by Müller-Dethlefs and Schlag.[42] This section will focus primarily on aspects of ZEKE-PFI which are important to its interpretation as a form of threshold photoelectron spectroscopy. Most importantly, *ZEKE-PFI relies on the continuity of oscillator strength across an ionization threshold in order to equate the population of bound Rydberg states, nl, just below threshold to the distribution of photoelectron partial waves, kl, just above threshold.* In this way, the distribution of cation and photoelectron final states determined by field ionization simulates rotationally-resolved threshold partial photoionization cross sections which usually cannot be obtained directly. Additional factors which can affect the analysis of ZEKE-PFI data are also considered below.

3.1.1 Line Positions from ZEKE-PFI

ZEKE-PFI is based on field ionization of high-n Rydberg states lying below the ionization threshold and the observed width and absolute energy position of spectral lines is dependent on the characteristics of the applied field. This is illustrated in Fig. 4 which gives the ZEKE-PFI line shapes as a function of applied pulsed field for one-photon ionization of Xe to its ground ${}^2P_{3/2}$ ionic state. These curves were taken with a small 50 mV DC offset (F_0) which is used to sweep very low energy prompt photoelectrons out of the ionization volume prior to the pulsed field (F). As expected, both the peak maximum and the onset move away (red-shift) from the zero-field ionization threshold as the pulsed field is increased. The blue side (high energy) of the line extending just above the zero-field threshold is essentially constant and reflects the bandwidth of VUV radiation. For the smallest pulsed fields the width of the ZEKE-PFI peak approaches the VUV bandwidth while for the highest fields the peak becomes flattened as field ionization extends further down into the Rydberg manifold. The data in Fig. 4 suggests that determining accurate line postions requires measurements of their field dependence and extrapolation to zero-field.

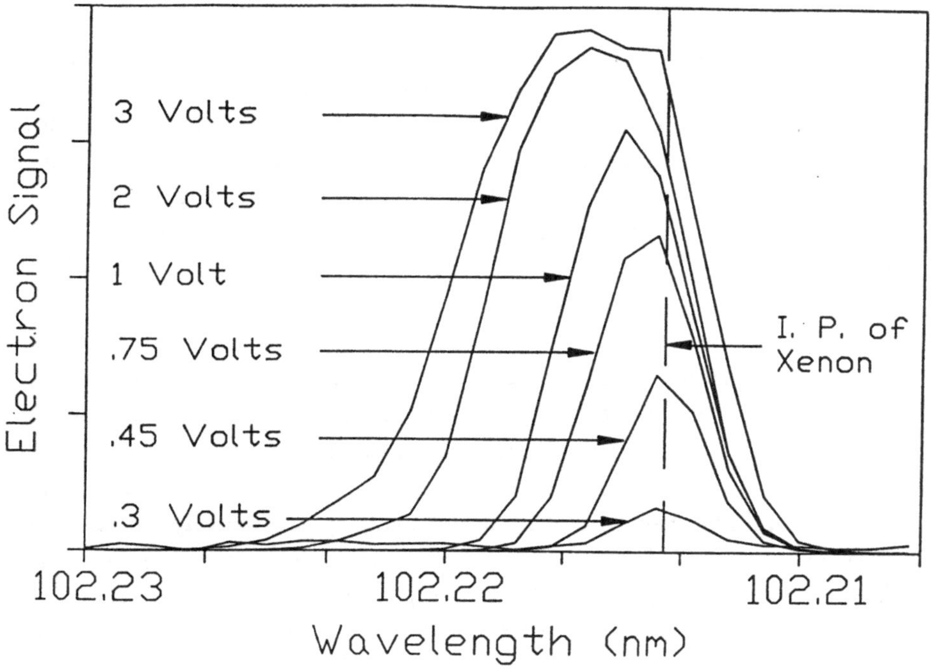

Figure 4: Pulsed field ionization spectra near the Xe^+ ground state ionization threshold ($^2P_{3/2}$) as a function of applied field.

Although this procedure is acceptable for sparse and well separated atomic thresholds, the larger number of lines (often overlapping) in cation rotational spectra make such a process impractical. In most cases, it is assumed that the ZEKE-PFI lines are Stark shifted by a uniform red-shift which is applied to the spectrum to place it on an absolute energy or wavelength scale. The energy shift of a cation threshold is given by $\delta = -6.1\sqrt{F}$ or $\delta \sim -4\sqrt{F}$ depending on whether field ionization takes place adiabatically or diabatically.[16] Here F is given in V/cm and δ in cm^{-1}. Adiabaticity refers to the avoided crossings of like m_l sublevels as the width of individual nl Stark manifolds increase with increasing applied field. The adiabatic limit corresponds to the classical saddle-point in the potential resulting from the Coulomb and applied electric field and is applicable to the case where ionization takes place in a DC field. For small pulsed fields with rapid onsets typical of most ZEKE-PFI experiments ($\sim 1 \times 10^8$ V/cm-sec), Chupka showed that field ionization should take place diabatically.[16] Chupka also found that the critical fields for diabatic ionization of different members of a Stark manifold are clustered about the limiting value $\delta \sim -4\sqrt{F}$ as $n \to \infty$. In a recent systematic study of threshold ionization of NO via (1+1′) REMPI, Pratt demonstrated both adiabatic and diabatic field ionization using a small DC field and an applied voltage pulse with a rise time ≤ 5 nsec, respectively.[44]

Because the diabatic shift can actually cover a range of values with $-4\sqrt{F}$ representing the most probable, the absolute accuracy of line positions obtained by applying a calculated diabatic red-shift are of limited accuracy. Combined with uncertainties in absolute wavelength calibration, ionization thresholds determined in this laboratory are usually no more accurate than ± 2 cm^{-1}. This level of accuracy is comparable to that obtained by Rydberg series extrapolation at relatively high n (≥ 20), but represents an improvement by a factor of 50 to 100 over conventional PES/TPES techniques. For most purposes, this accuracy is sufficient especially for the indirect determinations of thermochemical quantities, e.g. heats of formation, which often rely on ionization potentials and associated reaction cycles.[45] Relative accuracy within a ZEKE-PFI spectrum is limited only by the overall line width which is typically ≤ 1 cm^{-1} for small pulsed fields (≤ 0.5 V/cm) and a VUV bandwidth of ~ 0.7 cm^{-1}. In fact, rotational constants with an accuracy of ± 0.03 cm^{-1} have been derived from ZEKE-PFI spectra of the polyatomics NH_3, H_2CO and HCN.

The above discussion would imply that the width of the ZEKE-PFI lines would continue to increase as the applied field increases. In fact, several groups studying the vibronic structure of polyatomic molecules using (n+1′) ZEKE-PFI methods have obtained relatively narrow lines despite using pulsed fields as high as 300 V/cm ($4\sqrt{300} \sim 70$ cm^{-1}).[46] Recent studies by Pratt[44] and Zhang *et al.*[47] suggest that the width of the Rydberg state manifold contributing to the observed PFI signal results from a competiton between non-radiative decay and collisional deactivation which shorten the lifetime and efficient l and m_l mixing which can lead to substantial increases in lifetime. The rate of intramolecular decay scales as n^{-3} and is most likely responsible for limiting the low energy extent (red-shift) of ZEKE-PFI lines. In NO, rapid predissociation limits the observation of *np* Rydberg levels to

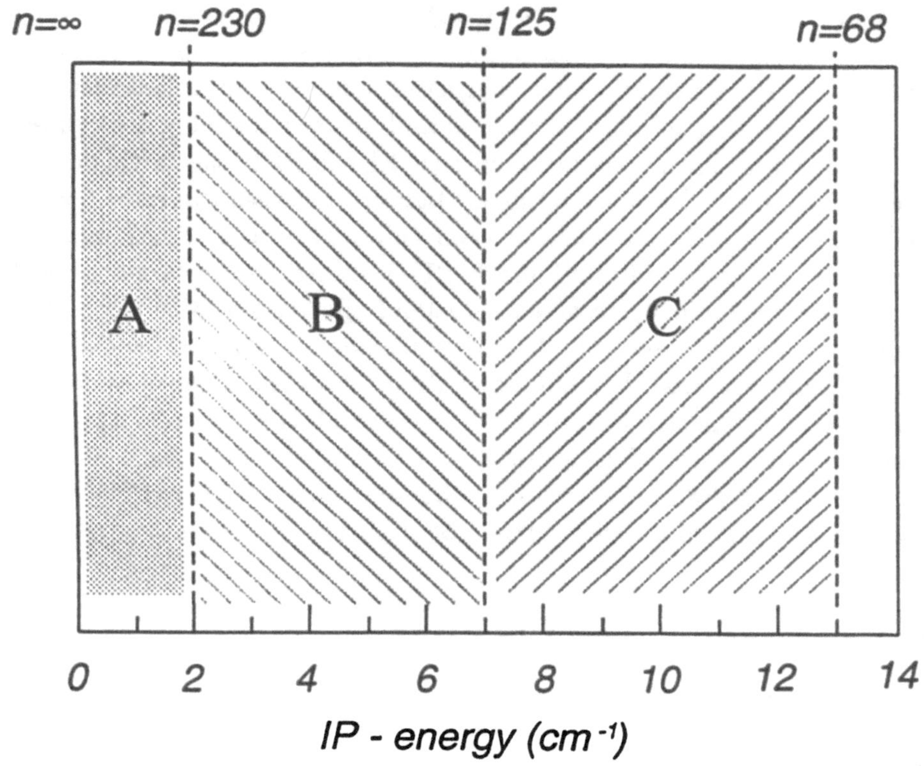

Figure 5: Schematic representation of the range of lifetime behavior for high-n Rydberg states for an applied pulsed field of 10 V/cm. The vertical dashed partitions at n = 68 and n = 125 are the approximate principal quantum numbers for Rydberg states at the field-induced Stark shift $(4\sqrt{F\ (V/cm)})$ and Inglis-Teller limit $(F\ (V/cm) = 1.71 \times 10^9 n^{-5})$, respectively. In region **A**, the very highest-n Rydberg states are lost due to long range collisions with other gas particles; in region **B**, Rydberg states are long lived ($\geq 1\mu$sec) due to l and m_l randomization induced by stray fields and/or interactions with surrounding ions; in region **C**, Rydberg states decay rapidly due to intra-molecular decay processes and are not detected in the typical ZEKE-PFI experiment.

those lying within 15–20 cm^{-1} of the ionization threshold.[44] Collisional loss of high-n Rydberg states via long-range ion-Rydberg and Rydberg-Rydberg inter-molecular interactions has the opposite effect and decreases the line width from the high energy (blue) side by depopulating the very highest-n Rydberg levels.[47] High-n depopulation was found to increase with increasing laser power which suggested *inter*molecular interactions between the increasing number of cations and excited Rydberg molecules. Hence, the observation of relatively narrow ZEKE-PFI lines with pulsed fields greater than 1 V/cm is a fortuitous balance between the rates for intramolecular decay, collisional deactivation and field induced Stark mixing. This situation is illustrated in Figure 5. The ability to use larger pulsed fields to obtain vibronically resolved threshold spectra has also led to the development of an ion-based PFI detection scheme. Developed by Zhu and Johnson,[48] mass analyzed threshold ionization (MATI) has the potential to provide ZEKE-PFI spectra at somewhat lower resolution (3 cm^{-1} has been reported) but with the added advantage of having a mass "tag." Experimental variants of the MATI technique have recently appeared[46,47] and the reader is referred to a review article by Johnson.[49]

3.1.2 Line Intensities from ZEKE-PFI

As noted at the beginning of this section, ZEKE-PFI can be considered a form of threshold photoelectron spectroscopy providing that the nl Rydberg state distribution excited just below threshold reflects the continuum kl partial wave distribution just above the ionization limit. Theoretically, the continuity of oscillator strength guarantees that the two distributions are smoothly connected as the excitation energy crosses the ionization threshold. From the standpoint of detecting high-n Rydberg states by field ionization, it is sufficient to show that the ZEKE-PFI line intensity accurately reflects the absorption oscillator strength to the manifold of Rydberg states initially photoexcited. Consequently, any process which results in a preferential depopulation of Rydberg states of a specfic l between the time of photoexcitation and pulsed field ionization will result in ZEKE-PFI intensities which are not representative of the photoexcitation process. Predissociation and autoionization of specific series are the most important differential loss mechanisms since they lead to molecular fragments or prompt photoelectrons which are not detected by ZEKE-PFI. The n^{-3} dependence of intramolecular decay rates suggests that processes such as predissociation can be minimized through the use of small fields, i.e. ≤ 1 V/cm, which ensures that only Rydberg levels with $n \geq 100$ are detected. Measurements on NO (Refs. [15,44]) as well as experience in this laboratory on several small molecules have shown that levels within $\sim$ 10 cm^{-1} of the ionization threshold have empirical lifetimes of several microseconds. These lifetimes are usually lower bounds since the Rydberg excited molecules moving with the molecular beam drift out of the field ionization volume if longer delays are used. Using a collinear excitation and molecular beam geometry, Zhang *et al.* recently demonstrated lifetimes as long as 25 μs in a (1+1′) ZEKE-PFI study of phenanthrene.[47]

In fact, simple n^{-3} scaling of known or estimated predissociation rates for low-n levels of simple molecules, e.g. NO, results in lifetimes for Rydberg states with $n \geq 150$ which are still too short (< 10 ns) to be observed with practical delay times. As shown in a recent analysis by Chupka, the observed lifetimes for high-n Rydberg states most likely result from rapid l and m_l mixing induced by small residual DC fields and long-range collisions with the surrounding prompt ions.[16] Stark mixing becomes quite facile at field strengths where adjacent Stark manifolds overlap (Inglis-Teller limit: $F(a.u.) = \frac{1}{3}n^{-5}$).[50] For Rydberg levels with $n = 150$ and small quantum defects, the Inglis-Teller field is only $\sim$ 0.02 V/cm . Residual fields of such magnitude are common in electron/ion spectrometers due to surface charging or contact potentials between lens elements or defining apertures. In addition, DC fields on the order of 0.01–0.05 V/m are typically used in ZEKE-PFI measurements to ensure the removal of low energy prompt photoelectrons from the ionization volume prior to pulsed field ionization. The rapid mixing of l and m_l for high-n Rydberg states can lead to increases in lifetime by a factor of n^2 (Ref. [16]) which is consistent with observations. The increase in lifetime is due in large part to the population of non-penetrating, high-l levels whose lifetimes with respect to intramolecular decay is extremely long.[50]

For high-n Rydberg levels lying above the Inglis-Teller limit, Stark mixing can also lead to nominally forbidden interactions with nearly degenerate, low-l predissociative or autoionizing states. These interactions have the effect of introducing a field-dependence to the ZEKE-PFI line intensities and line shapes. A particularly good example of field-induced interactions is the case of the 61d state of H_3 which undergoes strong mixing with an optically "forbidden" and strongly predissociated np interloper at field strengths as low as 0.2–0.4 V/cm.[50] Field-induced autoionization has been invoked to explain the disappearance of long-lived, high-l (f,g) Rydberg states near the ionization threshold of HCl in external fields of ≤ 1 V/cm.[51] According to Mahon *et al.*, field-induced autoionization involves a series of quadrupole couplings induced by the field in which l is conserved and the core changes by two units of rotational angular momenta ($\Delta N^+ = -2$).[52] Field-induced processes should be distinguished from normal (zero-field) rovibronic autoionization or predissociation in which Rydberg collisions with the core cause exchanges of electronic and rovibrational angular momentum and energy. Just as TPES is sensitive to, and often dominated by autoionization processes that yield near zero energy electrons, continuity arguments require that ZEKE-PFI line intensities reflect the bound-bound interactions below threshold that are responsible for autoionization or predissociation above threshold. One of the important advantages of ZEKE-PFI over TPES is that the delay between excitation and field ionization limits the detection of autoionization to only those states that lie isoenergetic with the narrow band of high-n Rydberg states initially photoexcited. The other unusual aspect of ZEKE-PFI is that the perturbing levels gain lifetime through their interaction with the high-n Rydberg states and only autoionize after the applied pulsed field depresses the ionization potential and the Stark manifold of states becomes a continuum. This forced-autoionization process has been described by Gilbert and Child[53] and accounts for the unusual rotational state distributions observed in the VUV/ZEKE-PFI spectra in HCl, NO, N_2O, and OH.[30,54,55,56] Combinations of forced and field-induced autoionization processes have also been succesfully invoked to explain the very recent REMPI/ZEKE-PFI spectrum of NO_2.[57]

It is clear from the above discussion that the final state distribution of Rydberg levels sampled by the pulsed field can be quite different than that initially photoexcited. The point to be emphasized, however, is that ZEKE-PFI is not sensitive to the evolution of the strongly mixed Stark levels since the pulsed field ionizes all energetically available Rydberg states uniformly. All that is required is that field induced l and m_l dephasizing not lead to preferential population loss. Again, the use of small pulsed fields minimizes differential loss mechanisms due to the inherent n^{-3} lifetime scaling and the high probability for field induced l, m_l mixing which can result in very long extrinsic lifetimes. In cases where field-induced autoionization or predissociation are suspected, it may be necessary to measure the ZEKE-PFI spectrum at several delays, pulsed field amplitudes and/or DC offset voltages.

4 Applications: Non-Linear Polyatomic Molecules

In generalizing the angular momentum and parity constraints for photoionization in non-linear polyatomic molecules, it is necessary to consider the additional quantum numbers required to specify individual rovibronic quantum states. In general, polyatomic molecules have three independent axes of rotation labeled in order of increasing moment of inertia ($I_a < I_b < I_c$). Consequently, there are three body-fixed projections of the total angular momentum (K_a, K_b, K_c), two of which are required in specifying a single rotational level, e.g. $|JK_aK_c\rangle$. In addition to conservation of the total angular momentum given by Eqn. (1), the projections of J (or N) on the molecule-fixed quantization axis are also conserved. The general form of this selection rule (not including spin) is given by

$$\Delta K_\eta = \mu + \lambda \tag{5}$$

where η represents the rotation axis of highest symmetry (body-fixed z axis) and λ and μ are the η-axis projections of the photoelectron orbital angular momentum and the photon angular momentum ($\mu = 1, 0, -1$).[58,59] For diatomic molecules, in which the rotational angular momentum (R) lies perpendicular to the internuclear axis, Eqn. (5) takes the simple form $\Delta\Lambda = \lambda, \lambda \pm 1$. In the bound-state spectroscopy of diatomics, transitions are characterized as parallel or perpendicular depending on whether the transition dipole lies along ($\Delta\Lambda = 0$) or is perpendicular to ($\Delta\Lambda = \pm 1$) the internuclear axis. Due to the uncoupling of the core and photoelectron angular momenta, photoionization transitions allow for a larger range of $\Delta\Lambda$ which are more conveniently classified as $\Delta\Lambda = even$ and $\Delta\Lambda = odd$. For symmetric tops, the subscript in Eqn. (6) is suppressed and K represents the projection onto the a axis for prolate tops ($I_a < I_b = I_c$) and the c axis for oblate tops ($I_a = I_b < I_c$). In this case, $\Delta K = even$ and $\Delta K = odd$ photoionization transitions have much the same meaning as $\Delta\Lambda$ in diatomic molecules. Unlike diatomic molecules, the rotational energy level spacings for symmetric tops includes energy terms proportional to both $J(J+1)$ and K^2, such that the rotational state distribution of the cation clearly reflects the *even* or *odd* character of $|\Delta K|$. For asymmetric tops, like H_2O, three types of rotational transitions are classified by the four combinations of even/odd for $\Delta K_a/\Delta K_c$. Both ΔK_a and ΔK_c are *odd* or *even* in Type B transitions, while type A transitions are characterized by $\Delta K_a = even$ and $\Delta K_c = odd$ and type C transitions by $\Delta K_a = odd$ and $\Delta K_c = even$. These different types refer to the axes along which the transition dipole has a non-zero projection, i.e. $\mu \parallel a$-axis in type A transitions. From the standpoint of parity, each unique specification of the transition dipole, i.e. $\Delta K = even/odd$, type A, type B, etc., is associated with a specific parity of the photoelectron, $l = odd$ or $l = even$. As will be seen below, complete K state resolution provides unique information concerning the photoionization dynamics of polyatomic molecules.

4.1 Asymmetric tops: H_2O

The non-linear triatomic hydrides H_2X (X ≡ C, N, O, S, CO) are planar asymmetric tops with C_{2v} symmetry. The outermost $1b_1$ orbital in NH_2, H_2O and H_2S is essentially a p-type atomic orbital, pointed out of the plane and centered on the heavy atom. The outermost $2b_2$ orbital of formaldehyde (H_2CO) lies in the molecular plane and is localized on the C-O moeity. The latter can be characterized as a d-type atomic orbital fixed at the center-of-mass. Given the atomic-like nature of these orbitals, one might expect atomic selection rules ($\Delta l = \pm 1$) to provide a reasonable description of photoionization in these molecules. In the case of H_2O, earlier *ab initio* static exchange and Schwinger variational calculations indicate dominant $kd\,(l = 2)$ photoionization continua, in general agreement with an atomic model.[60,61] More recently, Child and Jungen have applied multichannel quantum defect theory (MQDT) to photoexcitation of asymmetric tops and performed numerical calculations for $H_2O\ (1b_1)^{-1}$ near threshold ionization.[23] In these calculations, the authors assumed only $d \leftarrow p$ excitations from which they were able to obtain a plausible assignment for the high resolution H_2O photoionization spectrum obtained by Page *et al.*[62] Child and Jungen also showed that $l = 2$ photoexcitation leads only to type C rotational transitions, for which $\Delta K_a = odd$ and $\Delta K_c = even$.

The association of type C transitions with $l = 2$ continua is a general consequence of parity selection rules as illustrated for H_2O photoionization in Fig. 6. In this figure, the lowest rotational levels of the neutral X^1A_1 and ground $\tilde{X}^2B_1$ ionic states are shown along with their $|NK_aK_c\rangle$ assignments, rovibronic symmetries (Γ_{evr}) and parities (p_{evr}).[63] The symmetries are

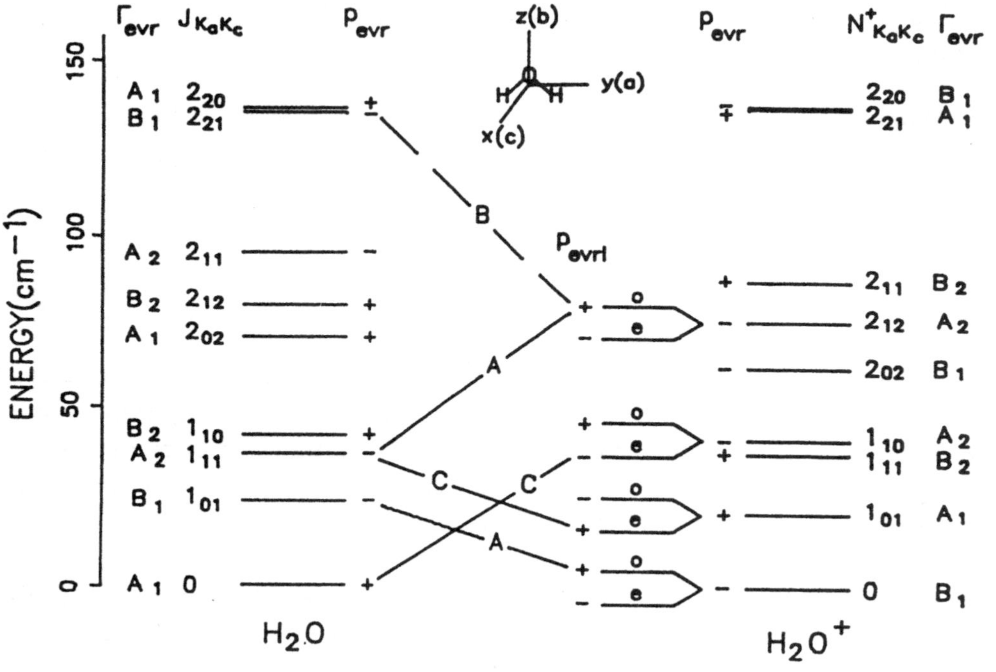

Figure 6: Level structure of the first few rotational levels of H_2O and H_2O^+. Overall rovibronic symmetries (Γ_{evr}) and parities (p_{evr}) are labeled for each level. Each level of the ion can be coupled to l = *even* (e) and l = *odd* (o) photoelectron partial waves to give an overall final state parity (p_{evrl}). A few examples of type A, B and C transitions are also shown. Type B transitions are forbidden by nuclear spin statistics.

appropriate to a near prolate top ($I_a < I_b \sim I_c$) with $a = y$, $b = z$ and $c = x$ (representation II of Zare).[64] The ion levels may couple to an l = *odd* or *even* photoelectron (shown as o and e) to give a complete Hund's case (d) parity (p_{evrl}). Using the general dipole selection rule for parity, $\oplus\ominus \leftrightarrow \ominus\oplus$, we find that l = *odd* continua accompany type A transitions (ΔK_a = *even* and ΔK_c = *odd*) and l = *even* continua are associated with type C transitions as noted above. Type B transitions, with both ΔK_a and ΔK_c *odd* or *even*, are strictly forbidden by nuclear spin statistics. A similar diagram describes H_2S photoionization, although the neutral and cation are closer to oblate tops.

The parity selection rules extracted from Fig. 6 have also been derived theoretically by Lee, *et al.* through the use of Eqn. (3), and general symmetry properties of asymmetric tops.[58,59] Specifically, they find

$$\Delta K_a + l = odd \tag{6}$$

which summarizes the parity restrictions for type A and type C transitions found above. Furthermore, they use Eqn. (6) to show that only type A and type C photoionization transitions are allowed independent of nuclear symmetry constraints. It is important to note that the type(s) of allowed transitions, e.g. type A, are specific to the asymmetric top case being studied. For formaldehyde, which is very nearly a prolate symmetric top ($I_a \ll I_b \sim I_c$), the C_2 axis is now identified with the $a\,(=z)$ top axis. In this case, the rovibronic symmetries (Γ_{evr}) and level structure differ significantly from that in Fig. 6, although the parities for any $|NK_aK_c\rangle$ state are the same. To see how this effects which rotational transitions are allowed, we use Eqn. (6) to calculate the allowed values of K_η which corresponds to K_a in H_2CO. Dipole-induced photoionization from the outer $2b_2$ molecular orbital leads to ka_1, ka_2 and kb_2 ionization continua which, combined with the symmetries of the dipole operator in C_{2v} symmetry ($a_1 \rightarrow \mu = 0, b_2 \rightarrow \mu = \pm 1$), gives $\Delta K_a = odd$. Consequently, only type C and type B ($\Delta K_a = odd$ and $\Delta K_c = odd$) ionization transitions are permitted, the latter being forbidden in NH_2 H_2O and H_2S. Parity selection rules, which are invariant to top classification, can now be summarized as $\Delta K_c + l = even$, in which type C and type B transitions remain coupled to $l = even$ and $l = odd$ continua, respectively.

Both the experimental ZEKE-PFI and calculated threshold spectra for the (000) level of H_2O^+ (Fig. 7) exhibit strong type C rotational transitions with a number of weaker yet well isolated lines which are assignable to type A transitions, e.g. $0_{00} \leftarrow 1_{01}$. According to the parity selection rule of Eqn. (6), these type A transitions arise from odd (almost pure p at threshold energies) wave contributions to the photoelectron matrix elements. These p waves of the ka_1 and kb_1 continua are entirely molecular in origin, since photoionization of the $1b_1$ orbital with its almost pure p (99.7 %) character

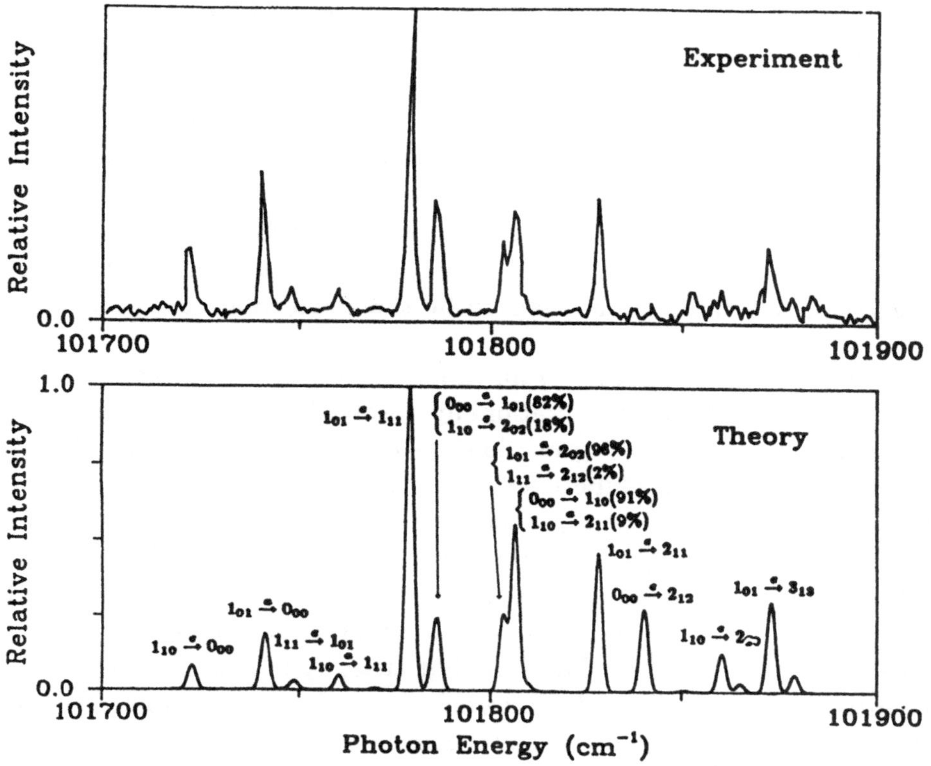

Figure 7: The measured (upper) and calculated (lower) ion rotational distributions for photoionization of the $1b_1$ orbital of the $X\,^1A_1$ ground state of jet-cooled H_2O. Rotational levels are labeled by the total core angular momentum without spin and the a and c axis projections, i.e. $N^+_{K_aK_c} \leftarrow J''_{K_aK_c}$. The a and c labels indicate type A and type C transitions, respectively.

should lead only to s and d (even) waves in an atomiclike picture. The ka_2 continuum makes almost no contributions to these type a transitions, since its p wave is forbidden and f or higher odd waves are negligible.[58,59]

The origin of type A transitions is not due to an inadequate description of the ground state, but results from $np - nd$ Rydberg series interactions induced by the small, but finite anisotropy of the H_2X^+ core potential. At close range (or small n) where the body-fixed C_{2v} symmetry components are well defined, the np and nd Rydberg series of the same λ (a_1 or b_1) can mix since l is not a good quantum number. At long range (or high-n), the Rydberg or photoelectron orbitals separate according to l, however, the short- to long-range fragmentation process results in p final states with non-zero oscillator strength. This l-uncoupling process is one-electron in nature and the $p-d$ interactions are simply a consequence of the short range electron-ion core interactions induced by the anisotropic core potential.

As noted earlier, Child and Jungen included only $d \leftarrow p$ excitations and thereby predicted that the rotationally resolved photoelectron spectrum would only include type C transitions as required by parity (Eqn. (6)).[23] In a subsequent paper, Gilbert and Child proposed that the "anomalous" type A transitions could be explained by introducing a dipole-induced mixing between the strongly allowed nd series at high-n and "forbidden", isoenergetic np levels converging to higher rotational limits.[53] In their model, the zero-field dipole-mixing would lead to observable type A transitions via forced autoionization.[53,52] The fact that these type A transitions are well accounted for by a "field-free" HF calculation indicates that forced autoionization is not responsible for the appearance of type A lines. However, the MQDT approach to $np - nd$ interactions mediated by long range dipolar coupling is quite appropriate. The non-zero dipole moment of H_2O^+ is simply a long range manifestation of the anisotropy of the cation potential. The essential physics are contained in both the HF and QDT approaches and the H_2O example highlights the multichannel nature of molecular photoionization.

4.2 Symmetric Tops: CH_3

The methyl radical and cation are oblate symmetric tops, belonging to the D_{3h} molecular point group. Dipole-allowed transitions from the $X\ ^2A_2''$ ground state of CH_3 occur to final states (ion core $(^1A_1')$ + photoelectron) of A_1' ("parallel" transition) and E'' ("perpendicular" transition) overall symmetry. These short-range states correspond to photoelectron continuum channels with ksa_1', kda_1' and kde'' symmetry, respectively. Transitions to

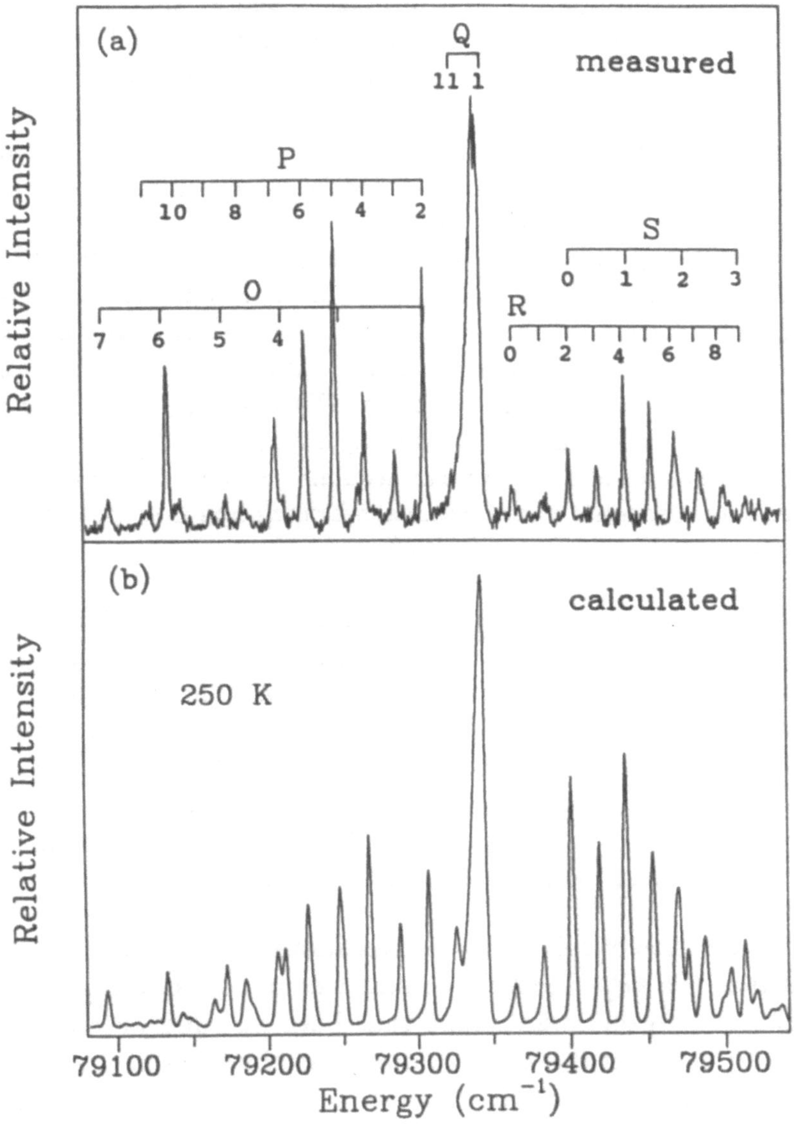

Figure 8: Rotationally resolved threshold photoelectron spectrum for one-photon ionization out of the a_2'' orbital of CH_3. Upper curve: experimental ZEKE-PFI spectrum. Lower curve: *ab initio* calculation of the threshold photoelectron spectrum using a photoelectron kinetic energy of 50 meV and assuming a ground state rotational temperature of 250 K.

p-type ($l = 1$) continua, which transform as a_2'' and e', are dipole-forbidden. Since the a_2'' HOMO is essentially an out-of-plane $2p$ orbital localized on the central carbon atom, these allowed final states are consistent with atomic selection rules ($\Delta l = \pm 1$). Furthermore, the center-of-charge and center-of-mass coincide for the CH_3^+ cation core and this nearly spherical potential (zero dipole moment) should result in nearly atomic-like photoionization.

The rotationally-resolved ZEKE-PFI spectrum for the methyl radical is shown in the upper panel of Fig. 8. Methyl radicals were produced by supersonic-jet flash pyrolysis of azomethane, CH_3NNCH_3, as described elesewhere.[65] The rotational structure is quite similar to that observed for the bound-bound $\beta, \delta \leftarrow X$ bands in CH_3 and CD_3 which are parallel electronic transitions ($\Delta K = 0$) with simple P, Q and R ($\Delta N = -1, 0, +1$) branches reminiscent of a diatomic molecule.[66] Due to the similarity of the ground and excited state structures, each branch line is comprised of many closely spaced $\Delta K = 0$ sub-bands.[66,67]

The bottom frame of Fig. 8 gives the calculated threshold photoelectron spectrum for one-photon ionization out of the a_2'' orbital of CH_3 assuming a ground state rotational temperature of 250 K. The dominant $\Delta K = 0$ structure is clearly reproduced with $\Delta K = \pm 1$ transitions predicted to be too weak to be observed. The relative branch intensities are in reasonable agreement with experiment with some notable exceptions at lower energy. The anomolously high intensity observed for the $O(6)$ branch line is most likely a result of near resonant rotational autoionization.[65]

The observation of only $\Delta K = 0$ transitions for photoionization of CH_3 suggests that the ka_1' "parallel" continua dominate at threshold. Application of Eqn. (5), however, shows that both ka_1' and ke'' continua lead to $\Delta K = 0$; values of the dipole-allowed combinations of μ and λ can be readily obtained from the D_{3h} character table. For photoionization transitions to ka_1' continua one obtains $\mu = \lambda = 0$, and for transitions to ke'' one obtains $\mu = \lambda = \pm 1$. This yields the result that for *both* ka_1' and ke'' continua, $\Delta K = 0, \pm 2$, while *odd* values of ΔK are forbidden. Rotational transitions with $\Delta K = \pm 2$ are predicted to make up less than 5% of the observed lines in the threshold spectrum. In addition, individual $\Delta K = 0$ rotational transitions between the $^2A_2''$ ground state and the $^1A_1'$ ionic state involve a change in total parity, i.e. $\oplus\ominus \leftrightarrow \ominus\oplus$.[68] Consequently, dipole transitions between the neutral and final (electron plus ion core) states require that l = *even*, confirming the $(s,d) \leftarrow p$ atomic nature of CH_3 $(a_2'')^{-1}$ photoionization. This result also means that $l = 3\,(kfa_1'$ and $kfe'')$ continua, which are formally allowed by Eqn. (5), are parity forbidden.

The physical significance of these results is that (1) the small anisotropy of the CH_3^+ core potential does not induce mixing of l = *even* and l = *odd* partial waves at short-range; (2) the classification of particular bound-to-continuum transitions as parallel or perpendicular is inappropriate. The latter is a result of l-uncoupling of the continuum as the photoelectron final states evolve from short- to long-range. In this fragmentation process (frame transformation), the molecular symmetry orbitals (ka_1' and ke'') do not have a one-to-one correspondence with a specific photoelectron angular momentum (l) at long-range. An analogous "decompostion" occurs for the asymmetric top molecules discussed earlier, where the body-fixed photoelectron symmetry orbitals contribute to all three types (A, B, C) of observed bound-to-continuum rotational transitions.[69]

5 Summary

In this review, we have tried to emphasize the the dynamical aspects of molecular photoionization as determined from rotationally resolved photoelectron spectroscopy. By characterizing the quantum state distribution of the cation, the degenerate photoelectron continuum is "resolved" into its partial wave distribution, kl. Coupled with general angular momentum and parity constaints, these experimental studies permit a detailed examination of the interaction of the photoexcited electron at both short and long range with respect to the molecular ion core. Short range interactions induced by the anisotropy of the ion core potential determine the eventual fragmentation channels observed by experiment. In the ZEKE-PFI technique, external fields can promote otherwise forbidden or "forced" processes which are unique to the high-n character of the photoexcited electron at long range. Measurements as a function of molecular symmetry, e.g. CH_3, H_2O and CH_2CO, provide a more "chemical" view of photoionization by using the photoelectron continuum to reflect the charge distribution of the molecule and its cation.

An area not explicitly covered here involves the influence of vibrational motion on molecular photoionization dynamics. Specifically, linear polyatomic cations such as CO_2^+, N_2O^+ and HCN^+ exhibit extensive vibronic structure associated with the $\tilde{X}\,^2\Pi$ ground electronic state.[30,70,71,72] The latter results from the coupling of vibrational angular momentum, $\mathcal{L}$, acompanying excitation of the ν_2 bend vibration (Renner-Teller coupling). Conventional PES/TPES has provided partially resolved vibronic spectra for these molecules, however, rotational resolution is required to make unambiquous symmetry assignments. In addition, rotational state distributions of the Renner-Teller components of different overall vibronic symmetry provide alternative probes of the photoelectron continuum through the use of angular momentum and parity selection rules.[71] Other vibronic interactions which would greatly benefit from rotational resolution involve the long standing problems of Jahn-Teller distortions in small molecules such as CH_4 and BH_3 as well as the complex vibronic structure associated with the bent-to-linear geometry change in the vibrationally excited $\tilde{A}\,^2A_1$ electronic states of H_2O^+ and H_2S^+.[73,74]

6 Acknowledgments

The author would like to thank R. G. Tonkyn, R. T. Wiedmann, E. R. Grant, K. Wang, M.-T. Lee, V. McKoy, J. A. Blush and P. C. Chen for their collaborative efforts in this work. This work was performed at Brookhaven National Laboratory and supported by the US Department of Energy, Office of Basic Energy Sciences under contract No. DE-AC02-76CH00016.

References

[1] L. R. Khundkar and A. H. Zewail, *Annu. Rev. Phys. Chem.* **41** 15 (1990) and references therein.

[2] G. E. Hall and P. L. Houston, *Ann. Rev. Phys. Chem.* **40** 375 (1989) and references therein.

[3] D. W. Turner, C. Baker, A. D. Baker and C. R. Brundle, *Molecular Photoelectron Spectroscopy*, Wiley-Interscience, London, 1970.

[4] K. Siegbahn, C. Nordling, G. Johansson, J, Hedman, P. F. Hedén, K. Hamrin, U. Gelius, T. Bergmark, L. O. Werme, R. Manne and Y. Baer, *ESCA Applied to Free Molecules*, North-Holland, Amsterdam, 1969.

[5] see for example: J. W. Rabalais, *Principles of Ultraviolet Photoelectron Spectroscopy*, Wiley-Interscience, New York, 1977; J. Berkowitz, *Photoabsorption, Photoionization and Photoelectron Spectroscopy*, Academic Press, New York, 1979.

[6] L. Åsbrink, *Chem. Phys. Lett.* **7** 549 (1970).

[7] Y. Morioka, S. Hara and M. Nakamura, *Phys. Rev. A* **22** 177 (1980).

[8] J. E. Pollard, D. J. Trevor, J. E. Ruett, Y. T. Lee and D. A. Shirley, *J. Chem. Phys.* **77** 34 (1982).

[9] D. Villarejo, R.R. Herm and M. G. Inghram, *J. Chem. Phys.* **46** 4495 (1967); T. Baer, W. B. Peatman and E. W. Schlag, *Chem. Phys. Lett.* **4** 243 (1969); R. Spohr, P. M. Guyon, W. A. Chupka and J. Berkowitz, *Rev. Sci. Instrum.* **42** 1872 (1971).

[10] I. Nenner and J. A. Beswick, in G. V. Marr (ed.), *Handbook on Synchrotron Radiation*, Vol. 2, North Holland, Amsterdam, 1987, p.355 and references therein.

[11] Müller-Dethlefs, M. Sander and E. W. Schlag, *Chem. Phys. Lett.* **112** 291 (1984).

[12] L.A.Chewter, K. Müller-Dethlefs and E. W. Schlag, *Chem. Phys. Lett.* **135** 219 (1987).

[13] L. A. Chewter, M. Sander, K. Müller-Dethlefs and E. W. Schlag, *J. Chem. Phys.* **86** 4737 (1987).

[14] W. Habenicht, R. Bauman, K. Müller-Dethlefs and E. W. Schlag, *Ber. Bunsenges. Physik. Chem.* **92** 414 (1988).

[15] G. Reiser, W. Habenicht, K. Müller-Dethlefs and E. W. Schlag, *Chem. Phys. Lett.* **152** 119 (1988).

[16] W. A. Chupka, *J. Chem. Phys.* **98** 4520 (1992).

[17] A. H. Kung, *Opt. Lett.* **8** 24 (1983).

[18] C. T. Rettner, E. E. Marinero, R. N. Zare and A. H. Kung, *J. Phys. Chem.* **88** 4459 (1984).

[19] R. H. Page, R. J. Larkin, A. H. Kung, Y. R. Shen and Y. T. Lee, *Rev. Sci. Instrum.* **58** 1616 (1987).

[20] E. Cromwell, T. Trickl, Y. T. Lee and A. H. Kung, *Rev. Sci. Instrum.* **60** 2888 (1989).

[21] C. H. Greene and Ch. Jungen, *Advances At. Mol. Phys.*, **21** 51 (1985) and references therein.

[22] P. A. Schultz, R. M. Mead, P. L. Jones and W. C. Lineberger, *J. Chem. Phys.*, **77**, 1153 (1982); P. A. Schulz, R. M. Mead, W. C. Lineberger, *Phys. Rev.*, **A27**, 2229 (1983).

[23] M. S. Child and Ch. Jungen, *J. Chem. Phys.* **93** 7756 (1990).

[24] A. D. Buckingham, B. J. Orr and J. M. Sichel, *Phil. Trans. Roy. Soc. Lond. A* **268** 147 (1970).

[25] R. G. Tonkyn, J. W. Winniczek and M. G. White, *Chem. Phys. Lett.* **164** 137 (1989).

[26] H. Lefebvre-Brion, *Chem. Phys. Lett.* **171** 377 (1990).

[27] S. N. Dixit and V. McKoy, *Chem. Phys. Lett.* **128** 49 (1986).

[28] J. Xie and R. N. Zare, *J. Chem. Phys.* **93** 3033 (1990).

[29] K. Wang and V. McKoy, *J. Chem. Phys.* **95** 4977 (1991).

[30] R. T. Wiedmann, E. R. Grant, R. G. Tonkyn and M. G. White, *J. Chem. Phys.* **95** 746 (1991).

[31] M. Braunstein, V. McKoy and S. N. Dixit, *J. Chem. Phys.* **96** 5726 (1992).

[32] K. Wang, J. A. Stephens and V. McKoy, *J. Phys. Chem.* **97** 9874 (1993).

[33] R. G. Tonkyn and M. G. White, *Rev. Sci. Instrum.* **60** 125 (1989).

[34] J. W. Hepburn, *Laser Techniques in Chemistry*, A. Myers and T. R. Rizzo, editors (John Wiley, New York), to be published.

[35] H. H. Fielding, T. P. Softley and F. Merkt, *Chem. Phys.* **155** 257 (1991).

[36] G. H. C. New and J. F. Ward, *Phys. Rev. Lett.* **19** 556 (1967).

[37] S. E. Harris and R. B. Miles, *Appl. Phys. Letters* **19** 385 (1971); A. H. Kung, J. F. Young and S. E. Harris, *Appl. Phys. Letters* **22** 301 (1973).

[38] R. Hilbig, G. Hilber, A. Lago, B. Wolff and R. Wallenstein, in R. G. Lerner (ed.), *Short Wavelength Coherent Radiation : Generation and Applications*, AIP, NY, 1986, p.382.

[39] see for example: J. L. Dehmer, D. Dill and A.C. Parr, in S. P. McGlynn, G. L. Findley and R. H. Heubner (ed.), *Photophysics and Photochemistry in the Vacuum Ultraviolet*, D. Reidel, Dordrecht, Holland, 1985, p.341.

[40] I. Drabrowski and G. Herzberg, *Can. J. Phys.* **52** 1110 (1975).

[41] K. Yoshino and D. E. Freeman, *J. Opt. Soc. Am. B* **2** 1268 (1985).

[42] K. Müller-Dethlefs and E. W. Schlag, *Ann. Rev. Phys. Chem.* **42** 109 (1991).

[43] W. Kong, D. Rodgers, J. W. Hepburn, K. Wang and V. McKoy, *J. Chem. Phys* **99** 3159 (1993).

[44] S. T. Pratt, *J. Chem. Phys.* **98** 9241 (1993).

[45] J. Berkowitz, G. B. Ellison and D. Gutman, *J. Phys. Chem.* **98** 2744 (1994).

[46] H. Krause and H. J. Neusser, *J. Chem. Phys.* **97** 5923 (1992).

[47] X. Zhang, J. M. Smith and J. L. Knee, *J. Chem. Phys.* **99** 3133 (1993).

[48] L. Zhu and P. M. Johnson, *J. Chem. Phys.* **94** 5769 (1991).

[49] P. M. Johnson and L. Zhu, *Int. J. Mass Spect. Ion Proc.*, **131**, 193 (1994).

[50] C. Bordas and H. Helm, *Phys. Rev. A* **45** 387 (1992).

[51] K. S. Haber, Y. Jiang, G. P. Bryant, E. R. Grant and H. Lefebvre-Brion, *Phys. Rev. A* **44** R5331 (1991).

[52] C. R. Mahon, G. R. Janik and T. F. Gallagher, *Phys. Rev. A* **41** 3746 (1990).

[53] R. D. Gilbert and M. S. Child, *Chem. Phys. Lett.* **187** 153 (1991).

[54] R. G. Tonkyn, R. T. Wiedmann and M. G. White, *J. Chem. Phys.* **96** 3696 (1992).

[55] R. T. Wiedmann, M. G. White, K. Wang and V. McKoy, *J. Chem. Phys.* **98** 7673 (1993).

[56] R. T. Wiedmann, R. G. Tonkyn, M. G. White, K. Wang and V. McKoy, *J. Chem. Phys.* **97** 768 (1992).

[57] G. P. Bryant, Y. Jiang, M. Martin and E. R. Grant, *J. Phys. Chem.* **96** 6875 (1992).

[58] M.-T. Lee, K. Wang, V. McKoy, R. G. Tonkyn, R. T. Wiedmann, E. R. Grant and M. G. White, *J. Chem. Phys.* **96** 7848 (1992).

[59] M.-T. Lee, K. Wang and V. McKoy, *J. Chem. Phys.* **97** 3108 (1992).

[60] G. H. F. Diercksen, W. P. Kraemer, T. N. Rescigno, C. F. Bender, B. V. McKoy, S. R. Langhoff and P. W. Langhoff, *J. Chem. Phys.* **76** 1043 (1982).

[61] L. E. Machado, L. M. Brescansin, M. A. P. Lima, M. Braunstein and V. McKoy, *J. Chem. Phys.* **92** 2362 (1990).

[62] R. H. Page, R. J. Larkin, Y. R. Shen and Y. T. Lee, *J. Chem. Phys.* **88** 2249 (1988).

[63] G. Herzberg, *Molecular Spectra and Molecular Structure, III. Electronic Spectra and Electronic Structure of Polyatomic Molecules*, (Van Nostrand, New York, 1966), p. 112.

[64] R. N. Zare, *Angular Momentum*, Wiley-Interscience, New York, 1988, p.268.

[65] J. A. Blush, P. C. Chen, R. T. Wiedmann and M. G. White, *J. Chem. Phys.* **98** 3557 (1993).

[66] G. Herzberg, *Proc. R. Soc. London Ser. A* **262** 291 (1961).

[67] P. Chen, S. D. Colson, W. A. Chupka and J. A. Berson, *J. Phys. Chem.* **90** 2319 (1986).

[68] G. Herzberg, *Molecular Spectra and Molecular Structure, II. Infrared and Raman Spectra of Polyatomic Molecules*, (Van Nostrand, New York, 1950).

[69] K. Wang, M.-T. Lee, V. McKoy, R. T. Wiedmann and M. G. White, *Chem. Phys. Lett.* **219** 397 (1994).

[70] F. Merkt, S. R. McKenzie, R. J. Rednall and T. P. Softley, *J. Chem. Phys.* **99** 8430 (1993).

[71] R. T. Wiedmann and M. G. White, to be published.

[72] R. T. Wiedmann and M. G. White, *J. Chem. Phys.* **102** xxx (1995).

[73] R. N. Dixon, G. Duxbury, J. W. Rabalais and L. Åsbrink, *Mol. Phys.* **31** 423 (1976).

[74] G. Duxbury, Ch. Jungen and J. Rostas, *Mol. Phys.* **48** 719 (1983).

MOLECULAR INTERACTIONS FROM THE DYNAMICAL BEHAVIOUR OF POLYATOMIC GASEOUS MIXTURES

F.A. GIANTURCO
Department of Chemistry
The University of Rome
Città Universitaria, 00185 Rome, Italy

Abstract

The microscopic determination of intermolecular potentials for simple systems is discussed in terms of the types of experimental data which can be employed to obtain such potential energy functions. The elementary dynamics which presides over internal energy transfers under dilute conditions is analysed and its power for the interpretation of experiments is discussed in relation to crossed beam data, transport coefficients and field-dependent properties of the mixtures. The quantum and classical approach are further shown to be valid under different temperature conditions and various, specific atom-diatomics examples are discussed.

1. Introduction

Intermolecular forces embrace all forms of matter and preside over an extremely broad variety of processes and phenomena of the natural world as we know it. Their role needs to be described and understood if we wish to determine the properties of simple systems such as gases, liquids and solids but especially of more complex, and often chemically more interesting, systems. These are neither the simple liquids nor the regular solids but rather a myriad of dissolved solute molecules, small molecular aggregates or macroscopic particles interacting in liquid or vapour. It is the forces in such systems that ultimately determine the behaviour and properties of everyday things and therefore touch on a very broad area of phenomena in physics, chemistry, chemical engineering and biology: all areas which have witnessed tremendous advances in the last 15 years or so [1-3].

Because of the wide range of topics existing within each of these very different disciplines, and because of the diversity of the phenomena which can help us to uncover the effects of intermolecular forces, one is necessarily limited to look into specific sectors of investigations where the manifestations of molecular interactions are viewed almost in isolation, in order to gain in depth what one misses in breath. It is important, however, not to loose sight of the forest because of the trees and therefore it is essential

E. Yurtsever (ed.), Frontiers of Chemical Dynamics, 79–105.

to be aware of the basic unity of the intermolecular forces at a microscopic level, while applying at the same time a very narrow approach in order to understand as deeply as possible their appearance in a specific area of investigation.

Intermolecular forces could be loosely classified into three categories. First, there are those which are purely electrostatic in origin and arise from the Coulomb forces between charges. The interaction between charges, permanent dipoles, quadrupoles, etc. fall in this category. Secondly, one can generally talk about polarisation forces which originate from the various dipole moments induced in the atoms and the molecules of the medium by the electric fields of nearby charges and permanent dipoles. All interactions in a solvent medium, for instance, involve polarisation effects. Third, there are forces that are quantum mechanical in nature and give rise to covalent or chemical bonding (including charge transfer interactions) or to repulsive steric or exchange interactions that balance the attractive forces at very short distances.

These three categories should be considered as neither rigid nor exhaustive, as often an unambiguous classification is not possible. There is, in fact a further type of force which, like gravity, acts between all atoms and molecules. This is variously known as dispersion forces, London forces, charge fluctuation forces and induced-dipole-induced-dipole forces. This last type of interaction, together with two of the previous forces (i.e. excluding electrostatic forces), makes up the third and perhaps most important contribution to the total Van der Waals (VdW) force between atoms and molecules. Because such forces are always present, they play a role in a host of important phenomena and will thus be the subject of our study in this chapter. More specifically, we shall see how we can glean information about VdW forces in gaseous mixtures (dilute mixtures) at the most fundamental level by analysing a variety of dynamical processes which are very directly related to the manifestation of such molecular interactions.

Our aim is to show that, in spite of their apparent simplicity, the accurate determination of intermolecular VdW forces requires a rather active interplay between a broad range of experimental data and an equally broad set of theoretical and computational models.

2. Representation of Potential Functions

As mentioned above, when we deal with neutral, apolar molecules, i.e. with molecules with small dipole moments, the dispersion energy becomes the most important of the attractive, long-range interactions. They are therefore completely determined by the permanent multipole moments and by the static, as well as the frequency-dependent, multipole polarisabilities of the partner monomers. Most of the current work on intermolecular interaction potentials is concerned with closed-shell molecules and they will also be the main object of the present overview. It is also worth noting, however, that the interactions between open-shell molecules are especially interesting since, as a direct consequence of the relation between the spin and the permutation symmetry of the electronic wavefunction [4], different couplings between non-zero spin states of interacting open-shell partners will lead to different exchange interactions and, ultimately, to different short-range behaviour of the potential functions. Some of these potentials may correspond to the rising of chemical bonds just as it occurs between open shell atoms. In other cases one finds the occurrence of very weak bonds of the Van der

Waals (VdW) type which, however, turn out to be spin-dependent as in the case of the $(O_2)_2$ dimers [5].

For most applications and tests of the quality of the selected potential surface it is practical to write it as much as possible in analytic form. This means that one can try to adopt a rather flexible approach whereby different regions of the intermolecular interaction are assigned different parametric expressions which could be obtained iteratively via some comparison with specific experiments. It is the aim of the present chapter to show that scattering experiments and scattering-related non-equilibrium properties are indeed very good sources for the accurate determination of various regions of overall phase space involved in defining the full range of action for the PES under study.

One of the most general representations of the interaction surface could be obtained via a generalized spherical expansion. In this approach, the orientational dependence (anisotropy) of the intermolecular potential between two arbitrary, non-linear molecules A and B is explicitely expressed in (symmetric top) free-rotor functions $D^{L_A}_{M_A K_A}(\Omega_A)^*$ for molecule A and $D^{L_B}_{M_B K_B}(\Omega_B)^*$ for molecule B. The Euler angles $\Omega_A = (\alpha_A, \beta_A, \gamma_A)$ and $\Omega_B = (\alpha_B, \beta_B, \gamma_B)$ describe the orientations of the molecules with respect to some arbitrary space-fixed (SF) coordinate reference system. The direction of vector **R**, which connects the center of mas (c.o.m.) of molecule A to that of molecule B, is given with respect to that SF frame via the polar angles (Θ, Φ). The ensuing intermolecular potential can therefore be expressed by the following expansion over orthonormal functions [6]

$$V(\mathbf{R}, \mathbf{q}_A, q_B, \Omega_A, \Omega_B) = \sum_{L_A K_A L_B K_B L} v^{K_A K_B}_{L_A L_B L}(R, q_A, q_B) \cdot \tag{1}$$

$$\cdot A^{K_A K_B}_{L_A L_B L}(\Omega_A, \Omega_B, \Theta, \Phi)$$

where one has defined the angular, complete basis set as

$$A^{K_A K_B}_{L_A L_B L}(\Omega_A, \Omega_B, \Theta, \Phi) = \sum_{M_A M_B M} \begin{pmatrix} L_A & L_B & L \\ M_A & M_B & M \end{pmatrix} D^{L_A}_{M_A K_A}(\Omega_A)^* \cdot \tag{2}$$

$$\cdot D^{L_B}_{M_B K_B}(\Omega_B)^* C^L_M(\Theta, \Phi)$$

The new functions $C^L_M(\Theta, \Phi)$ are spherical harmonics in the Racah normalisation and the expression in the large round brackets is a 3-j symbol [7]. The unknown quantities are now the expansion coefficients $v^{K_A K_B}_{L_A L_B L}$ which depend on the radial distance R between the molecules, and on the radial internal coordinates for each of the partner molecules, q_A and q_B. Any intermolecular pair potential can therefore be expressed to any desired accuracy, over the whole range of interest of the above variables, whenever a sufficient number of such expansion coefficients is given for the system at hand. For molecular partners which exhibit well-defined equilibrium structures (i.e. for what we shall call nearly-rigid molecules) it is often expedient to make a Taylor expansion, for each of the coefficients obtained in eq.(1), about the equlibrium values of the internal coordinates q_A and q_B. However, only for very simple systems it has been possible to reliably obtain the further coefficients of such Taylor expansions [8]. In most cases, therefore one makes use of the rigid-molecule approximation, i.e. one replaces q_A and q_B by their equlibrium values or one assumes the usual Born-Oppenheimer separation between internal and external molecular coordinates, thereby replacing q_A and q_B by their values averaged over some ground-state vibrational wavefunction of each of the involved partners. Such an approach would produce what are often called vibrationally adiabatic expansion coefficients in eq. (1) [9].

In some specific cases and with a specific choice of the coordinate frame of reference the general form of the spherical expansion (1) can be simplified considerably. If one chooses, in fact, a body-fixed (BF) frame of reference with its z-axis along the vector $\underline{R}$, one may make use of the fact that $C_M^L(0,0) = \delta_{M,O}$ and therefore simplify the coefficients of eq.(2). For linear molecules the free-rotor functions are simply given by spherical harmonics

$$D_{MO}^L(\alpha,\beta,\gamma)^* = C_M^L(\beta,\alpha) = \left[\frac{4\pi}{2L+1}\right]^{1/2} Y_M^L(\beta,\alpha) \tag{3}$$

Furthermore, if either A or B is an atom in an S electronic state, one can use $C_o^o(\beta,\alpha) = 1$ and one can therefore obtain the well-known Legendre expansion for atom-diatom potentials

$$V(\mathbf{R},q_A,\beta_A) = \sum_{L_A} V_{L_A}(R,q_A) P_{L_A}(\cos\beta_A) \tag{4}$$

The coordinate q_A is now treated as a scalar quantity and the angle β_A is the internal angle between $\hat{R}$ and $\hat{q}_A$. Alternatively, the spherical harmonics expansion for an atom-molecule potential

$$V(\mathbf{R},q_A,\beta_A,\gamma_A) = \sum_{L_A K_A} V_{L_A K_A}(R,q_A) C_{K_A}^{L_A}(\beta_A,\gamma_A) \tag{5}$$

Once a specific frame of reference is chosen and the potential is written as some specific sum over expansion coefficients, then one is left with the determination of the analytic functions that are used to describe such coefficients. In principle, each of them can be obtained numerically by an angular quadrature over the initial, full PES. For example, if we consider the atom-molecule potential of eq. (5) as given for the specific example of a target molecule with C_{3v} symmetry (e.g. the NH_3 rigid-rotor), then eq. (5) can be written as

$$V(\mathbf{R},\vartheta,\varphi) = \sum_{\lambda \geq \mu \geq o} V_{\lambda\mu}(R) C_{\mu}^{\lambda}(\vartheta,\varphi) \tag{6}$$

where

$$C_{\mu}^{\lambda}(\vartheta,\varphi) = \frac{1}{\left(1+\delta_{\mu o}\right)} \left\{ Y_{\lambda}^{\mu}(\vartheta,\varphi) + (-)^{\mu} Y_{\lambda}^{-\mu}(\vartheta,\varphi) \right\} \tag{7}$$

and λ is the L_A index with respect to a BF frame where (ϑ,φ) are the polar angles of the vector R within that frame. The λ-projection onto the z-axis of the molecular figure axis is now symmetry-restricted to be given by $\mu=3n$, with n positive integer. Furthermore, if the actual values of the intermolecular potential on the l.h.s. of eq. (6) are given over a grid of variable values, then the required coefficients of the symmetry-adapted multipolar expansion (SAME) can be obtained by numerical quadratures

$$V_{\lambda\mu}(R) = 2\int_{o}^{\pi} d\varphi \int_{-1}^{1} d(\cos\vartheta) V(\mathbf{R},\vartheta,\varphi) C_{\mu}^{\lambda}(\vartheta,\varphi) \tag{8}$$

The further problem of fitting the numerical results with some sort of parameter-dependent analytic expression is a well-known task if one wants to increasingly optimize the overall PES by comparisons with existing experiments and by successive adjustments of the involved parameters [10,11]. One of the most widespread approximations, one which is most commonly used for Montecarlo calculations of equilibrium structures, is to write it as a sum of atom-atom potentials which are assumed to be isotropic, i.e. to depend only on the distance R_{ab} between the atom a in the molecule A and atom b in molecule B. It is in turn written very often as a parameter-dependent Lennard-Jones (LJ) function and applied to the study of condensed phases [12] or to the equilibrium structures of complex VdW systems [13] for which only experimental bulk data are available. The reason why such simple potential are so popular in the study of the condensed phases is that they contain relatively few parameters while they still describe (albeit implicitely) the anisotropy of the intermolecular potentials and they even model, in a qualitative way, its dependence on the internal molecular coordinates. Furthermore, they are often believed to be transferable which signifies that the same atom-atom interaction parameters can be used for the same pairs of atoms in different molecular environment. The main drawback of such potentials, however, is that their parameters are usually obtained from properties which are related to a fairly limited range of interatomic distances and therefore it is not possible to use them outside such domains, which is what is often required when dynamical properties in the gas phase are employed to test the intermolecular forces [14].

In the multipolar expansions discussed above, on the other hand, one is left with the apparently simpler problem of selecting a functional form of the radial coefficients and to obtain one which can be employed and tested over a wide range of R distances between the two collision partners. Here again a possibly simple approach is to select exponential functions to describe the short-range (SR) repulsive region:

$$V_{AB}^{SR}(R,\Omega) = \exp\left[-\alpha^{SR}(\Omega)(R-\rho(\Omega))\right] \tag{9}$$

in which the 'slope' parameter $\alpha(\Omega)$ and the 'range' parameter $\rho(\Omega)$ are assumed to be functions of the molecular orientations, Ω_A and Ω_B. Such functions may be then expanded in terms of the angular basis functions of eq.s (2), (4) or 5), thereby occurring non-linearly in the potentials. As a result of this, the ensuing expansions may need fewer terms to converge.

In the case of weak interactions between neutral species, the VdW potentials required usually exhibit an attractive well which depends on the overall anisotropy of the full PES, $\varepsilon(\Omega)$, which can be treated as the previous parameters and included in the more general form of a Morse potential [15]

$$V_{AB}^{M}(R,\Omega) = \varepsilon(\Omega)\left\{1-\exp\left[-\alpha_M(R-\rho(\Omega))\right]\right\}^2 \tag{10}$$

which is connected with the SR part via a spline function on the inner region of the well.

The outer part of the interaction that describes the dispersion forces between the molecular partners is in turn introduced via long-range (LR) coefficients that are made to depend on orientational effects:

$$V_{AB}^{LR}(R,\Omega) = -\frac{C_6(\Omega)}{R^6} - \frac{C_8(\Omega)}{R^8} - \frac{C_{10}(\Omega)}{R^{10}} - \frac{C_{12}(\Omega)}{R^{12}} \tag{11}$$

and which are then connected with the region of the well via another set of spline function coefficients [10].

In conclusion, the multiparameter description mentioned above is a very flexible model that allows one to separately improve on the parameter choice by selecting, as we shall see below, different dynamical properties or different values of collision energy for which one compares calculations with existing experiments [14,15]. The acronym used for the above parametrisation of the expansion coefficients in eq. (1) is that of ESMSV, or E(xponential)-S(pline)-M(orse)-S(pline)-V(an der Waals) [17]. One can further expand on the multiparameter treatment and employ a set of Morse functions, with different 'range' and 'size' parameters each, to describe the repulsive wall, the well region and the onset of the attractive part of the potential [18], respectively. The use of three different functions in the inner region and the intermediate range of distances allows for greater flexibility in adjusting the parameter values to specific features which can be obtained from scattering experiments (see below) [15] and are further connected to the LR form of eq. (11) through spline functions either at each of the Ω values selected to describe the interaction or directly within the radial coefficients of expansion (4). The ensuing full potential is usually called the M3SV functional representation of the atom-molecule potential energy functions [19].

It should have become clear from the above discussion that one can basically have two different approaches to the evaluation of interaction potentials between molecular systems:

(i) to evaluate specific values of the potential on the l.h.s. of eq. (1), over an extensive grid of relative distances and of internal molecular coordinates, using ab initio methods and striving to attain as high an accuracy as possible in getting the largest feasible amount of grid points [20]. This approach is obviously computationally very intensive and can only be applied to relatively simple systems, and mostly to those which have stronger interactions than the weaker dispersion interactions [21]. Its use in obtaining the weaker VdW potentials between neutral partners has only been partly successful, in fact, and often over a rather limited range of relative geometries [22]. It is often considered a sort of 'brute force' approach which allows one to test to its limit the capabilities of the broad variety of Many-Body Perturbation Theory (MBPT) treatments which are available today. We will not further discuss this specific approach in the present chapter but rather examine in detail what possibilites there are for testing its results vis à vis specific experimental data in the domain of dynamical effects from the features of the given PES.
A variant of the direct calculations mentioned above is to carry out the spherical expansion of eq.(1) and then to obtain some of the needed expansion coefficients from various types of information which are partly determined from ab initio calculations, especially for the evaluation of the contributions to the asymptotic R^{-n} coefficients of the potential where they are completely determined by monomer properties. The computation of molecular multiple moments is today rather straightforward, both at the Self-Consistent Field (SCF) level and with the further inclusion of electron correlation effects. On the other hand, the calculation of accurate frequency-dependent polarisabilities is still not a routine job with ab initio methods [23,24] and therefore the direct evaluation of expansion coefficients in eq.(1) is not yet a viable possibility for interaction forces between complicated molecular monomers;

(ii) the other alternative approach, as partly mentioned in the previous discussion, is to generate either side of eq.(1), i.e. either the full PES as function of molecular internal coordinates and of relative distances and orientations or the numerous spherical expansion coefficients on the r.h.s. of that equation, by using parameter-dependent functions where the selection of the parameters is guided by three different possible schemes that we shall list below. It is evident that the latter approach relies heavily of the capability of such schemes to provide as unique a set of parameters as possible from either the analysis of available data or from specific theoretical models.

The first scheme is to combine ab-initio data with a model for the overall behaviour of the expansion coefficients, thereby obtaining parameters from that model treatment. Chief examples of such an approach have been the Hartree-Fock-plus-Damping (HFD) model [25] and the Tang-Toennies (TT) damping model [26], although other empirical procedures have been suggested along similar lines [27]. The main idea behind such methods is to provide the SR part of the coefficients (or of the full interaction) from SCF-Hartree Fock (HF) calculations that are inherently simple in nature and to then obtain the LR behaviour from R^{-n} expansion coefficients. The intermediate region which often includes the well region is then obtained through empirical damping formulae which usually require no additional parameters. Clearly one can then use the coefficients of the spherical expansion, or of the simplified expansions (4) and (5), which have been obtained with such method to determine the needed parameters in a functional representation of the ESMSV or of the M3SV variety.

The second possible scheme to determine the parameters is, obviously, to adjust any of the most commonly used functional forms to an extensive set of ab initio points obtained from quantum chemical calculations. Provided that a sufficiently large number of points be available, one can fix quite accurately the various 'range' parameters, exponential parameters or 'slope' parameters and the well depth parameters. However, such an approach is essentially based on having solved the problem through the (i) procedure of before and wanting now to test the PES thus obtained by using a flexible, functional representation of the intermolecular potentials. It therefore suffers from the same limitations and difficulties of any fully ab initio treatment and has been used only over a very limited number of cases [8].

The third possible scheme, and the one which has been used most often, is to start off with a pre-determined functional form which guarantees some flexibility in its set of parameters and to then adjust the parameters to a broad variety of experiments, either of spectroscopic nature for the internal molecular structures and for the bound state region of the clusters [28,29] or from scattering experiments which sample a large range of intermolecular distances and mostly deal with the continuum states of the cluster system [30,31]. This latter possibility essentially means that one needs to develop a closer link between the various feature of the intermolecular potentials and the dynamical observables which appear in the different experiments. Such a procedure has been employed under the general name of multiproperty analysis of a given representation for a specific potential energy function which one wants to determine [32,33] and will be the main object of the present study.

What one is required to do is to examine more in detail at least two groups of experiments which are related to dynamical behaviour: the microscopic dynamics of single collision events as brought out in high resolution molecular beam experiments and its connection with the statistical behaviour of non-equilibrium gases as it appears in the measurements of transport and relaxation coefficients of molecular mixtures. The following Sections 3 and 4 will briefly discuss those two aspects, respectively, while Section 5 will present some specific example.

3. The Single Collision Processes

As mentioned before, molecular beam scattering data can yield valuable information on the intermolecular potentials. The well-known techniques used for atomic collisions [32] can be applied to molecular systems and provide total cross sections, summed over the initial and final internal states of the partner molecules [14].

More sophisticated techniques are also available for measuring state-to-state inelastic cross sections between specifically selected molecular partners, either by time-of-flight techniques [34] or by using state-selective detection methods [35]. Under favourable conditions, elastic and inelastic differential cross sections (DCS) can be directly inverted to test specific parts of an anisotropic potential [36]. However, such an inversion requires very high quality experimental data and data over a fairly wide range of relative collision energies: as a result, it has only been used in relatively few cases. In general, one can say that total cross sections (TCS) provide information mainly on the spherical part of the interaction potential, whereas inelastic cross sections, integral and differential, provide information on the anisotropy. Nevertheless, some information on the orientational features of the intermolecular potential can be obtained from the damping of the quantum oscillations in either the DCS or the integral cross sections (ICS). For collisions with small reduced mass of the cluster (e.g. the systems which involve He atoms as one of the partners) the differential cross sections are dominated by diffraction oscillations, and the damping of these is principally sensitive to the anisotropy of the repulsive part of the PES. When larger reduced masses are involved, the diffraction oscillations (even at low collision energies) are more difficult to resolve and the rainbow oscillations become more important for the determination of the potential parameters [37]. These oscillations are also damped by the potential anisotropy, but in this case the damping is most sensitive to the anisotropy of the attractive region [38].

Since the scattering dynamics of the elastic and inelastic processes are inextricably connected during the single collision event, a quantum treatment of the physics involved (to extract the relevant scattering amplitudes) starts with writing down the total Hamiltonian that contains in it the internal Hamiltonians for the monomer partner molecules [39]. The simplest situation is the one where one partner is a structureless atom and the other a diatomic molecule; they are considered to interact through a single PES and in a given initial state $|\alpha\rangle$, with total energy $E_\alpha = 2\mu k_\alpha^2 + \varepsilon_\alpha$, with $\underline{k}_\alpha$ being the relative wavevector and ε_α the molecular internal energy content.

$$H = \frac{\hbar^2}{2\mu R} \cdot \frac{\partial^2}{\partial R^2} \mathbf{R} + \frac{\hat{L}^2}{2\mu R^2} + H_M + V(R, r, \gamma) \qquad (12)$$

where μ is the reduced mass of the system, $\hat{L}$ the relative orbital (transitional) angular momentum operator and H_M internal Hamiltonian of the isolated molecule. Here $\mathbf{R}, \mathbf{r}$ are the vector quantities giving the atom-molecule c.m. distance and the molecular bond distance respectively, while γ represents their relative angle of orientation: $\arccos\gamma = \hat{R} \cdot \hat{r}$

$$H_M = \frac{\hbar^2}{2mr} \frac{\partial^2 \mathbf{r}}{\partial r^2} + \frac{\hat{j}^2}{2mr^2} + U(r) \qquad (13)$$

The $\hat{j}$ operator now refers to the molecular rotations and U(r) is the ground-state, B.O. potential energy curve for the diatomic molecule. The unperturbed states of interest here are given by

$$(\varepsilon'_\alpha - H_M)\Phi_{\alpha'} = 0 \qquad (14)$$

where the corresponding molecular eigenfunctions Φ'_α, constitute a convenient basis for representing the total wavefunction Ψ_α for the scattering process

$$\Psi_{\alpha}(\mathbf{R},\mathbf{r}) = \sum_{\alpha'} \Phi_{\alpha'}(\mathbf{r}) F_{\alpha\alpha'}(\mathbf{R}) \tag{15}$$

with the unknown expansion coefficients representing the relative translational part of the wavefunction given in its initial state $|\alpha>$ and outgoing in the final channel $|\alpha'>$. Since the expansion is over an infinite set, the channel $|\alpha'>$ physically describe both the energetically accessible (open) final channels and the inaccessible (closed) final channels, depending on the selected $\underline{k}_{\alpha}$ value for the scattering event.

Substituting expansion (15) into eq.(16)

$$(E_{\alpha} - H)\Psi_{\alpha} = 0 \tag{16}$$

gives

$$\sum_{\alpha'} \Phi_{\alpha'}(\mathbf{r}) \left\{ \frac{\hbar^2}{2\mu R} \frac{\partial^2}{\partial R^2} + E_{\alpha} - \varepsilon_{\alpha'} - \frac{\hat{L}^2}{2\mu R^2} \right\} F_{\alpha\alpha'}(\mathbf{R}) = \tag{17}$$
$$= V(R, r, \gamma)\Psi_{\alpha}(\mathbf{R}, \mathbf{r})$$

Multiplying on the left by $\Phi^{*}_{\alpha'}(\underline{r})$ and integrating over $\underline{r}$, gives the following equation:

$$\left\{ \frac{\hbar^2}{2\mu} \frac{\partial^2}{\partial R^2} + E_{\alpha} - \varepsilon_{\alpha'} - \frac{\underline{L}^2}{2\mu R^2} \right\} F_{\alpha\alpha'}(\underline{R}) = \sum_{\alpha''} F_{\alpha\alpha''}(\underline{R}) U_{\alpha\alpha''}(\underline{R}) \tag{18}$$

where:

$$U_{\alpha'\alpha''}(\underline{R}) = \int \phi^{*}_{\alpha''}(\underline{r}) U(r, R, \gamma)\ \phi_{\alpha'}(\underline{r}) d\underline{r} \tag{19}$$

corresponds to a coupling matrix element between the PES and the rotovibrational, unperturbated, states of the target molecule. From eq.(18) one sees that such states are all coupled together by the potential and affect the specific form of each collision exit channel $F_{\alpha\alpha'}(R)$.

Given that a certain number of states must be retained in the eq.(18) according to the energy range of the process considered and to the strength of the coupling elements of eq.(19), the first task is to reduce the number of the coupled equations that need to be solved by a judicious choice of the functions used in eq.(15). In case of simple rotors, the angular parts of (15) are chosen to be eigenfunctions of the total angular momentum $\hat{\mathbf{J}} = \hat{\mathbf{l}} + \hat{\mathbf{j}}$, which leads to a block-diagonalisation of the matrix **U** of (19) and solutions are then seeked for only one of these blocks, for each value of $\hat{\mathbf{J}}$. The final coupled equations then become 1D coupled equations where the angular variables have been separated out through standard angular momentum coupling algebra [39]:

$$\frac{d^2}{dR^2}\mathbf{u}(R) + \left\{\mathbf{K} - \mathbf{W}(R) - \mathbf{L}(\mathbf{L}+\mathbf{I})R^{-2}\right\}\mathbf{u}(R) = \mathbf{O} \quad (20)$$

where **u** is a column vector, **K** and **L** constant diagonal matrices and **W** is the coupling matrix. The elements of **L** are given by non-negative integers l_i which label the relative orbital angular momentum and **I** is the unit matrix. The dimension of the vector space is N and controls the size of the required coupled equations (CC) needed to attain convergence on the individual elements of the scattering **S**-matrix [39].

One immediately sees that the range of action of the PES controls the size and strength of the matrix **W** and therefore is affecting the form of the solution vectors. They are in turn producing the required phaseshifts which go into the scattering S-matrix that finally yields the cross sections [39]. Even when only vibrational and rotational states are considered, the vector space N is often very large, although at low collision energies the actual number of open channels may be rather limited. It therefore follows that, for the test of the PES which experiments at relatively low energies, rather few vibrational functions may be needed but a major problem is still given by the number of rotational levels. Because of the (2j+1) degeneracy associated with the energy levels of the molecular rotor, the number of coupled equation that need to be solved increases rapidly with the maximum value of j included in the basis set used in the expansion [15]. As the computational time required to solve the coupled equations depends on the cube of the number in the set, the task becomes rapidly very difficult even at quite low J values. Consequently, a large number of approximate methods have been developed in the last few years, in an attempt to reduce solution time without significant loss of accuracy [39,40].

One of the most popular ways of markedly reducing the space of the u vectors is the so called Infinite-Order-Sudden-Approximation (IOSA) [39,41], whereby two basic simplifications are induced in the diagonal matrices **K** and **L**. The first one is the centrifugal sudden approximation (CS) which replaces the angular momentum matrix by a prechosen 1 value, l_o:

$$\mathbf{L}(\mathbf{L}+\mathbf{I}) \approx \hbar^2 l_o(l_o+1) \quad (21a)$$

and the second one is the energy sudden (ES) approximation which simplifies the wavevector matrix by selecting one reference energy level for all coupled channels:

$$\mathbf{K} \approx k_o \quad (21b)$$

For a rigid rotor, i.e. disregarding for the moment the molecular vibrations, the above replacements result in an uncoupled set of ordinary differential equations for each value of the relative angle γ in the coupling potential of eq.(12):

$$\left\{\frac{d^2}{dR^2} + k_o^2 - \frac{l_o(l_o+1)}{r^2} - \frac{2\mu}{\hbar^2}V(R;\gamma)\right\}u_{l_o}(R,\gamma) = 0 \quad (22)$$

The angle dependent radial wavefunctions u_{l_o} (R;γ) are obtained by solving eq.(22), for each of the contributing l_o values and for each selected γ value, subject to the usual scattering boundary conditions

$$g_{l^o}(R;\gamma) \underset{R\to\infty}{\approx} k_o^{1/2}\left\{e^{-i(kR-l_o\pi/2)} - e^{-\left[\left(kR-l_o\pi/2+2i\eta_{l_o}(\gamma)\right)\right]}\right\} \quad (23a)$$

and

$$g_{l_o}(0;\gamma) = 0 \quad (23b)$$

The phaseshift is obtained for each fixed orientation and produces in turn an angle dependent scattering amplitude

$$f(\gamma|\theta) = \frac{i}{2k_o}\sum_{l=o}^{\infty}(2l+1)\left[1-e^{2i\eta_l(\gamma)}\right]P_l(\cos\theta) \quad (24)$$

It can then be shown [39] that the TDCS is simply given here by the weighted orientation average

$$T(\theta|k_o) = \frac{1}{2}\int_{-1}^{1} I(\gamma,\theta)\; d\cos\gamma \quad (25)$$

of the effective 'central field' DCS, at each γ, obtained from eq.(24) for each chosen value of the wavevector k_o

$$I(\gamma|\theta) = \frac{1}{4k^2}\left\{\left[\sum_l (2l+1)sin[2\eta_l(\gamma)]P_l(\cos\theta)\right]^2 + \right. \quad (26)$$

$$\left. + \left[\sum_l (2l+1)\cos[2\eta_l(\gamma)-1]P_l(\cos\theta)\right]^2\right\}$$

Moreover, the degeneracy-averaged, state-to-state DCS can be determined [39] from the transition out of j=0

$$I^{DA}(j \leftarrow j|\theta) = \sum_{j''=|j-j'|}^{j+j'} C^2(j,j'',j',0,0,0)^2 I(j'' \leftarrow 0|\theta) \qquad (27)$$

where

$$I(j' \leftarrow j|\vartheta) = |F_{j'}(\vartheta)|^2 \qquad (28)$$

and the coefficients on the r.h.s. of eq.(28) are obtained from a multipolar expansion of the orientation-dependent scattering amplitude of eq. (24)

$$f(\gamma|\vartheta) = \sum_L F_L(\vartheta)P_L(\cos\gamma) \qquad (29)$$

at the chosen value of k_o for the collision energy [41].

The above simplification of the dynamics has been found to provide, at least for VdW systems involving weak anisotropic intermolecular forces [42], a fairly reliable way of evaluating the range and steepness parameters of the spherical part of the potential function and a rather realistic starting point for estimates of the lower anisotropy coefficients in expansions (4) and (5). It is usually considered valid when two conditions are met:

(i) The relative kinetic energy is large compared to the rotational energy spacing, i.e. the ES is acceptable;

(ii) Rotational transitions are dominated by relatively small values of the orbital angular momentum and the centrifugal barrier is therefore less important than the main potential in the regions of the relevant rutning points (CS approximation).

In most collision experiments in molecular beams that involve Van der Waals (VdW) systems, the above conditions have been tested for many cases [30,31,19,15,43] and found to be satisfied at the energy considered. Thus, the state-to-state DCS of eq.(27) can be easily computed and used to test the anisotropic part of the PES. This method of testing is of course more reliable than the one based only on quenching effects in TDCS, as discussed in the previous Section, since in the latter case the desired anisotropy is directly obtained from experimental cross sections.

$$I^{DA}(j \leftarrow j|\theta) = \sum_{j''=|j-j'|}^{j+j'} C^2(j,j'',j',0,0,0)^2 I(j'' \leftarrow 0|\theta) \qquad (27)$$

where

$$I(j' \leftarrow j|\vartheta) = |F_{j'}(\vartheta)|^2 \qquad (28)$$

and the coefficients on the r.h.s. of eq.(28) are obtained from a multipolar expansion of the orientation-dependent scattering amplitude of eq. (24)

$$f(\gamma|\vartheta) = \sum_L F_L(\vartheta) P_L(\cos\gamma) \tag{29}$$

at the chosen value of k_o for the collision energy [41].

The above simplification of the dynamics has been found to provide, at least for VdW systems involving weak anisotropic intermolecular forces [42], a fairly reliable way of evaluating the range and steepness parameters of the spherical part of the potential function and a rather realistic starting point for estimates of the lower anisotropy coefficients in expansions (4) and (5). It is usually considered valid when two conditions are met:

(i) The relative kinetic energy is large compared to the rotational energy spacing, i.e. the ES is acceptable;

(ii) Rotational transitions are dominated by relatively small values of the orbital angular momentum and the centrifugal barrier is therefore less important than the main potential in the regions of the relevant rutning points (CS approximation).

In most collision experiments in molecular beams that involve Van der Waals (VdW) systems, the above conditions have been tested for many cases [30,31,19,15,43] and found to be satisfied at the energy considered. Thus, the state-to-state DCS of eq.(27) can be easily computed and used to test the anisotropic part of the PES. This method of testing is of course more reliable than the one based only on quenching effects in TDCS, as discussed in the previous Section, since in the latter case the desired anisotropy is directly obtained from experimental cross sections.

4. The Non-equilibrium Bulk Properties

The kinetic theory of dilute gases establishes that the transport coefficients are related to the forces between pairs of molecules and therefore provides observables which carry information on molecular interactions [6]. The molecular diffusion coefficient, the viscosity coefficient in pure gases and in gaseous mixtures, the thermal conductivity, the thermal diffusion and the virial coefficients, all quantify the difficulty for the transport of mass, momentum and energy in a molecular gas subjected to gradients of concentration, velocity or temperature [44]. Because in a diluted gas this transport is achieved by motion of the molecules, the nature of the collision that the molecules undergo influences the difficulty of transport. Furthermore, since the outcome of the binary collisions is determined by the form of the intermolecular potential, the latter is going to influence the transport properties in a rather complicated but specific way and therefore the study and the calculation of such properties can provide a useful approach to the accurate evaluation of intermolecular forces.

The fundamental kinetic theory for transport coefficients can be formulated, however, in a way which involves only a set of well defined collision integrals [6,44] in addition to molecular masses, temperature and pressure in the gaseous mixture. These collision integrals represent variously weighted, energy-averaged cross sections for the binary encounters between molecules in the dilute gas.

The most significant features of the above theory therefore is that each of the transport coefficients in the gas mixture can be expressed in terms of well defined integrals over the relevant PES for each of the possible binary encounters in the system.

Thus, provided that the relevant PES is either known or estimated from other sources, it becomes possible to calculate the transport properties of a gas consisting of such molecules to any desired degree of accuracy at any temperature.

The great improvement, over the last ten years or so, that has occurred on the quality of the experimental determination of the various transport coefficients, and the corresponding increased efficiency of the computational tools used to generate collision integrals, is now making possible to test different potential surface against the outcome of experiments over rather wide ranges of temperature. In the following we briefly summarize the theoretical formulation and give some specific examples on how effective the various properties are in testing potential functions. A more detailed treatment of the theory can be found in a recent, well documented monograph on the subject [45].

The necessary generalized collision integrals are defined by [6,45]:

$$\Omega_{ij}^{(n,s)} = \frac{1}{2}\left(\frac{K_B T}{2\pi\mu}\right)^{1/2} \frac{1}{(K_B T)^{s+2}} \int d\zeta \zeta^{s+1} e^{-\zeta/K_B T} Q^{(n)}(\zeta) \tag{30}$$

where i and j refer to the collision partners and the $Q^{(n)}(\zeta)$ are the generalized integral cross sections:

$$\begin{aligned} \Omega^{(n)}(\zeta) &= Z_r^{-1} \sum_{j=0}^{\infty} (2j+1) e^{-\varepsilon_j/K_B T} \times \\ &\times \sum_{j'=0}^{\infty} \int d(\cos\theta) I^{DA}(j' \leftarrow j\,\theta) \Phi_n(\zeta,\theta) \end{aligned} \tag{31}$$

where the degeneracy-averaged differential cross sections of eq.(27) are used and:

$$\Phi_1(\zeta,\theta) = 1 - (\zeta'/\zeta)^{1/2} \cos\theta \tag{32a}$$

$$\Phi_2(\zeta,\theta) = \left(\frac{\zeta'}{\zeta}\right)\left(1 - \cos^2\theta\right) + \frac{1}{3}\left(1 - \frac{\zeta'}{\zeta}\right)^2 \tag{32b}$$

In the expressions $\zeta = \hbar^2 k^2 / 2\mu$ and $\zeta' = \zeta - (\varepsilon_{j'} - \varepsilon_j)$ are the incident and final relative kinetic energies, and Z_r is the rotational partition function for the partner molecules in an atom (ion)-molecule mixture:

$$Z_r = \sum_{j=0}^{\infty} (2j+1)\, e^{-\varepsilon_j/K_B T} \tag{33}$$

and ε_j is the rotational energy of the isolated molecule. If one assumes, consistently with the IOS approximation discussed before, that $\zeta = \zeta'$, then the Φ_n became independent of energy and are given by:

$$\Phi_n(\zeta, \theta) \simeq \Phi_n(\theta) = 1 - \cos^n \theta \tag{34}$$

If one uses the above prescription in eq.s (32), then the generalized integral cross sections are considerably simplified as the final and initial rotational states can be summed over analytically to give:

$$Q^{(n)}(\zeta) = \frac{1}{2}\int_{-1}^{1} Q^{(n)}(\zeta;\gamma)d(\cos\gamma) \tag{35}$$

which in turn contains the following generalized IOS, angle-dependent cross section [45]:

$$Q^{(n)} = \frac{4\pi}{k^2}\sum_l \sum_j C_{lj}^{(n)} sin^2\left[\eta_{l+j}(\gamma) - \eta_l(\gamma)\right] \tag{36}$$

where the phaseshifts are those appearing in eq.(26) and the coupling coefficients on the r.h.s. of (36) are given by

$$C_{lj}^{(n)} = \sum_{k=j}^{n} \frac{2^k(2k+1)n![(n+k)/2]!(2l+1)(2l+2j+1)}{(n+k+1)![(n-k)/2]!(2l+1)(2l+2j+1)} \cdot \tag{37}$$

$$\cdot\, C^2(k,l,k+l;\ 0,0,0)$$

Thus, IOSA cross sections producing the $\eta(\gamma)$ can be directly, and rapidly, used to produce the relevant collision integrals. Only recently, however, has the IOS approximation been tested against rigorous calculations [46-48] and has been proved to be accurate, at least for the He-N_2 case, within 1-2%. The IOS scheme is therefore the one quantum approach that has been recently most frequently employed for the calculation of the above integrals in order to gain some knowledge on the quality of the PES that is being used.

The first order diffusion coefficient, physically describing transport of mass in the gaseous mixture along a concentration gradient, is given by [6]:

$$D_{ij}^{(I)} = \frac{3K_BT}{16N\mu\Omega_{ij}^{(1,1)}} \tag{38}$$

where N is now the total number density (molecules/cm^3) of the gas. One can also go further [44] and define the second-order approximation which is given by:

$$D_{12}^{(2)} = D_{12}^{(1)}\left(1 - \Delta_{12}\right)^{-1} \tag{39}$$

where the Δ_{12} are defined in the literature as function of masses and of higher order Ω integrals [44]. In the cases currently tested, the corrections are often very small at the temperature examined [49] but obviously depend very markedly on the system at hand.

Another quantity of interest is given by the first order interaction viscosity in the mixture, which corresponds to the transport of particle momentum along a velocity gradient. It could be considered as the simplified viscosity of a model gas of molecules with mass equal to twice the reduced mass of the species in the mixture, between which the acting forces are given by the interaction potential one wishes to evaluate

$$\eta_{ij}^{(1)} = \frac{5\ K_B T}{8\ \Omega_{ij}^{(2,2)}} \tag{40}$$

The first order mixture viscosity of the binary mixture depends more explicitely on the concentration [44]

$$\eta_{mix}^{(1)} = \frac{H_{22}x_1^2 - 2H_{12}x_1x_2 + H_{11}x_2^2}{H_{11}\cdot H_{22} - H_{12}^2} \tag{41}$$

where the x_i's are the molar fractions of the two species and the H_{ij} are complicated functions of mole fractions, interaction viscosity and the viscosity coefficients of the component pure species [6].

Two further quantities of interest are the thermal conductivity, λ_{12}, and the thermal diffusion coefficient, α_T. The former is given by the following equation

$$\lambda_{12} = \frac{75\ K_B^2 T}{64\ \mu_{12}}\frac{1}{\Omega_{ij}^{(2,2)}} \tag{42}$$

its simple physical meaning is that of a kinetic energy transport along a temperature gradient and, as one can see from eq.(40) is rather directly related to the same collision integral appearing in the interaction viscosity [49].

The thermal diffusion factor, α_T, physically describes the extent of the partial separation of an initially uniform gas mixture which takes place when the mixture is subjected to a temperature gradient. Its phenomenological expression is given by

$$\frac{\partial x_1}{\partial \mathbf{R}} = -\alpha_T x_1 x_2 \frac{\partial}{\partial \mathbf{R}}(\ln T) \tag{43}$$

here **R** is the molecular vector position in real space, αT is the factor one is looking for and the x_i's are the usual molar fractions of the species present in the mixture. Since, to a first order approximation, the αT coefficients are identically zero [6], the thermal diffusion is a property that appears only when second-order corrections in the Chapman-Cowling expansion are explicitely included [44]. The corresponding coefficients are given by [49]

$$\alpha_T = (6C^*_{12} - 5)\frac{S_1 x_1 - S_2 x_2}{Q_{11}x_2 + Q_{12}x_1 x_2 + Q_{22}x^2}$$

The coefficients C, S and Q are all quantities of a similar nature as the H_{ij} integrals introduced in eq. (41) and have all been explicitely given before [6,49] in the recent literature.

5. Discussion of Some Examples

In the previous Sections we have endeavoured to show that the use of multichannel scattering theory, in its exact, Close-Coupled (CC) formulation and in its various dynamical decoupling approximations, CS and IOS, allows one to produce from a given functional form of the intermolecular potential specific angular distributions, elastic and inelastic, total integral cross sections and state-to-state partial integral cross sections over a broad range of collision energies that sample different region of the intermolecular forces.

Such computed quantities also appear in the formulation of the generalized collision integrals that are the basic microscopic quantities which control and explain the behaviour of the non-equilibrium properties of the gaseous mixture, thus linking in a rather direct way the outcome of the dynamics with the statistical observables of a molecular ensemble.

The comparison with experiments can be carried out, therefore, at two different levels:

(i) at the 'elementary' level involving single collisions between individual partners as they are measured in crossed molecular beam experiments. If the resolution of the experiments allows it, then the amount of inelastic processes can be gauged vis à vis the size of the angular distributions for state-to-state transitions [37,50] and therefore the correct anisotropy of the full interaction could be modified to better reproduce the experimental findings.

(ii) at a more macroscopic level, by evaluating at various level of approximation the involved collision integrals and generalized cross sections and then compare their temperature behaviour with a wide variety of bulk properties for transport and relaxation coefficients of the type we have briefly discussed in the previous Section. It is always important, in such comparisons, to know experimentally a broad variety of properties and also to have them available over a broad range of temperature values, since the latter set of data imply that the underlying scattering quantities are obtained over a correspondingly broad range of relative energies, hence they sample a wider interval of the interaction potential range of action.

An interesting example is shown by the recent results for the He-CO full anisotropic potential that has been first obtained from adjustment to molecular beam experiments only [51] and employed then to compute rigorously a broad range of transport and relaxation cross sections [38]. Since the test indicated that modifications were still needed in the potential function, it was altered to optimize its agreement with diffusion and viscosity coefficients [11]. Table 1 shows an example of the results obtained with the original molecular-beam-extracted potential at different level of approximation in the employed dynamics.

TABLE 1. Computed diffusion, interaction viscosity and mixture viscosity coefficients for He-CO, using the PES optimized to molecular beam experiments [51]. The experiments are from ref. [11]

		η_{mix}			η_{AB}			$D_{AB}^{(0)}$	
T/K	x(CO)[a]	IOS	(CC+CT)[b]	Expt	IOS	(CC+CT)[b]	Expt	(CC+CT)[b]	IOS
298	0.3766	18.92	18.89	19.42	13.69	13.63	0.684	0.666	0.654
	0.6442	18.36	18.34	18.66					
323	0.3766	20.03	20.00	20.58	14.44	14.38	0.785	0.762	0.749
	0.6442	19.48	19.46	19.83					
373	0.3766	22.17	22.14	22.75	15.89	15.82	0.007	0.970	0.955
	0.6442	21.65	1.63	20.02					
423	0.376	24.19	24.1	24.85	17.29	1.21	1.23	1.20	1.182
	0.6442	23.67	23.64	24.08					
473	0.3766	26.12	26.07	26.83	18.62	18.55	1.49	1.45	1.427
	0.6442	25.56	25.53	26.01					

[a] x(CO) = mole fraction of CO in the mixture.
[b] with quantal Z_{rot} partition function.

The IOS calculations are those described in the previous Section, while the exact calculations involve both quantum (CC) and classical (CT) evaluations of the relevant collision integrals [11].

If one now scales the repulsive 'range' parameter and the well position of the spherical coefficient in eq.(4), one finds that the transport properties are affected quite markedly even if the changes are never bigger than about 1% of the initial values of such quantities. A specific examples with IOS calculations is shown in Figure 1, where the relative percentage error, Δ%, defined as: $(D_{calc}-D_{expt})/D_{expt}$, is shown for nine different modifications of the interaction potential [11]. One clearly sees that the results marked by open circles in the bottom part correspond to the best agreement with the measured data, while the original PES optimized to the beam data is given by the open circles at the center: it shows a much larger relative error than the one using the optimized PES.

If one now goes back to the original molecular beam experiments and evaluates angular distributions with the new PES, one finds that the individual, state-to-state cross sections are indeed modified, as is shown by the exact, CC calculations of Figure 2. The

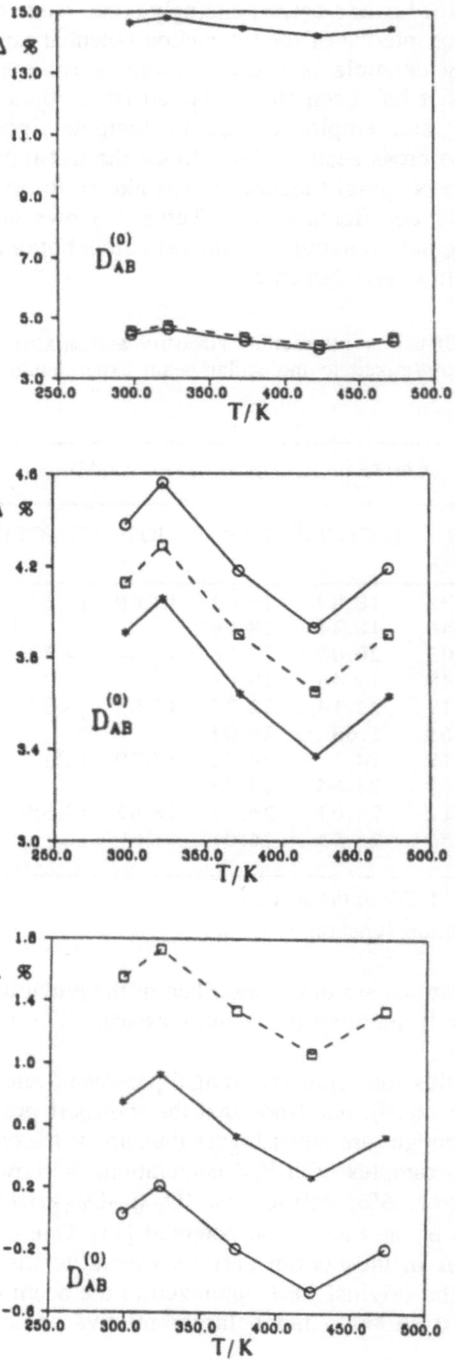

Fig. 1. IOS computed diffusion coefficients with different scaling of the PES from beam experiments [11]. See text for meaning of symbols.

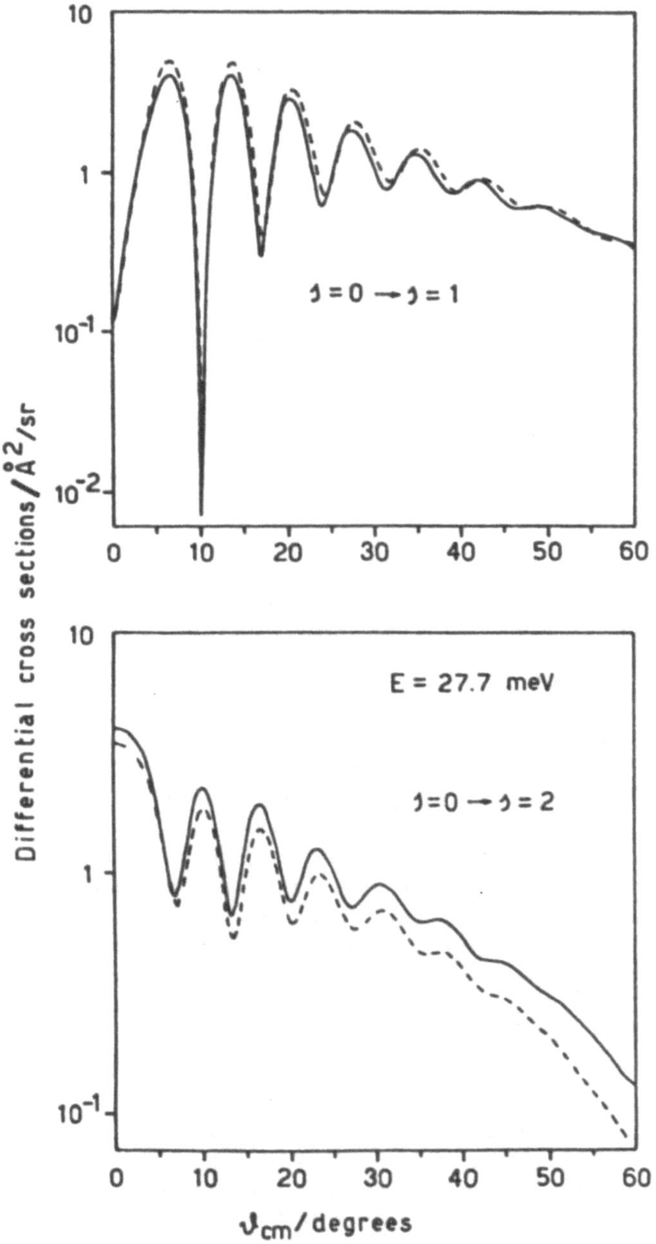

Figure 2. State-to-state inelastic differential cross-sections computed for the He-CO system at 27.7 meV of collision energy. Solid line, close-coupling (CC) calculations using the molecular beam potential of ref. (51), dashed line, CC calculations using the PES optimized as in Fig. 1 top, j=0→j=1 transition; bottom j=0→j=2 transition.

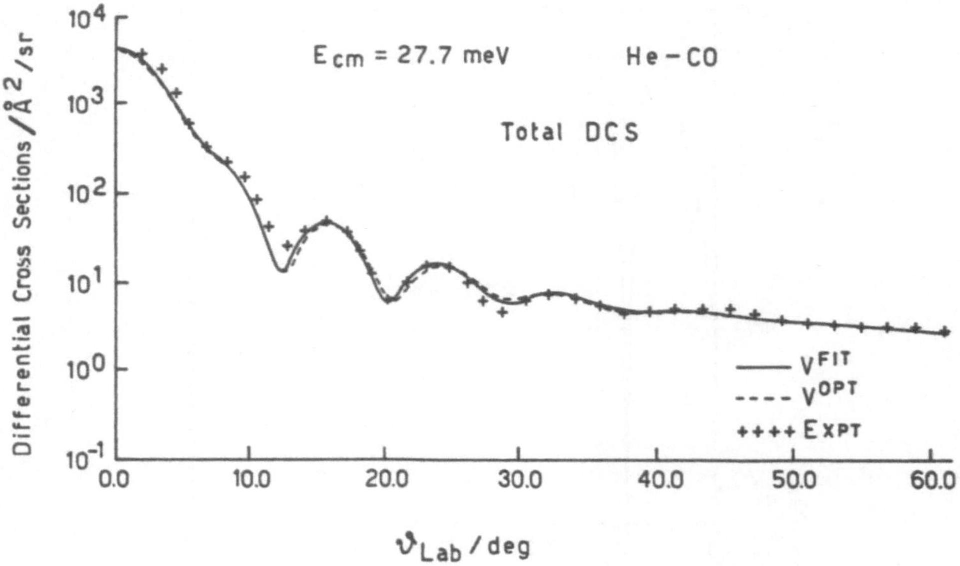

Figure 3. Close-coupled calculations for the total differential cross-sections at 27.7 meV collision energy. Solid line, results with original PES of ref. [51]. Dashed line, results with new, optimized PES. Crossed, experimental data from [52].

modified PES clearly shows that the fast oscillations in the angular distributions are affected by changes of the well depth position, even if its anisotropy remains unchanged. If one now sums over the known population of rotational states in the beam experiments [51] and calculates all the relevant elastic and inelastic state-to-state cross sections, one can finally see the effects of the small modifications in the 'size' of the well and of the repulsive wall in the spherical component of expansion (4) suggested by the optimized results for diffusion coefficients. Such effects are pictorially reported in Figure 3, together with the experimental total differential cross sections, summed over all rotationally inelastic processes which occur in the high-resolution, crossed beam experiments involving the present He-CO system [52].

One clearly sees from the comparison that the optimized PES V_{OPT}, follows as closely as the previous selection, V_{fit}, the experimental data. Its small changes, however, are enough to improve dramatically the agreement between computed and measured diffusion coefficients, as shown in the previous Figure 1.

One of the points which we have tried to make in the present analysis is to stress the importance of selecting the most appropriate dynamical approximation for the system under study in order to truly test the quality of the intermolecular function in relation to the experimental findings for dynamical properties. An example of differences which can be attributed solely to the dynamical approximation employed is shown in Figure 4, by some recent calculations on the He-CH_4 system [53]. We present there the relative percentage errors with respect to experimental data obtained with IOS calculations of the mixture viscosity coefficients (marked as stars in the Figure). We also show the relative percentage errors of the IOS data with respect to CS calculations (open circles). One observes that the IOS comparison with experiments yield results which indicate that the repulsive 'range' of the interaction employed is possible too small, since it produces collision integrals that are slightly too small, hence larger viscosity coefficients with respect to measurements. On the other hand, the CS calculations gives smaller viscosity coefficients that are essentially well within the error bars of the experiments [54]: thus, the suggested PES for this system is tested more accurately when CS calculations are employed, in spite of the already good qualitative behaviour of the IOS results.

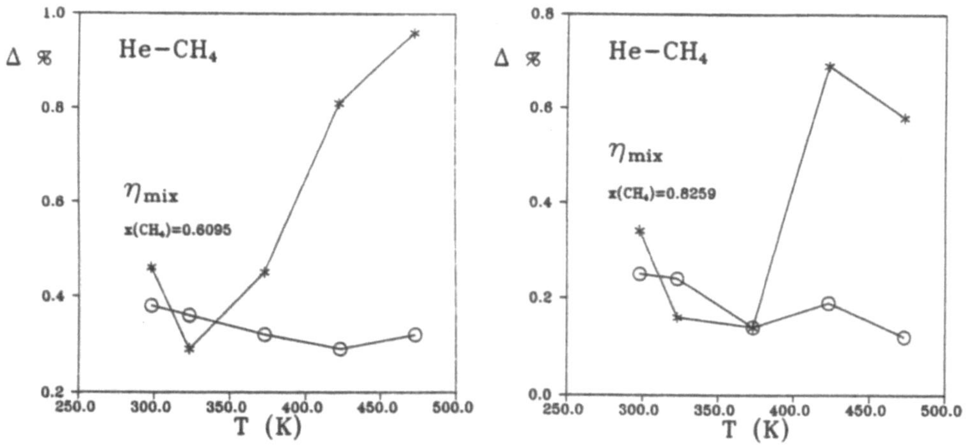

Figure 4. Computed mixture viscosity coefficients, as a function of temperature, for He-CH_4 mixtures. The left part of the figure refers to a CH_4 molar fraction of 0.6095, while the right part refers to x(CH_4)= 0.8259.

In conclusion, we have shown that the determination of intermolecular forces from the dynamical behaviour of gaseous mixtures with the species under study can be carried out to a very high level of accuracy and can be employed to test a broad range of the various regions of the full potential energy function. In particular, once the quantum or the classical treatments of the dynamics can be specifically tested with the results of microscopic collisional events, i.e. under the single-collision conditions of high resolution crossed beam experiments, the bulk properties which closely depend on it, i.e. the non-equilibrium features given by transport and relaxation coefficients can also be profitably employed to test the global quality of a selected potential energy function.

Such an approach becomes often the only viable approach since the full evaluation with quantum chemical techniques of a large grid of values in eq. (1), at chemical accuracy level, is still out of range for many systems of current interest, e.g. the ones given before as examples, and has proven to be reliable only for very simple partners and over a rather limited range of relative geometries.

Acknowledgements

The calculations presented here were carried out with the financial support of the Italian Nat. Research Council (CNR) and of the Italian Ministry of Science, Education and Technology (MURST). Very profitable discussions with Dr. S. Serna, N. Sanna and E. Buonomo are also gratefully acknowledged.

References

1. Hobza, P. and Zahradnik, R. (1988) *Intermolecular Complexes*, Elsevier, Amsterdam
2. Buckingham, A.D., Fowler, P.W. and Huston, J.M. (1988) *Theoretical Studies of Van der Waals Molecules and Intermolecular Forces*, Chem. Rev.**88**, 963-988
3. Israelachvili, J. (1992) *Intermolecular and Surface Forces*, Academic Press, London
4. Masten, F.A., Klein, D.J., Foyt, D.C. (1971) J. Phys. Chem., **75**, 1866-78
5. Warmer, P.E.S., van der Avoird, A. (1984) J. Chem. Phys. **81**, 1929-38
6. Maitland, G.C., Rigby, M., Smith, E..B., Wakeham, W.A. (1981) *Intermolecular Forces: Their Origin and Determination*, Clarendon Press, Oxford
7. Brink, D.M., Satchler, G.R. (1975) *Angular Momentum*, Clarendon Press, Oxford
8. e.g. see: Schwenke, W.D., Walch, S.P. and Taylor, P.R. (1993) J. Chem. Phys. **98**, 4738-44
9. Le Roy, R.J. and Carley, J.S. (1980) Adv. Chem. Phys. **42**, 353-78
10. Sathyamurthy, N. (1985) *Computetional Fitting of ab initio Potential Energy Surfaces*, Comp. Phys. Rep. **3**, 1-70
11. Gianturco, F.A., Sanna, N. and Serna, S. (1994) Mol. Phys. **81**, 421-46
12. Pertsin, A.J., Kitaigordsky, A.I. (1987) *The Atom-Atom Potential Method for Organic Molecular Solids*, Springer, Berlin
13. Claverie, P. (1987) in: Pullman B (ed.) *Intermolecular Interactions: from Diatomics to Biopolymers*, Wiley, New York
14. e.g. see: Gianturco, F.A. (1989) in: *New Theoretical Concepts for Understanding Organic Reactions*, Bertràn J. and Csizmadia, I.G., Eds., Kluwer Academic Publ., Amsterdam, pg 257-89

15. Faubel, M., Kohl, K.H., Toennies, J.P. and Gianturco, F.A. (1983) J. Chem. Phys. **78**, 5629-35
16. Meath, W.J. and Aziz, R.A. (1984) Mol. Phys. **52**, 225-36
17. Beneventi, L., Casavecchia, P. and Volpi, G.B. (1986) J. Chem. Phys. **85**, 7011-24
18. Pack, R.T. (1978) Chem. Phys. Lett. **55**, 197-211
19. Gianturco, F.A. and Palma, A. (1985) J. Phys. B **18**, L519-24
20. van der Avoid, A., Wormer, P.E.S., Mulder, F., Berns, R.M. (1980) Topics in Curr. Chem. **93**,1-28
21. Gianturco, F.A. and Schneider, F. (1991) Adv. Chem. Phys. **82**, 135-174
22. Williams, H.L., Szalewicz, K., Jeziorski, B., Moszynski, R., Rybak, S. (1993) J. Chem. Phys. **98** 1279-92
23. Kumar, A., Meath, W.J. (1985) Mol. Phys. **54**, 823-35
24. Mormer, P.E.S., Rijks, W. (1986) Phys. Rev. **A33**, 2928-39
25. e.g. see: Douketis, C., Scoles, G., Marchetti, S., Zen, M. and Thakkar, A.J. (1982) J. Chem. Phys. **76**, 3057-69
26. Tang, K.T., Toennies, J.P. (1984) J. Chem. Phys. **80**, 3726-39
27. Knowles, P.J., Meath, W.J. (1987) Mol. Phys. **60**, 1143-57
28. Tennyson, J., van der Avoird (1984) Chem. Phys. Lett. **105**, 49-58
29. Le Roy, R.J., Hutson, J.M. (1987) J. Chem. Phys. **86**, 827-49
30. Gianturco, F.A., Venanzi, M., Candori, R., Pirani, F., Vecchiocattivi, F., Lee, M.S., Dickinson, A.S. (1986) Chem. Phys. **109**, 417-29
31. Battaglia, F., Gianturco, F.A., Casavecchia, P., Pirani, F., Vecchiocattivi, F. (1982) Faraday disc. Chem. Soc. **73**, 257-69
32. e.g. see: Faubel, M., Toennies, J.P. (1977) Adv. At. Mol. Phys. **13**, 262-94
33. Candori, R., Pirani, F., Vecchiocattivi, G., Gianturco, F.A., Lamanna, U.T., Petrella, G. (1985) Chem. Phys. **92**, 345-67
34. Faubel, M. (1983) Adv. At. Mol. Phys. **19**, 345-77
35. e.g. see: Bergmann, K., Hefter, U., Witt, J. (1980) **72**, 4777-89
36. Buck, U. (1986) Comp. Phys. Rep. **5**, 1-48
37. Gianturco, F.A., Venanzi, M., Faubel, M. (1989) J. Chem. Phys. **90**, 2639-50
38. Gianturco, F.A., Sanna, N., Serna, S. (1993) J. Chem. Phys. **98**, 3833-45
39. Gianturco, F.A. (1979) *The Transfer of Molecular Energy by Collisions*, Springer Verlag, Berlin
40. e.g. see: Bernstein, R.B., Ed. (1979) *Atom-Molecule Collision Theory: a Guide to Experimentalists*, Plenum Press, New York
41. Parker, G.A., Rock, R.T. (1978) J. Chem. Phys. **68**, 1585-97
42. Gianturco, F.A., Venanzi, M., Dickinson, A.S. (1989) Mol. Phys. **66**, 563-78
43. Gianturco, F.A., Palma, A., Venanzi, M. (1985) Mol. Phys. **56**, 399-411
44. Chapman, S. and Cowling, T.G. (1970) *The Mathematical Theory of non-uniform Gases*, Cambridge University Press, Cambridge
45. Mc Court, F.R.W., Beenakker, J.J.M., Köhler, W.E., Kusscer, I. (1990) *Non-equilibrium Phenomena in Polyatomic Gases*, Clarendon Press, Oxford
46. Gianturco, F.A., Sanna, N., Serna, S. (1991) Molec. Phys. **74**, 1071-86
47. Gianturco, F.A., Sanna, N., Serna, S. (1992) J. Chem. Phys. **97**, 6720-33
48. Gianturco, F.A., Sanna, N., Serna, S. (1993) Molec. Phys. **78**, 1015-28
49. Gianturco, F.A., Venanzi, M. (1989) J. Chem. Phys. **9**, 2525-41
50. Gianturco, F.A., Palma, A., and Sanna, N. (1992) Chem. Phys. **158**, 77-89
51. Dilling, W. (1985) Ph D Thesis, University of Göttingen, Germany

52. Faubel, M., Kohl, K.H. and Toennies, J.P. (1980) J. Chem. Phys. **73**, 250-67
53. Gianturco, F.A., Serna, S. (1994) J. Chem. Phys. **100**, 4316-23
54. Hellemans, J.M., Kestin, J., Ro, S.T. (1973) Physica **65**, 376-89.

RESONANCES IN MOLECULAR DYNAMICS : *CONCEPTS AND APPLICATIONS*

O. ATABEK

Laboratoire de Photophysique Moléculaire du CNRS,
Université Paris-Sud, 91405 Orsay, France

INTRODUCTION

The concept of *resonance* is present in all fields of physics. From wind-induced oscillations that make a bridge to collapse, to musical instruments like flutes or trumpets where the size or shape of the cavity in which air enters is changed to fulfill appropriate conditions, resonances may have, in human life, destructive consequences but they can also serve as sources of pleasure.

An intense activity has recently been devoted to this concept in molecular physics, primarily due to the use of experimental devices, such as atomic or molecular beams and short intense laser pulses, which allow for a considerable improvement in the quality of the information obtained from a study of collisions or half-collisions involving atoms, molecules and photons. The measured lifetime, energy and angular distributions require the consideration of resonances mediating the processes. Resonances play a major part in the detailed understanding of photochemical and energy transfer mechanisms where they occur as more or less sharp structures of the various cross sections or inelastic transition probabilities. Their position (energy) and lifetime (proportional to the inverse of their width) reflect fundamental properties of the system. In reactive scattering they may be considered as direct probes of the so-called "transition state", but due to the summation over many partial waves, each leading to slightly different energy positions, the sharp structures are most likely washed out. In photodissociation on the other hand, the molecule is initially in a well prepared (single total angular momentum) bound state and resonances are expected to be more clearly attributed to the peaks, dips or even more complicated profiles of the photon absorption spectrum.

There has simultaneously been considerable progress toward a rigourous theory of resonant states which are described by solutions of the wave equation under a special type of boundary conditions. Here again, molecular physics with its large

E. Yurtsever (ed.), Frontiers of Chemical Dynamics, 107–129.

variety of interacting entities and force fields provides many opportunities of applying these new tools. A favourable circumstance in this undertaking is the existence, in quantum molecular physics and quantum chemistry, of a tradition of high numerical accuracy in the manipulation of large basis sets and the integration of coupled differential equations.

In the following we will concentrate on recent theoretical advances on resonance mediated molecular (heavy particle) motion in a few atom system. The wave equation of such a system possess solutions corresponding to either bound or scattering states. In a bound state the particles are constrained to remain in a finite region of configuration space, whereas the scattering states allow for the description of collision processes. Actually the situation is not as clear-cut as this. All excited bound states are subjected to radiative decay. The bound character may also appear as the result of an approximation and the system may, in fact, dissociate by the loss of electrons (ionization) or of atomic and molecular fragments (dissociation). Even the ground state is not stable since the collision with a photon may lead to dissociation. Thus bound states are just zeroth order view of decaying states, that is *resonances*.

In terms of potential energy curves accomodating them, two classes of resonances have been considered, namely shape or Feshbach. In the first category, the resonance results from a quasi bound level decaying through the potential barrier of a unique potential energy curve with a tunneling lifetime determined by the height and thickness of the barrier and the fragment reduced mass. A Feshbach resonance arises from the interaction of, at least, two potential energy curves, a quasi-bound level of one of them being embedded in the continuum of the other. The lifetime is then determined by the strenght of the potential coupling and the interfragment motion in the inner region.

The aim of the present paper is to identify properly the resonances in terms of an accurate definition characterizing their positions, lifetimes and wavefunctions and then to relate them to the interpretation of molecular processes, such as photon absorption and emission, dissociation, energy and angular distributions in the fragments. Which definition for a resonance (intuitive or formal approaches) and the real or complex energy characterizations, are the concerns of Section II. Section III is devoted to some of the more commonly used methods for calculating resonance energies and wavefunctions. Finally, Section IV provides a thorough illustration of the H_2^+ multiphoton intense field dissociation in terms of laser induced resonances.

II.- REAL AND COMPLEX ENERGY CHARACTERISATIONS

Several definitions characterizing resonances exist, which make use of criteria, physical or mathematical, to be fulfilled. Some of them relate a resonance with the presence of a long-lived quasi-bound state of a composite system. The intuitive and physical picture is a system, which for a particular range of energies, is temporarily trapped within a finite spatial region. Provided the lifetimes of the states of the composite system are long enough, the analysis of the wavefunction may lead to some recipes supporting a quantitative characterization of resonances. These recipes are based on the study of model problems described by *real energies*. An intrinsic and formal definition for a resonance has a different starting point. The emphasis is put on a specific resonance eigenfunction which is the solution of a Schrödinger equation with particular boundary conditions involving a pure outgoing behaviour and leading to a *complex energy* eigenvalue. Let us examine, in more detail, these different characterizations.

i) Real energy characterizations

The mathematical model which is referred to is a single discrete state $|\phi_p\rangle$ of energy $E_p^{(0)}$ interacting with an unbounded continuum of states $|\phi_{E'}\rangle$ (energy E'), through a smoothly varying coupling V [1]. The scattering eigenvector of this system is given by the Lippmann-Schwinger equation :

$$|\psi_E\rangle = |\phi_E\rangle + G^+(E)\, V\, |\phi_E\rangle \qquad (1)$$

where $G^+(E) = \lim_{\varepsilon \to o} (E + i\varepsilon - H)^{-1}$ is the resolvent and H the Hamiltonian. For some energies E, the scattering wavefunction shows specific behaviors which are retained as being characteristic for a resonance :

a) At short distances (inner region), a measure of the amplitude of $|\Psi_E\rangle$ can merely be obtained as its projection on $|\phi_p\rangle$. A simple calculation, involving the evaluation of the $\langle\phi_p | G^+(E) | \phi_p\rangle$ matrix element of the resolvent leads to :

$$|\langle\phi_p|\psi_E\rangle|^2 = |\langle\phi_p | G^+(E) | \phi_p\rangle \langle\phi_p | V | \phi_E\rangle|^2 = \frac{1}{\pi}\frac{\Gamma_p}{(E - E_p)^2 + \Gamma_p^2} \qquad (2)$$

i.e., a Lorentzian centered at an energy E_p given by

$$E_p = E_p^{(0)} + P\int dE' \frac{|\langle\phi_p|V|\phi_{E'}\rangle|^2}{E_p - E'} \qquad (3)$$

(P indicating a principal part integral) with a half width at half maximum (HWHM) Γ_p :

$$\Gamma_p = \pi\, |\langle\phi_p | V | \phi_E\rangle|^2 \qquad (4)$$

The amplitude of the scattering wavefunction has a sharp variation in a small energy region around E_p which is associated with the resonance position (or energy).

In terms of the time evolution of an initial wavepacket prepared on the $|\phi_p\rangle$ state :

$$| \Phi(t=0) \rangle = | \phi_p \rangle \tag{5}$$

one has

$$| \Phi(t) \rangle = -\frac{1}{2i\pi} \int_{-\infty}^{+\infty} dE' \, e^{-iE't/\hbar} G^+(E') |\Phi(t=0)\rangle \tag{6}$$

The probability, at time t, to find the system in its initial state, is

$$P_p(t) = |\langle \phi_p | \Phi(t) \rangle|^2 = \frac{1}{4\pi^2} \left| \int dE' \, e^{-iE't/\hbar} \langle \phi_p | G^+(E') | \phi_p \rangle \right|^2 \tag{7}$$

The evaluation of this integral by Cauchy's residue formula, leads to the description of a decaying (resonance) system :

$$P_p(t) = \exp(-2\Gamma_p t / \hbar) \tag{8}$$

with a lifetime $\hbar/2\Gamma_p$. Γ_p is the so-called width of the resonance.

b) At large distances (asymptotic region), the scattering wave function $|\Psi_E\rangle$ accumulates a phase shift $\Delta(E)$ with respect to the incident wave $|\phi_E\rangle$. More precisely if the incident wave asymptotically behaves as :

$$\langle R | \phi_E \rangle \xrightarrow[R\to\infty]{} A \sin [k(E) R + \delta] \tag{9}$$

then

$$\langle R | \psi_E \rangle \xrightarrow[R\to\infty]{} A \sin [k(E) R + \delta + \Delta(E)] \tag{10}$$

(R being an interfragment distance).

The scattering amplitude S(E), defined as the ratio of the amplitude of the outgoing wave (i.e. $e^{+i(kR+\delta)}$) to that of the ingoing wave (i.e. $e^{-i(kR+\delta)}$) is given by [1] :

$$S(E) = e^{2i\Delta(E)} = \frac{E - E_p - i\Gamma_p}{E - E_p + i\Gamma_p} \tag{11}$$

The energy dependance, in the vicinity of the resonance, for the real (ReS) and imaginary (ImS) parts of this scattering amplitude, is indicated on figs. (1a) and (1b). ReS as a function of E has a near constant unit value except close to the resonance position E_p where a well shape is observed with a HWHM Γ_p. ImS shows a near constant null value if an oscillation of amplitude ±1 in the neighbourhood of the resonance is taken apart, the distance between the extrema being $2\Gamma_p$.

The phase shift itself shows a very typical behaviour given by the Breit-Wigner formula [1] :

$$\Delta(E) = \tan^{-1} [-\Gamma p / (E-Ep)] \qquad (12)$$

Eq.(12) is extensively applied in resonance characterisation and means that the phase shift increases by π in passing through the resonance energy E_p, with a rate of increase determined by Γ_p as indicated in fig (1c).

Finally, the most intuitive characterization of a resonance relies upon the collisional time delay $\tau(E)$. By Taylor expanding the phase shift around the incident wave number k_o :

$$\Delta(k) = \Delta(k_o) + \frac{\partial\Delta}{\partial k}\Big|_{k_o}(k - k_o) + \ldots \qquad (13)$$

and retaining the first two terms, one observes that the scattering wave :

$$\Psi_E(R) \xrightarrow[R\to\infty]{} A \sin [k(R + \frac{\partial\Delta}{\partial k}\Big|_{k_o}) + \eta) \qquad (14)$$

is shifted by a distance $\bar{R} = \frac{\partial\Delta}{\partial k}\Big|_{k_o}$ with respect to the incident wave.

Expressing the k-derivative in terms of an E-derivative and introducing the velocity of the incident wave :

$$v_o = \frac{\hbar k_o}{\mu} \ ; \ (\mu \text{ being a reduced mass}) \qquad (15)$$

one gets the time delay due to a collision process [2] :

$$\tau(E) = \frac{\bar{R}}{V_o} = \hbar\frac{\partial\Delta}{\partial E} = \hbar\,\frac{\Gamma_p}{(E - E_p)^2 + \Gamma_p^2} \qquad (16)$$

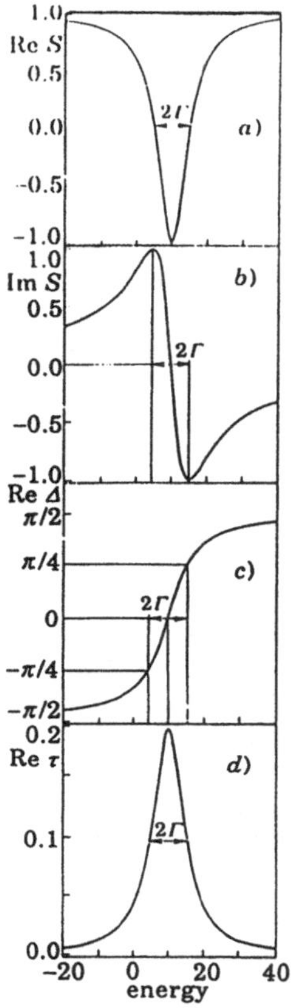

Figure 1. Scattering functions on the real energy axis : a) scattering amplitude (ReS); b) scattering amplitude (ImS) ; c) Phase shift (ReΔ) ; d) time delay (Reτ) [8].

i.e. the Lorentzian, illustrated on fig.(1d), centered on the resonance, with a HWHM, Γ_p. This is the additional time it takes for a wavepacket to move accross the interaction region as compared to free motion.

In summary, the scattering wavefunction when passing through the resonance, has a short range amplitude with a sharp variation, its asymptotic phase shift as a function of energy presents a π-jump and its collisional time delay rises sharply.

As far as a physical system bears some similarities with the present model, a real energy characterization of its resonance is possible by obtaining more or less accurately its energy and width from the S-like behaviour of the phase shift. Resonances with very large widths are difficult to characterize, specially in cases of overlapping (widths of the order of magnitude of energy differences), leading to the breakdown of the model involving a unique, or isolated resonance) [see for instance 3].

ii) Complex energy characterizations

We start from the observation that in the asymptotic region where molecular potentials are constant, $\psi_E(R)$ is proportional to a combination of two exponentials (Eq. 10) :

$$\psi_E(R) \underset{R \to \infty}{\propto} e^{-ikR} + S(E)\, e^{+ikR} \tag{17}$$

The scattering amplitude S (or scattering matrix in the multidimensional case) defined by (Eq. 11) is of modulus unity, since the phase shift Δ is real. We can, however, try to look for complex values of E (or k) such that the outgoing term dominates over the ingoing one. An extreme situation is a value k_p of k which is a pole of S. A class of these poles are found to be of the form $k_p = k_o - ik_1$ with k_o and k_1 positive. A possible way to reach these poles is to analyse a wavefunction solution of a Schrödinger equation, regular at the origin, and asymptotically purely outgoing. Such boundary conditions can only be fulfilled by quantized complex eigenenergies :

$$E_p - i\Gamma_p = \frac{\hbar^2}{2\mu}k_p^2 = \frac{\hbar^2}{2\mu}(k_o^2 - k_1^2) - i\frac{\hbar^2}{\mu}k_o k_1 \tag{18}$$

which constitute the Gamow-Siegert definition of a resonance.

The complex energy characterization leads to its own intuitive picture of a resonance, in terms of a decaying state. This is clear from :

i) either the time evolution where the complex energy

$$e^{-iEt/\hbar} = e^{-iE_p t/\hbar}\, e^{-\Gamma_p t/\hbar} \tag{19}$$

results into a damping factor ;

ii) or the exponentially diverging asymptotic behavior :

$$e^{ikR} = e^{ik_o R}\, e^{+k_1 R} \tag{20}$$

illustrating the observation of the density (or current) at distance R which corresponds to breaking events occured in the past at the origin, where there was more activity in the source [4].

This formal definition of the resonance is mathematically clear and presents uniform range of validity. By using it, one can obtain a complete set of resonance positions whatever the lifetimes are, which can then be used for example in the reconstruction of the scattering matrix at any energy [5]. In particular, reliable results can be obtained for very large resonances with energies well above potential barriers [6] or even in more unexpected situations corresponding to purely repulsive force-fields [5].

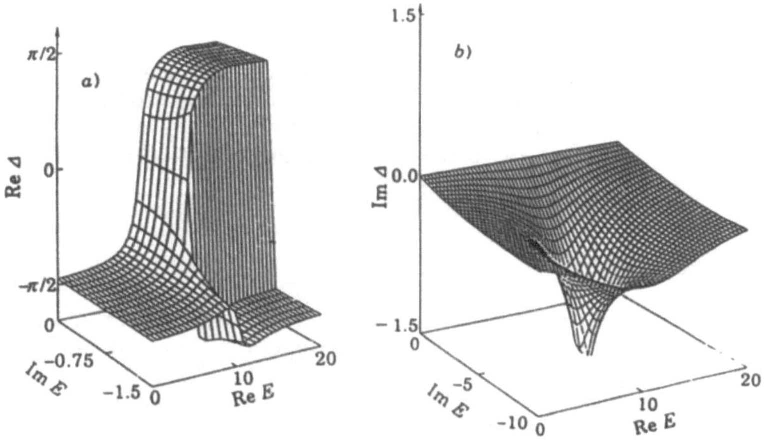

Figure 2. Phase shift ReΔ (a) and ImΔ (b) in the complex energy plane [8].

The relation between the two characterisations of resonances (either real or complex energy) is a question that has been addressed in the literature in different contexts. In some particular one-dimensional model cases, on pure mathematical grounds, it has been found that, even for small widths, there is no one-to-one relation between the shift of the phase of the wavefunction and the presence of a pole in the scattering amplitude [7]. A recent study, based on the behavior of the phase shift Δ (Eq. 12) analytically continued in the complex energy plane around the pole of the scattering amplitude, leads to the following interesting results [8] :

i) The real part of Δ, as plotted on fig. (2a) in a 3D representation, is analysed by cuts parallel to the real axis, i.e. ImE = constant. Closer a cut is to the discontinuity associated with the pole of S, sharper is the variation around the pole position. More unexpectedly, ReΔ is nonsymmetric with respect to a cut defined by ImE = - Γ_p. A Heaviside-type behavior is observed at ImE = - Γ_p with an infinitely sharp π-jump for ReΔ, in contrast with what happens on the real E-axis, where the same jump may occur, for large Γ_p on a large energy scale such that overlapping effects may wash out any specificity due to a resonance.

ii) The resonance pole shows up only in the imaginary part of Δ. As is clear from fig.(2b), ImΔ has a well-localized variation close to the pole position.

These observations not only help understanding the behaviors of the scattering amplitude, the phase shift and the time delay for real energies, but also provide an accurate definition of the resonance in terms of the phase shift and therefore a possible accurate computational scheme.

III.- COMPUTATIONAL METHODS

There is an increasing variety of methods, to calculate the energies (and widths) of resonant states in a time independent multidimensional quantum frame, which can roughly be classified into two categories, following real or complex energy characterisations. In both categories we can distinguish between basis set or grid methods.

1 - Real energy methods

The real energy context leads to procedures, which are rather numerical recipes having a limited range of validity with accuracies depending on resonance parameters (position and lifetime). The maximum amplitude localisation of the resonant wavefunction in the inner part of the potential can be exploited in so-called stabilization techniques which are actually connected with square integrable basis set methods. The box quantization procedure deals with the quantized energies of a system obtained by inclosing it in a box [9]. The scattering wavefunction is expanded on a basis of sine functions presenting nodes at both edges of the box with variationaly obtained expansion coefficients. The (real) eigenvalues of the Hamiltonian are plotted as a function of the box size L. The presence of a resonant state in some energy range manifests itself by a "stable" (nearly L-independent) eigenvalue : the inner localisation of the corresponding wavefunction is such that, a change in the box size which is large enough, hardly modifies its energy. As for the nearby scattering states their "unstable" eigenvalues are decreasing when the box size is larger indicating an increase in the density of states of the quasi-continuum with L. The avoided crossings between the networks of "stable" and "unstable" curves are analysed to give the position (region of minimum slope) and the width of the resonance. The simplest way to reach the width is to apply the Fermi golden rule :

$$\Gamma_p = 2\,\pi\rho\,V^2 \qquad (21)$$

where V is half of the energy separation at the avoided crossing and ρ the density of quasi scattering states. The stabilization procedure leads to good achievements for narrow resonances, otherwise the localization is less pronounced and the possibility of resonance overlapping makes the technique unexploitable [10]. Box quantization treats the quantum system in a global manner with all eigenenergies provided by the diagonalization of a matrix without the need for any resonance searching procedure.

The π-jump of the phase shift in the vicinity of the resonance energy is the basis for one of the most popular computational tools belonging to grid methods. The regular scattering wave solution is obtained from the numerical integration of the radial Schrödinger equation, starting by a null value at the origin. The phase shift Δ results from the comparison of the asymptotic behavior of this solution and the standard form given by (Eq. 10). The procedure is repeated for different energies to get the S-like behavior of the phase shift, from which the position and the width of the resonance is extracted using (Eq. 12). Here again accuracies are limited by the overlapping of neighbouring resonances. As for very narrow resonances, the energy range for the π-jump is so sharp that a trial energy may be very hard to find.

2 - Complex energy methods

The complex energy frame deals with the poles of the scattering matrix. There are some rare cases for which S(k) is known analytically such that the resonance energies are immediatly available. There is also the possibility to define approximations (semiclassical treatments [11]) yielding analytical forms of S(k). The standard situation remains the obtention of the poles of S(k) from the Gamow-Siegert resonance wavefunction, with a technical difficulty related to it, namely its asymptotic divergent behavior (Eq. 20) which renders the use of square integrable basis set expansions unappropriate. A very efficient and by now very popular device to modify this behavior is that of complex rotation, amounting to the use of $\rho\, e^{i\theta}$ instead of R in the wave equation [12]. By writing the wavenumber $k = K\, e^{-i\beta}$ (K and β positive), the asymptotic resonance wavefunction

$$e^{ikR} = \exp\,[iK\rho\cos(\theta - \beta)]\,\exp\,[-\,K\rho\sin(\theta - \beta)] \qquad (22)$$

goes to zero if θ exceeds β, just as for a bound state. Thus all the expertise developed in the study of bound states can be applied to the study of resonant states [see for instance 12]. In particular, this property of the complex rotation of transforming the resonance wavefunction into a localizable one, justifies the use of square integrable basis set methods for its calculation [13]. An alternative is to introduce a rotated complex coordinate into the Schrödinger equation and solve it in a way analogous to that prescribed for bound states using grid methods [14].

The basis set methods consist in expanding the resonant wavefunction in a basis of localized functions (like harmonic oscillators, gaussian functions, etc...). The solution of the wave equation is replaced by the diagonalization of the complex rotated Hamiltonian matrix. Due to troncation effects (finite dimension of the matrix), all eigenvalues depend on the rotation angle θ . It is however observed that some of them pause at some particular values of θ corresponding to the fulfillment of the condition $\partial E/\partial\theta = 0$. They are identified with the resonant states. The main limitation of such methods is related with the variational estimation of the energy, mainly depending on the ability of the form and number of basis set functions to describe the resonant wavefunction [15].

The most powerfull and widely used tool for calculating resonances seems to be the use of grid methods which are well adapted to multidimendional cases treated by the coupled channel approach [14,16]. The degrees of freedom of the system are separated into two types : an interfragment radial motion (coordinate R) and all other motions (described by internal variables of the fragments, angular variables of the interfragment axis, electronic and field variables... etc collectively denoted by r). It is only the R-motion which is dynamically considered, all others being treated in terms of square integrable basis set functions $\psi_j(r)$ $j=1,2....$ The complete wave function is written in the form :

$$\Psi(r;R) = \sum_j \chi_j(R)\psi_j(r) \tag{23}$$

with unknown R-dependant functions $\chi_j(R)$. Introduction of (Eq. 23) into the Schrödinger equation leads, after left multiplication by $\psi_j(r)$ and r-integration, to the coupled differential equations :

$$\left[-\frac{d^2}{dR^2} + W_{ii}(R) - E\right]\chi_i(R) = -\sum_{j\neq i} W_{ij}(R)\chi_j(R) \tag{24}$$

where

$$W_{ij}(R) = \left\langle \psi_i(r) \middle| H(r,R) \middle| \psi_j(r) \right\rangle_r \tag{25}$$

One term in the series, resulting into a single Schrödinger equation, may sometimes provide a reasonably accurate description of the dynamics. Retention of more than one term leads to a multichannel approach, each channel being labelled i, with a channel potential $W_{ii}(R)$ and interchannel couplings $W_{ij}(R)$. In an N-channel situation it is possible to build N independent sets of functions $\chi_i(R)$ obeying (Eq. 24) and satisfying some prescribed boundary conditions either for small R or large R values. These sets are conveniently grouped into a matrix $\underline{\underline{U}}$ (R) and any wavefunction corresponds to an appropriate combination of the columns of it. Numerical unaccuracies related with possible loss of independency within the basis of functions can be avoided by propagating ratio of $\underline{\underline{U}}$ matrices defined following Fox by [17] :

$$\underline{\underline{P}}^{o}(R) = \underline{\underline{U}}^{o}(R-h) \;/\; \underline{\underline{U}}^{o}(R) \tag{26}$$

h being the step-size on the grid R, which can be propagated outward with :

$$\underline{\underline{P}}^{o}(R+h) = [\,2\,\underline{\underline{\beta}}\,(R) - \underline{\underline{\alpha}}(R-h)\;\underline{\underline{P}}^{o}(R)\,]^{-1}\;\underline{\underline{\alpha}}\,(R+h) \tag{27}$$

and

$$\underline{\underline{P}}^{i}(R) = \underline{\underline{U}}^{i}(R+h) \;/\; \underline{\underline{U}}^{i}(R) \tag{28}$$

which can be propagated inward with :

$$\underset{\approx}{P}^{i}(R\text{-}h) = \left[2\underset{\approx}{\beta}(R) - \underset{\approx}{\alpha}(R+h)\underset{\approx}{P}^{i}(R)\right]^{-1} \underset{\approx}{\alpha}(R-h) \tag{29}$$

$\underset{\approx}{\alpha}$ and $\underset{\approx}{\beta}$ are the Numerov matrices of elements

$$\alpha_{ij}(R) = h\left\{\delta_{ij} + h^2/12\left[E - W_{ij}(R)\right]\right\} \tag{30}$$

$$\beta_{ij}(R) = h\left\{\delta_{ij} - 5h^2/12\left[E - W_{ij}(R)\right]\right\} \tag{31}$$

Each propagation carries a boundary-value information. If the integration goes from R_o to R_M, we must choose $\underset{\approx}{P}^{o}(R_o+h)$ and $\underset{\approx}{P}^{i}(R_M\text{-}h)$ from the expected behavior of the wavefunction at the end points. Continuity of the wavefunction and of its derivative at an arbitrarily given matching point R_m requires :

$$\det \left| \underset{\approx}{P}^{i}(R_m) - \underset{\approx}{P}^{o}(R_m+h)^{-1} \right| = 0 \tag{32}$$

i.e., a quantization condition which can only be fulfilled for discrete energies (real or complex).

The regularity at the origin imposes the boundary-value for the outward integration, namely :

$$\underset{\approx}{P}^{o}(R_o + h) = 0 \tag{33}$$

As for the inward propagation the Gamow-Siegert resonance definition is such that, all closed channels start from zero and all open channels have a pure outgoing behavior. It can be shown that, without the complex rotation of the coordinate, such a choice may lead to numerical unstabilities and that, in a potential free region the algorithm will rather built the ingoing e^{-ikR} function even when starting from the outgoing e^{ikR} [18]. An additional difficulty is related with the choice of off diagonal elements of $\underset{\approx}{P}^{i}(R_M\text{-}h)$ in cases for instance where off diagonal elements of the potential matrix dominate the diagonal ones for large interfragment separation (cases of the so-called persistent effects, atomic static Stark effect, etc...). All these limitations and unaccuracies can be avoided by the complex rotation technique, which leads, following (Eq. 22), to :

$$\underset{\approx}{P}^{i}(R_M\text{-}\ h) = 0 \tag{34}$$

It is worthwhile noting that, within the frame of the complex rotation technique this choice (Eq. 34) is even not critical. A propagation in the potential free asymptotic region will automatically build the outgoing e^{+ikR} [18]. This can be understood by referring to the asymptotic wavefunction with complex wavenumber of the form $k_o - ik_1$:

$$\chi(R) \underset{R\to\infty}{\to} A(k)\, e^{-ik_oR}\, e^{-k_1R} + B(k)\, e^{+ik_oR}\, e^{+k_1R} \tag{35}$$

Resonance states correspond to choosing k in such a way that only the outgoing term is left. This means trying to detect the vanishing of the ingoing component ($A(k) = 0$), which is normally dominated (because of the factor e^{+k_1R}) by the diverging outgoing component. In other words when propagating inwards, it is the unwanted ingoing component which dominates and after a number of propagation steps the (outgoing) memory of the end point is lost. After complex rotation $R = \rho\, e^{i\theta}$, the wave function becomes :

$$\chi(\rho e^{i\theta}) \to A(k)e^{-iK\rho\cos(\theta-\beta)}e^{K\rho\sin(\theta-\beta)} + B(k)e^{+iK\rho\cos(\theta-\beta)}e^{-K\rho\sin(\theta-\beta)} \tag{36}$$

If $\sin(\theta-\beta) > 0$, the unwanted component is increasing so that its detection is made easier. The outgoing component, on the other hand, is decreasing as $\rho \to \infty$; that is in an inward propagation it is precisely this outgoing component which is automatically built [18].

The complex energy grid methods are very powerful and accurate. One may for instance calculate the resonances of a repulsive exponential potential, a case which would be hard to approach in terms of constructive interference effects [5]. One may also prove in this way that to a given potential there generally corresponds a rich spectrum of resonant states with, for some of them, imaginary parts exceeding real parts thus going far beyond a simple intuitive picture [6]. Such resonances, although not as useful as those of the more common type, should be taken into account when attempting the reconstruction of the scattering matrix from its poles [5].

In multidimensional systems the study of resonances is important both for the evaluation of lifetimes and for the calculation of the branching ratios of different channels. The outcome of the fragmentation process may involve several channels (several internal states of separating species are energetically accessible and can experimentally be discriminated by measuring their relative populations) such that the decomposition is characterized not only by a total rate Γ_p but also by a set of partial rates (or partial widths γ_i, which are the residues of de scattering matrix element S_{ij} [19] :

$$\left|S_{ij}(E)\right|^2 = \frac{4\gamma_i\gamma_j}{(E-E_p)^2+\Gamma_p^2}\ ;\ \Gamma_p = \sum_j \gamma_j \tag{37}$$

The problem of partial widths, widely discussed in the literature [20], can accurately be solved from the asymptotic amplitudes of the resonance wavefunctions calculated using grid methods. It is worthwile noting that for a single-channel complex-energy Schrödinger equation $H\chi = E\chi$, one formally has :

$$E = -\frac{\chi''}{\chi} + V \tag{38}$$

(double-dash meaning second derivative). V being assumed real and the eigenenergy E = $E_p - i\Gamma_p$ already provided, the relation :

$$\Gamma_P = \mathrm{Im}(\chi''/\chi) = \mathrm{Im}\left[\chi^*(R)\chi''(R)\right] / |\chi(R)|^2 \tag{39}$$

is directly obtained multiplying numerator and denominator by the conjugate χ^* (R). Integration of (Eq. 39) results into a flux expression :

$$\Gamma_P = \mathrm{Im}\left[\chi^*(R)\chi'(R)\right] / \int_0^R |\chi(R')|^2 dR' \tag{40}$$

both Eqs. (39) and (40) are R-invariant [21].

The multichannel generalisation of (Eq. 40) involves straightforward algebra and gives [21] :

$$\Gamma_P = \sum_j \mathrm{Im}\left[\chi_j^*(R)\chi_j'(R)\right] / \sum_j \int_0^R |\chi_j(R')|^2 dR' \tag{41}$$

Provided that asymptotic channels are decoupled, a partial width γ_i accounting for the decay of the resonance into the corresponding channel i, is associated to the asymptotic flux in this channel :

$$\gamma_i = \mathrm{Im}\left[\chi_i^*(R)\chi_i'(R)\right] / \sum_j \int_0^R |\chi_j(R')|^2 dR' \tag{42}$$

For the case of closed channels the above expression turns out to be vanishingly small.

Finally, the determination of partial widths relies upon an accurate calculation of the wavefunction channel components. At the matching position R_m the channel functions can be obtained by solving the homogeneous set of equations :

$$\left[\underset{\approx}{P}^i(R_m) - \underset{\approx}{P}^o(R_m + h)^{-1}\right]\underset{\sim}{U}(R_m) = 0 \tag{43}$$

where $\underset{\sim}{U}(R_m)$ denotes the solution column vector with components $\chi_j(R_m)$. The channel functions at the other grid points could in principle be constructed by making use of the relations [21] :

$$\underset{\sim}{U}(R_m + h) = \underset{\approx}{P}^o(R_m + h)\underset{\sim}{U}(R_m) \quad \text{and} \quad \underset{\sim}{U}(R_m - h) = [\ \underset{\approx}{P}^i(R_m - h)\ ]^{-1}\underset{\sim}{U}(R_m) \tag{44}$$

However, the divergence of the wavefunction in classically forbidden regions (small R for all channels or large R for closed channels) is unavoidable. A much more satisfactory computation method is based on the following fact : once the iterative method giving the resonance energy has converged, throughout the entire grid we have :

$$\left[\underset{\approx}{P}^{o}(R_m+h)\right]^{-1}=\underset{\approx}{P}^{i}(R_m) \quad \text{and} \quad \left[\underset{\approx}{P}^{i}(R_m-h)\right]^{-1}=\underset{\approx}{P}^{o}(R_m) \tag{45}$$

Provided that all these matrices have been stored, $\underset{\approx}{P}^{i}$ can be used to perform *outward* propagation of the vector solution at R_m, while $\underset{\approx}{P}^{o}$ is used for *inward* propagation. This procedure omits completely the unstabilities which are observed when $\underset{\approx}{P}^{o}$ and $\underset{\approx}{P}^{i}$ are used to perform outward and inward propagations, respectively [21].

IV.- AN ILLUSTRATIVE EXAMPLE : LASER INDUCED RESONANCES IN THE MULTIPHOTON DISSOCIATION OF H_2^+

Photodissociation is one of the dynamical processes, where the resonance concept is crucial for the understanding and the interpretation of the underlying mechanisms. In this context the multiphoton dissociation of H_2^+ interacting with an intense electromagnetic (laser) field, provides a thorough illustration. A large amount of experimental and theoretical effort has recently been directed to the detailed analysis of this simple molecular ion prepared by a five-photon absorption from the X $^1\Sigma_g^+$ ground state of H_2. An additional sixth, photon causes the dissociation of H_2^+ involving, within good accuracy, only two Born-Oppenheimer electronic states $1s\sigma_g$ and $2p\sigma_u$ (briefly denoted by g and u) :

$$H_2^+(1s\sigma_g, v=0) \xrightarrow{\hbar\omega} H_2^+(2p\sigma_u) \longrightarrow H^+ + H(1s) \tag{46}$$

It is precisely this last step which is under consideration in this section. Fig. (3a) displays the potential energy curves g (attractive) and u (repulsive) as a function of the internuclear distance R, in the absence of any laser field. When a continuous wave (cw) laser (of frequency ω and intensity I) is switched on, the potential energy curves are said the be "dressed" by the photon : the ground state potential after absorbtion of the photon energy $\hbar\omega$ presents a curve crossing with the excited state potential, as indicated on fig. (3b). The v=0 vibrational level of g (of energy $E_{g,v=0}$) is no more a bound state ; it is now embedded in the nuclear continuum of the excited u state to which it is coupled by the radiative (electromagnetic) interaction :

$$V(R)=\mu(R).\xi=\mu(R)\sqrt{I} \tag{47}$$

where $\mu(R)$ is the field induced transition dipole moment between g and u, and ξ the field amplitude. This is a traditional picture of a Feshbach resonance with a characteristic decay time (or width).

Three different approaches have to be considered, with respect to the field intensity, when the calculation of this resonance is attempted. Following the analysis of the previous Section, these are connected with the possibility of observing isolated or overlapping resonances. Quantitatively, one has to compare the strength of the radiative coupling given by the Rabi frequency ($\omega_R = 1/\hbar\mu\sqrt{I}$) to the local vibrational level spacing ω_v in the ground electronic state.

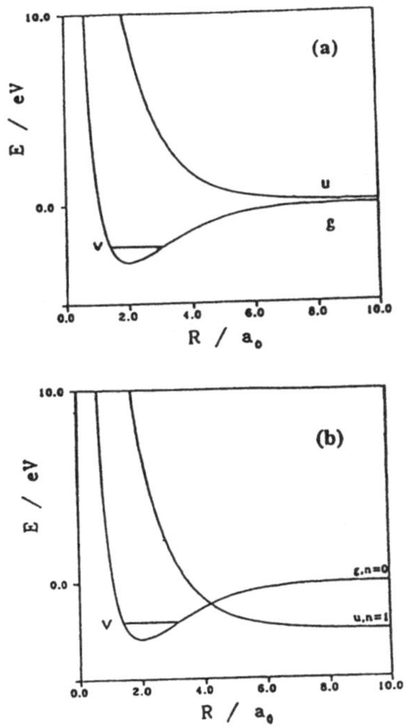

Figure 3. Potential energy curves for H_2^+ ground g and first excited u states. (a). In the absence of the laser field, (b) dressed by the field.

i) Weak field regime

As far as $\omega_R \ll \omega_v$, the resonance can be considered to be isolated. This situation holds for H_2^+, for which the vibrational spacing is large ($\omega_v \approx 2000$ cm^{-1}), up to intensities of the order of 10^{11} W/cm^2. In addition for such intensities the lowest order perturbative approach is appropriate. Fano's formula (Eq. 11) for the pole of the scattering amplitude can now be written as [22] :

$$E - (E_{g,v=0} + \hbar\omega) - F_{ug}(E) + i\pi\left|V_{ug}(E)\right|^2 = 0 \qquad (48)$$

where

$$V_{ug}(E) = \sqrt{I}\,\left\langle \chi^{(0)}_{u,E}(R)\,|\mu(R)|\,\chi^{(o)}_{g,v=0}(R)\right\rangle_R \qquad (49)$$

and

$$F_{ug}(E) = \mathcal{P}\int dE' \frac{|V_{ug}(E')|^2}{E - E'} \qquad (50)$$

$\chi^{(o)}_{g,v=0}(R)$ and $\chi^{(o)}_{u,E}(R)$ being the field-free wavefunctions associated with the ground vibrational level and the energy-normalized nuclear continuum of the excited state at energy E. The zeroth order perturbation leads to $E = E_{g,v=0} + \hbar\omega$ (i.e. the field dressing) and the lowest (second) order width is obtained from :

$$\Gamma_{v=o} = \pi\left|V_{ug}(E)\right|^2 \delta(E - E_{g,v=0} - \hbar\omega) \qquad (51)$$

The level shift is given at the same order of approximation by :

$$F_{v=0} = \int dE' \frac{|V_{ug}(E')|^2}{E_{g,v=0} + \hbar\omega - E'} \tag{52}$$

Two remarks are in order :

1- The position and the width of the resonance are linear functions of the field intensity ;
2- The width of the resonance correspond to the Fermi golden rule approximation for the absorption cross section $\sigma_g(\omega)$, calculated using the first order perturbation in the Born expansion of the transition matrix :

$$T = V + VG_oV + VG_oV\, G_oV + \ldots.. \tag{53}$$

G_o being the zeroth order resolvent and V the radiative interaction. This cross section, proportional to the square of the matrix element of T between the initial $\chi_{g,v=0}$ and the final $\chi_{u,E}$ states, is given by :

$$\sigma_g(\omega) = \frac{4\pi^2\omega}{cI}\left|V_{ug}(E)\right|^2 \delta(E - E_{g,v=0} - \hbar\omega) \tag{54}$$

ii) Intermediate field regime

When $\omega_R \approx \omega_V$ (intensity of the order of 10^{12} W/cm^2) the indirect coupling of successive vibrational levels of the ground state g, via the continuum u is large enough for the neighbouring resonance overlapping to occur. A single pole perturbative approach (Eq. 48) to the scattering matrix is no more appropriate. Which can still be retained, as an approximation, is a single photon scheme : the molecule is supposed not to absorbe more than one photon consecutively. Within the coupled equation frame, the total (molecule plus field) eigenket is expanded on the dressed electron-field channel kets, $\psi_{g\text{ (or u)}}(r)|\vec{k}\,\vec{e}\text{ (or vac)}\rangle$

$$|\Psi(r,R)\rangle = \psi_g(r)\chi_g(R)|\vec{k}\,\vec{e}\rangle + \psi_u(r)\chi_u(R)|\text{vac}\rangle \tag{55}$$

with unknown functions $\chi_{g\text{ (or u)}}(R)$. $|\vec{k}\,\vec{e}\rangle$and $|\text{vac}\rangle$ designate the state of the electromagnetic field with either one photon of wavevector $\vec{k}$ (frequancy ω) and polarization $\vec{e}$, on zero photon, respectively. The Schrödinger equation written for $|\Psi(r,R)\rangle$, after appropriate left multiplication by $\psi_{g(u)}(r)|\vec{k}\,\vec{e}\text{ (vac)}\rangle$, and integration over electronic and field variables leads to the set of coupled equations [22]:

$$\left[-\frac{d^2}{dR^2}+V_g(R)+\hbar\omega-E\right]\chi_g(R)+V_{gu}(R)\chi_u(R)=0 \tag{56a}$$

$$\left[-\frac{d^2}{dR^2}+V_u(R)-E\right]\chi_u(R)+V_{ug}(R)\chi_g(R)=0 \tag{56b}$$

The resonances we are looking for are obtained by solving these equations with the complex rotation technique and boundary conditions appropriate to Siegert-Gamow condition. For the lowest resonance we are considering, the components $\chi_{g\,(\text{or}\,u)}$ (R) correspond, in the limiting case of $V_{gu}(R) \to 0$, to the zeroth order $\chi^{(0)}_{g,v=0}$ (R) and $\chi^{(0)}_{u,E_g}$ (R) defined in (Eq. 49), but for intermediate fields, the strength of the radiative coupling is such that the vibrational labelling v is completely lost. The variations of resonance parameters (position and width) with respect to the increasing field intensity, for several wavelengths choosen close to the maximum of the absorption line shape of H_2^+, are illustrated on Fig. (4). Two points can be emphazised :

1- the resonance behavior is no more linear with the laser intensity ;
2- as far as the resonance overlapping is not the dominant process, the observable, i.e. the absorption cross section can still be defined with a formula involving but one resonance [23] :

$$\sigma_g(\omega)=\frac{4\pi\omega}{c}\frac{\Gamma}{I} \tag{57}$$

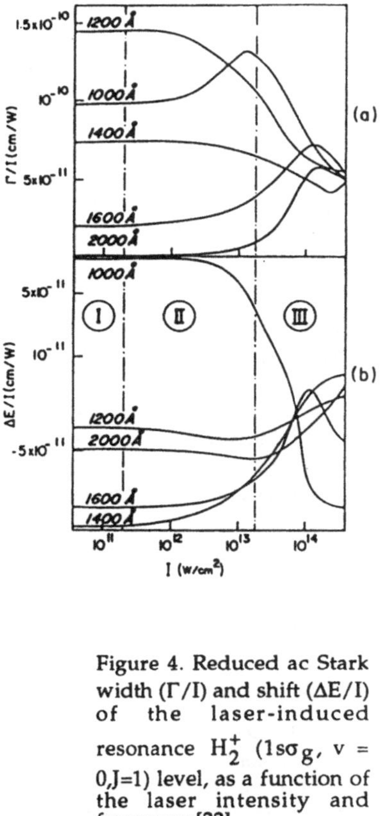

Figure 4. Reduced ac Stark width (Γ/I) and shift (ΔE/I) of the laser-induced resonance H_2^+ ($1s\sigma_g$, v = 0,J=1) level, as a function of the laser intensity and frequency [22].

the corresponding non linearities, related with those of Γ, being interpreted by the higher order radiative couplings building up the Born series of the transition operator (Eq. 53).

iii) Strong field regime

Situations for which $\omega_R >> \omega_V$ (occuring for H_2^+ at intensities larger than 10^{13} W/cm^2) correspond to the strong field regime, where not only perturbative approaches (as the one which is addressed to in Eq. (53)) are not valid, but also a single photon approximation can not be retained. One has to take into account the possibility of successive absorption or emission of several photons. The corresponding set of coupled equations are written for unknown wavefunction components bearing two labels : one (g or u) for the electronic degree of freedom, and another n for the field [22] :

$$\left[-\frac{d^2}{dR^2}+V_g(R)+(n+1)\hbar\omega-E\right]\chi_{g,n+1}(R)+V_{gu}(R)[\ \chi_{u,n}(R)+\chi_{u,n+2}(R)\ ]=0 \qquad (58a)$$

(for any n)

$$\left[-\frac{d^2}{dR^2}+V_u(R)+n\hbar\omega-E\right]\chi_{u,n}(R)+V_{ug}(R)[\ \chi_{g,n+1}(R)+\chi_{g,n-1}(R)\ ]=0 \qquad (58b)$$

(for any n)

The corresponding dressed potential energy curves are indicated on fig. (5). The coupling scheme involves two selection rules ; $g\leftrightarrow u$ and $\Delta_n = \pm 1$, such that channel $|g,n=0\rangle$ is coupled to $|u,n=1\rangle$ by the absorption of one photon and to $|u, n=-1\rangle$, by the emission of one photon. $|u,n=1\rangle$ in turn is coupled to $|g,n=0\rangle$ by the emission of one photon and to $|g,n=2\rangle$ by the absorption of one photon. In other words, starting from $|g,n=0\rangle$, one can successively reach $|u,n=1\rangle$ by absorbing one photon, $|g,n=2\rangle$ by absorbing two photons, $|u,n=3\rangle$ by absorbing three photon, etc...

The analysis can be summarized up as following :

1 - Resonance parameters for several wavelengths close to the maximum of the H_2^+ absorption lineshape, are gathered on fig. (4) as a function of the field intensity. High non-linearities are observed.

2 - Due to large resonance overlapping the absorption cross section cannot be represented by a unique resonance, one has to consider a more subtle reconstruction scheme using all resonances [24]. Results concerning the variations of the absorption lineshapes for three typical field strengths are displayed on fig. (6). A broadening and blue-shift of the maximum is observed. A tendancy for a cut-off frequency followed by a plateau behavior seems to be confirmed at high intensities [22,25].

3 - The multiphoton nature of the process puts the emphasis on the evaluation of partial widths or branching ratios towards the different channels of the dissociation. This constitutes the key for the interpretation of the final kinetic energy distribution of

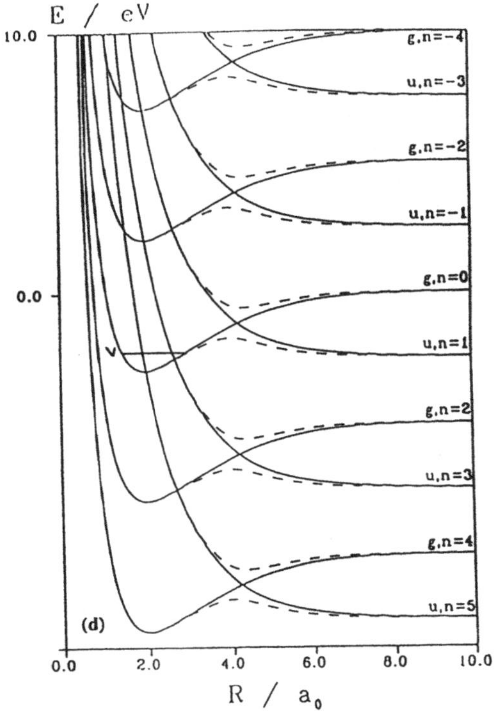

Figure 5. Dressed potential energy curves for H_2^+ ($\lambda = 330$ nm, $I = 10^{13}$ W/cm^2). Solid line : diabatic frame ; dotted line : adiabatic frame

the fragments. The formalism which is used derives from the asymptotic analysis of the resonance wavefunction components (Eq. 42). A difficulty is related with the increasing nature of the induced dipole moment as a function of R, for the homonuclear ion H_2^+ [26]. In the electric field gauge, the radiative coupling (Eq. 47), is said to be persistent at large R, which renders unphysical the definition of coupled channels [21]. This difficulty can be overwhelmed by referring to adiabatic channels diagonalizing the radiative coupling [21]. For an infinite number of channels differing asymptotically by a photon frequency $\hbar\omega$, the diabatic and adiabatic frames are identical at large R where molecular forces are constant. This allows for a partial fluxes calculation based on an expression :

$$\gamma_i = \mathrm{Im}\left[\tilde{\chi}_i^*(R)\tilde{\chi}_i'(R)\right] / \sum_j \int_0^R \left|\tilde{\chi}_j(R')\right|^2 dR' \tag{59}$$

formally identical to that of (Eq. 42) where the diabatic channel components χ_i of the wavefunction are merely replaced by their adiabatic counterparts $\tilde{\chi}_i$ [21]. The photon channel pathways followed during the fragmentation are determined in terms of these partial fluxes. Multiphoton branching ratios are obtained as the large R limit of the γ_i's.

The calculations are done at a wavelength (λ = 330 nm) for which the single photon absorption is very much unlikely. A multiphoton mechanism prevails with a leading three photon absorption step. Fig. (7) illustrates, for a typical laser intensity of the intermediate regime ($I = 10^{12}$ W/cm^2), the way in which the initial flux is shared among the different channels as the dissociation evolves and the fragments move apart. The upper pannel shows the adiabatic potentials involved in the leading three photon process. They are labelled 0,1,2 and 3 denoting also the net amount of absorbed photons as the system dissociates into the corresponding thresholds. The lower pannel shows the partial flux γ_i of each of the open channels as a function of R. As long as the fragments are close to the origin the entire flux is concentrated in the close channel 1. Once the intersection with the dissociative channel 2 is reached, due to very favorable nuclear wavefunction overlapping at the equilibrium position (~ 2 bohr), the three-photon absorption mechanism takes place, the flux is abruptly transferred into channel 2 under the effect of indirect nonadiabatic coupling. Another flux rearrengement takes place at the crossing region (3.5 bohr) where non-adiabatic couplings mix them again between channels 2 and 3. The physical interpretation is a forth photon emission during the dissociation, having as a consequence a breaking down of the velocity of the fragments. Fig. (8) gives the branching ratios (calculated as asymptotic partial fluxes) for dissociation into the single, double and triple adiabatic thresholds, for all intensities. For low laser intensity and for the wavelength which is studied, the total dissociation probability of the system is very small ($\Gamma \sim 10^{-10}$ cm^{-1}). Actually, the photodissociation process starts when the laser intensity is high enough ($I \sim 10^{11}$W/cm^{-2}), with the simultaneous absorption of three photons. This refers to a situation (Above Threshold Dissociation [26,27]), where the molecule absorbs more photons than the minimum number needed for dissociation to occur. During the evolution of the dissociation process itself, the radiative interaction may lead, at higher laser intensities ($I \sim 10^{13}$W/cm^2) to additional photon exchanges ; an emission producing the rise of the double photon branching ratio. Finally, for very strong fields ($I > 10^{13}$W/cm^2), again the single photon branching ratio is dominating. The interpretation of this process is still an open question [28]. It seems that it can be envisionned either as an adiabatic potential barrier lowering (Bond Softening [26]) or a five photon : three photons mechanism simultaneously absorbed followed by two emissions [29].

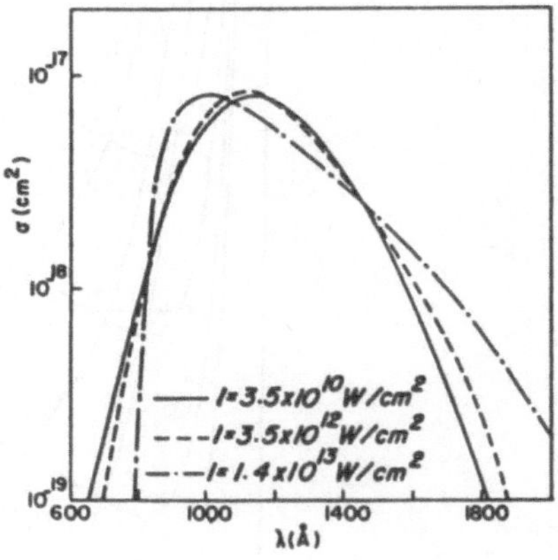

Figure 6. Line shape for the photodissociation processs of the (v = 0, J=1) level of H_2^+ ground state, for different laser intensity [22]

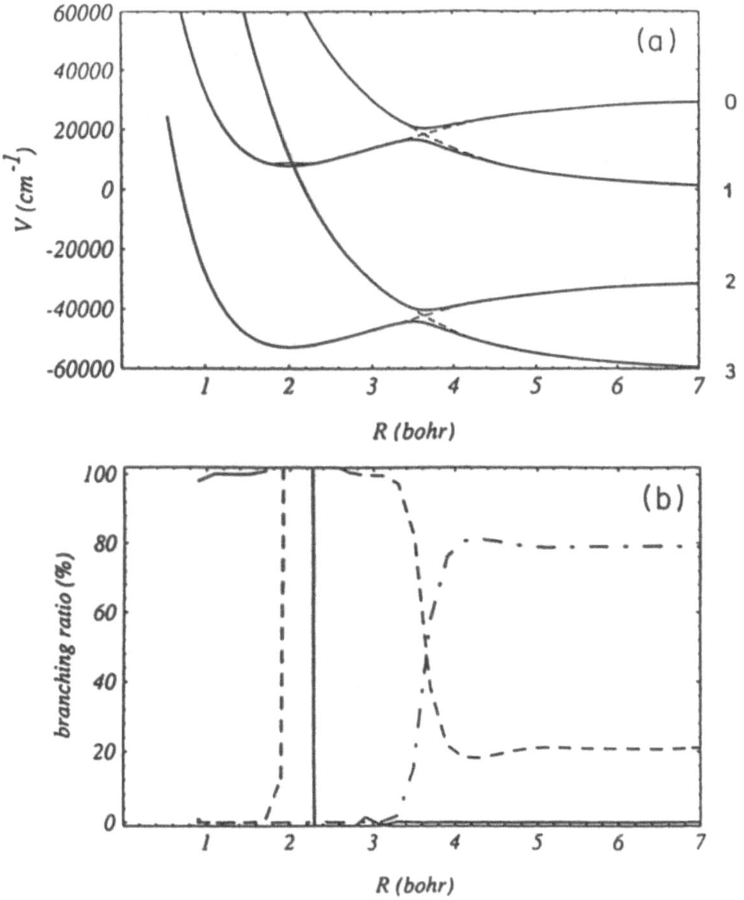

Figure 7 - Adiabatic potentials of the two main Floquet blocks involved in the photodissociation process (panel (a)) and branching ratios (%) for dissociation in these potentials as a function of the interfragment distance (b) for $I = 10^{12}$ W/cm^2 (intermediate intensity regime). In panel (a), the dashed lines denote the corresponding diabatic channels. In panel (b), the branching ratios (%) for dissociation into channels 1,2 and 3 are indicated by solid, dashed, and dotted-dashed lines, respectively [21]

As a word of conclusion, H_2^+ intense field multiphoton dissociation thoroughly illustrates the use of the resonance concept as a tool of understanding and thus controlling the underlying non linear optical processes. A formal complex energy definition and the introduction of the complex rotation of the coordinate in a close coupling scheme, provide the basis for very accurate calculations of total dissociation rates and partial fluxes analysis in extreme situations of persistent asymptotic couplings.

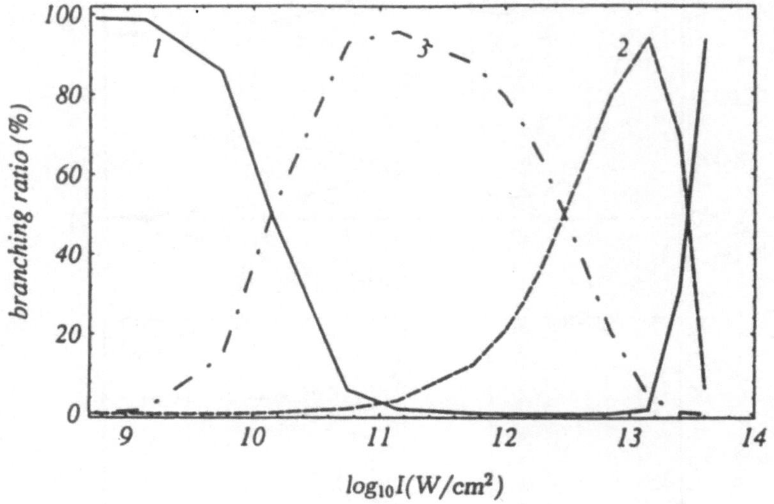

Figure 8 - Branching ratios (%) of the ground dissociating resonances, accounting for net absorption processes of one (solid line), two (dashed line), or three (dotted-dashed line) photons, as a function of increasing intensity in a semilogarithmic scale. I is measured in W/cm^2 [21]

REFERENCES

[1] U. Fano, Phys. Rev. 124, 1866 (1961)
[2] R.G. Newton, Scattering Theory of Waves and Particles, Springer-Verlag, New-York, N.Y. 1982
[3] O. Atabek, R. Lefebvre, M. Garcia-Sucre, J. Gometz-Llorente and H. Taylor, Inter. J. Quantum Chem. 49, 211 (1991)
[4] R.A. Bain, J.N. Bardsley, B.R. Junker and C.V. Sukumar, J. Phys. B7, 2189 (1974)
[5] O. Atabek, R. Lefebvre and M. Jacon, J. Phys. B15, 2689 (1982)
[6] O. Atabek and R. Lefebvre, Chem. Phys. Lett. 84, 233 (1981)
[7] L. Fonda, G.C. Ghirardi and A. Rimini, Rep. Progr. Phys. 41, 587 (1978)
[8] G. Raseev and O. Atabek, Il Nuovo Cimento, 107B, 463 (1992)
[9] U. Hazi and H.S. Taylor, Phys. Rev. A1, 1109 (1970) ; C.H. Maier, L.S. Cederbaum and W. Domcke, J. Phys. B13, L119 (1980)
[10] R. Lefebvre, J. Phys. Chem. 89, 4201 (1985)
[11] A.D. Bandrauk and O. Atabek, J. Phys. Chem. 91, 6469 (1987)
[12] see for instance, W.P. Reinhardt, Ann. Rev. Phys. Chem. 33, 223 (1982)
[13] e.g. Special Issue of Int. J. Quantum Chem. 14 (1978)
[14] O. Atabek and R. Lefebvre, Phys. Rev. A22, 1817 (1970) ; ibid Chem. Phys. 56, 195 (1981)
[15] O. Atabek, R. Lefebvre and A. Requena, Mol. Phys. 40, 1107 (1980)
[16] R.G. Gordon, J. Chem. Phys. 51, 14 (1969) ; B.R. Johnson, J. Chem. Phys. 69, 4678 (1978)
[17] L. Fox, The Numerical Solution of Two-point Boundary Value Problems in Ordinary Differential Equations, Oxford, U.P., London (1957)
[18] M. Chrysos, R. Lefebvre and O. Atabek, J. Phys. B27, 3005 (1994)
[19] O. Atabek and R. Lefebvre, J. Chem. Phys. 67,4983 (1977)
[20] C.J. Ashton, M.S. Child and J.M. Hutson, J. Chem. Phys. 78, 4025 (1983) ; S. C. Tucker and D.G. Truhlar, J. Chem. Phys. 88, 3667 (1988)
[21] M. Chrysos, O. Atabek and R. Lefebvre, Phys. Rev. A48, 3845 (1993) ; ibid 3855 (1993)
[22] X. He, O. Atabek and A. Giusti-Suzor, Phys. Rev. A42, 1585 (1990)
[23] S.I. Chu, J. Chem. Phys. 75, 2215 (1981)
[24] G. Jolicard and O. Atabek, J. Chem. Phys. 93, 4750 (1990)
[25] A. Keller, R. Numico and O. Atabek (unpublished)
[26] A.D. Bandrauk, E.Constant and J.M. Gauthier, J. Phys. (France) II, 1, 1033 (1991)
[27] A. Giusti-Suzor, X. He, O. Atabek and F.H. Mies, Phys. Rev. Lett. 64, 515 (1990)
[28] O. Atabek, Int. J. Quantum Chem. (in press)
[29] S. Miret-Artès and O. Atabek, Phys. Rev. A49, 1502 (1994).

MOLECULES IN LASER FIELDS

A.D. BANDRAUK
Laboratoire de Chimie Théorique
Faculté des sciences
Université de Sherbrooke, Que, J1K 2R1, Canada

1. Introduction

Current laser technology is capable of generating laser fields from the IR to visible wavelength regions in the form of well-tailored sequences of pulses with controllable phase and envelopes (pulse shape) [1-2]. Such pulses can be used for the efficient preparation of ensembles of atoms or molecules in specific states. This is of considerable interest not only in spectroscopy but also in studies of chemical dynamics [2-4]. Furthermore short pulses allow one to attain electric field strengths $\mathcal{E}$ (V cm^{-1}) or equivalently laser intensities I (W/cm^2) = $c\mathcal{E}^2/8\pi$ (c = velocity of light) which are comparable or greater than atomic fields. An important consequence of the progress in this area is that one can greatly enhance radiative transition rates and even ionize molecules. Clearly strong field science and short pulse science are two fields which will become increasingly intertwined. Efficient rapid excitation requires increasingly higher intensities as can be seen from the simple example of a resonantly driven two-level system [6-7]. The population of the upper state is given by the transition probability formula from level 1 to level 2

$$P_{12}(t) = \sin^2(\omega_R t/2) \quad , \quad \omega_R = d\mathcal{E}/\hbar \quad , \tag{1}$$

where ω_R, the Rabi frequency or radiative transition rate can be calculated from the expression [2],

$$\omega_R(\mathrm{cm}^{-1}) = 1.17 \times 10^{-3} \left[I(\mathrm{W/cm}^2)\right]^{1/2} d(ea_0) \quad . \tag{2}$$

d is the dipole transition moment in atomic units (a_o = 1 bohr = 0.529 x 10^{-8} cm). Therefore intensities of at least 10^{10} W/cm^2 are required to achieve subpicosecond (ps) radiative excitation rates per atomic unit of transition moment. As an example, the time for a single coherent radiative transition is given from equation (1) by $\tau = \pi/\omega_R$, then for τ = 1 ps = 10^{-12} sec, one requires $\omega_R \simeq 100$ cm^{-1}, or intensities well above 10^{10} W/cm^2 for transition moments less than ea_o, 1 atomic unit.

Driving atoms or molecules at intensities exceeding 10^{13} W/cm^2 produces a new phenomenon at long wavelength (e.g. IR), called Keldysh tunneling ionization [8]. Using a static tunneling ionization model, one can obtain an ionization threshold law for the ionization of atoms and molecules alike, [9-10],

E. Yurtsever (ed.), Frontiers of Chemical Dynamics, 131–150.

$$I_t(W/cm^2) = 4 \times 10^9 \, (IP(ev))^4 / Z^2 \tag{3}$$

where IP is the ionization potential and Z is the charge of the resulting ion. Thus for the neutral HF molecule which has IP = 16 ev and Z = 1, we one obtains readily $I_t = 2.6 \times 10^{14}$ W/cm^2. This law has been recently verified experimentally for HCl [10] and has been found to apply also in exact calculations of ionization rates of H_2^+ [11].

We recall that the atomic unit of the electric field and the equivalent laser field intensity are given by [2],

$$E_o = e/a_o^2 = 5.14 \times 10^9 \, V/cm \quad ; \quad I_o = \frac{cE_o^2}{8\pi} = 3.54 \times 10^{16} W/cm^2 \tag{4}$$

Equations (2), (4) serve as conversion factors in addition to reminding us that for intensities approaching the limits (4), one has to deal with nonperturbative radiative transition rates which can compete with and exceed IVR rates [4-5]. The implication is therefore that at higher intensities, coherences become important on subpicosecond time scales. Such coherences can induce new highly nonlinear optical phenomena such as laser-induced avoided crossings in the nuclear motion. These were first described in a dressed-molecule representation [2], [12-15]. This unusual phenomenon has now been detected experimentally [16-18]. High order radiative coherences can also be used to control vibrational excitations [4-5] as well as controlling high order harmonic generation [11], [19].

In the present chapter we will therefore investigate the fundamental properties of electromagnetic fields in relation to various computational methods, i.e., time independent and time dependent, which can be used to describe theoretically multiphoton transitions in molecules [2], [20]. The time-independent approach leads via the dressed molecule representation based on quantum electrodynamics to time-independent coupled equations and is applicable to long pulses or plane wave fields. The time dependent approach leads to full space and time dependent coupled equations [2], [21] which are most appropriate for short and intense laser pulses. Examples of published computational results will be cited in order to illustrate the use of these methods to describe both perturbatively and nonperturbatively multiphoton transitions in molecules.

2. Classical Electrodynamics, and Coherent States

Maxwell's equations show that one can derive consistently electrodynamics from a vector potential $\vec{A}(\vec{r},t)$ via a Lagrangean or Hamiltonian formalism [2], [22-25]. In free space, the vector potential which is a solution of a hyperbolic partial differential equation can be expressed as freely propagating plane waves:

$$\vec{A}(\vec{r},t) = A_o \hat{\varepsilon} \cos(\vec{k}\cdot\vec{r} - \omega t + \phi) \tag{5}$$

where $\hat{\varepsilon}$ is the unit polarization vector, ϕ is an arbitrary phase, k is the wave number $k = \frac{w}{c} = 2\pi/\lambda$, ω is the angular for a given wavelength λ. The solutions (5) are obtained by applying the Coulomb (nomelativistic) gauge condition $\vec{\nabla}\cdot\vec{A} = 0$ to the complete set of Maxwell equations. This gives rise to the transversality condition: $\vec{\nabla}\cdot\vec{A} = -2A_0\,\vec{k}\cdot\hat{\varepsilon}\sin(\vec{k}\cdot\vec{r}-\omega t+\phi) = 0$ from which $\vec{k}\cdot\hat{\varepsilon} = 0$ implies two possible polarizations $\hat{\varepsilon}$ mutually orthogonal to the propagation vector $\vec{k}$. The electric field $\vec{\mathcal{E}}(r,t)$ is then defined as the canonically conjugate momentum variable of $\vec{A}$ if the latter is considered as a generalized coordinate,

$$\vec{\mathcal{E}}(\vec{r},t) = -\frac{1}{c}\frac{\partial\vec{A}}{\partial t} = +\frac{\omega}{c}A_0\,\hat{\varepsilon}\sin(\vec{k}\cdot\vec{r}-\omega t+\phi) \tag{6}$$

Due to the nonexistence of magnetic charges (monopoles) one obtains from Maxwell's equations that the magnetic field satisfy $\vec{B} = \vec{\nabla}\times\vec{A}$, so that from (5) one obtains

$$\vec{B} = \vec{\nabla}\times\vec{A} = -A_0\,(\vec{k}\times\hat{\mathcal{E}})\sin(\vec{k}\cdot\vec{r}-\omega t+\phi)\ . \tag{7}$$

One notes that $\vec{\mathcal{E}}$ and $\vec{B}$ are exactly in phase but the two are mutually orthogonal as well as to the propagation direction of $\vec{k}$.

The average energy is defined classically via the expression

$$E = \int d^3r\left[\int_0^\tau \frac{\frac{(\vec{\mathcal{E}}^2+\vec{B}^2)}{8\pi}}{\tau}\,dt\right] = \frac{V\omega^2A_0^2}{8\pi c^2} \tag{8}$$

after averaging over a volume V and one cycle $\tau = 2\pi/\omega$. The result (8) is obtained after replacing the average of $\sin^2(k\cdot r-\omega t+\phi)$ by ½ and using the relation $(\hat{k}\times\hat{\varepsilon})^2 = \omega^2/c^2$.

In the quantum field theory of electrodynamics, $\vec{\mathcal{E}}$ and $\vec{A}$ become operators $\hat{\mathcal{E}}$ and $\hat{A}$ which are related through the quantum commutation relations of canonically conjugate variables. Thus for x and p, one has in quantum mechanics $[\hat{x},\hat{p}] = i\hbar$, then similarly because of the classical canonical relation (6) between $\vec{A}$ and $\vec{\mathcal{E}}$, the Heisenberg quantum commutator becomes:

$$\left[\hat{A}/c\,,\,\hat{\mathcal{E}}\right] = i\hbar \tag{9}$$

This leads to quantization of the radiation field with the ensuing concept of photons, i.e., the number of elementary excitations of the electromagnetic field, as we will show below. The field operators $\hat{\mathcal{E}}$ and $\hat{A}$ can be expanded as linear superpositions of the field quantum states of photon number n and energy $n\hbar\omega$. We can thus equate (8) to the average energy $N\hbar\omega$, where N is an average photon number, with the result that the vector potential amplitudes A_o and electric field amplitudes $\mathcal{E}_o$ become:

$$A_o = 2c\left(\frac{2\pi\hbar N}{\omega V}\right)^{1/2} \; ; \; \mathcal{E}_o = + \frac{\omega}{c} A_o = + 2\left(\frac{2\pi\hbar\omega N}{V}\right)^{1/2} . \tag{10}$$

This leads to new expressions for the total electric field1 $\vec{\mathcal{E}}$, equation (6) as:

$$\bar{\mathcal{E}}(\bar{r},t) = i\hat{\varepsilon}\left(\frac{2\pi\hbar\omega N}{V}\right)^{1/2}\left[e^{i(\bar{k}\cdot\bar{r}-\omega t+\phi)} - c\cdot c\cdot\right] \tag{11}$$

where c·c = complex conjugate. The time average of (11) leads from (10) to $\mathcal{E}_o^2 = 8\pi\hbar\omega N / V$ so that intensities I are related to N as:

$$I = \frac{c\mathcal{E}_o^2}{8\pi} = \frac{N\hbar\omega}{V} c \tag{12}$$

Equation (12) is fundamental in relating intensities to photon numbers. It is readily seen that I equals the energy density times the velocity, thus corresponding to a flux of energy. The units of I follow from (12) clearly as joules S^{-1} cm^{-2} or Watts/cm^2. As an example, for a CO_2 laser (ω= 1000 cm^{-1}), at an intensity of 10^{13} W/cm^2, the photon density N/V is ~ 10^{29} cm^3, so that N is indeed a very large number.

We now return to the quantum description of the fields $\bar{A}$ and $\bar{\mathcal{E}}$. We write these as operators:

$$\hat{A}(\bar{r},t) = c\left(\frac{2\pi\hbar}{\omega V}\right)^{1/2} \hat{\varepsilon}\left[\hat{a}e^{i(\bar{k}\cdot\bar{r}-\omega t)} + \hat{a}^+ e^{-i(\bar{k}\cdot\bar{r}-\omega t)}\right], \tag{13}$$

$$\hat{\mathcal{E}}(\bar{r},t) = -i\left(\frac{2\pi\hbar\omega}{V}\right)^{1/2} \hat{\varepsilon}\left[\hat{a}e^{i(\bar{k}\cdot\bar{r}-\omega t)} - \hat{a}^+ e^{-i(\bar{k}\cdot\bar{r}-\omega t)}\right], \tag{14}$$

where $\hat{a}$ and $\hat{a}^+$ are operators. In view of the quantum Heisenberg commutation relation (9), we have that for equal space and time (x = x', t - t') for $\hat{a}$ and $\hat{a}^+$ the well known annihilation and creation operators for bosons (i.e., photons).

$$[\hat{a},\hat{a}^+] = \hat{a}\hat{a}^+ - \hat{a}^+\hat{a} = 1 \; . \tag{15}$$

In quantum mechanics, operators define representations or Hilbert spaces which span eigenstates of these same operators. Thus we look for eigenstates $|\alpha\rangle$ of $\hat{a}$ and its transpose $\hat{a}^+$, i.e.

$$\hat{a}|\alpha\rangle = \alpha|\alpha\rangle \ , \ \langle a|\hat{a}^{+} = \alpha^{*}\langle\alpha| \ , \ \alpha = |\alpha|\ell^{i\phi} \ . \tag{16}$$

Then the average of the operator $\hat{\mathcal{E}}$ ($\bar{r}$,t) becomes in this space,

$$\langle\alpha|\hat{\mathcal{E}}(r,t)|\alpha\rangle = 2|\alpha|\left(\frac{2\pi\hbar\omega}{V}\right)^{1/2}\hat{\varepsilon}\sin\left(\bar{k}\cdot\bar{r}-\omega t+\phi\right) \ . \tag{17}$$

one therefore recovers the classical wave, equation (11) provided the amplitude

$$|\alpha| = N^{1/2} \tag{18}$$

Alternatively one can define an electric field operator with positive and negative frequency components:

$$\hat{\mathcal{E}} = \hat{\mathcal{E}}^{+} + \hat{\mathcal{E}}^{-} \ , \tag{19}$$

$$\hat{\mathcal{E}}^{-} = -i\left(\frac{2\pi\hbar\omega}{V}\right)^{1/2}\hat{a}\hat{\varepsilon}e^{-i(\bar{k}\cdot\bar{r}-\omega t)} \ , \tag{20}$$

$$\hat{\mathcal{E}}^{+} = (\hat{\mathcal{E}}^{+})^{+} = i\left(\frac{2\pi\hbar\omega}{V}\right)^{1/2}\hat{a}^{+}\hat{\varepsilon}\,e^{i(\bar{k}\cdot\bar{r}-\omega t)} \ , \tag{21}$$

$$\mathcal{E}^{+}|\alpha\rangle = \hat{\varepsilon}\frac{\mathcal{E}_{o}}{2}e^{i\phi}|\alpha\rangle \ , \ \langle\alpha|\mathcal{E}^{-} = \hat{\varepsilon}\frac{\mathcal{E}_{o}}{2}e^{-i\phi}\langle\alpha| \ . \tag{22}$$

$\mathcal{E}_o$ has already been defined in equation (10) in terms of the average photon number N. Equations (19) to (22) emphasize the fact that it is possible to define electric field operators and their eigenstates, with eigenvalues equal to the classical field amplitudes. These are called coherent states, i.e. quantum states of the electromagnetic field whose eigenvalues are the coherent classical fields, states with well defined amplitude $\mathcal{E}_o$, frequency ω and phase ϕ.

The coherent states $|\alpha\rangle$ can be shown to be minimum uncertainty states and are expressible as linear superpositions of the number states $|n\rangle$, i.e. the photon states, [26-27],

$$|\alpha\rangle = e^{-|\alpha|^{2}/2}\sum_{n=0}^{\infty}\left(\frac{\alpha^{n}}{(n!)^{1/2}}\right)|n\rangle \ . \tag{23}$$

Equation (23) shows that the photon number distribution in a coherent state is given by the Poisson distribution law:

$$P_n = |\langle n|\alpha\rangle|^2 = \frac{|\alpha|^{2n}}{n!} e^{-|\alpha|^2} , \qquad (24)$$

which has its maximum value at $|\alpha|^2 = N$, the most probable photon number. The variance of the photon distribution, i.e. $(\langle n^2\rangle - \langle n\rangle^2)^{1/2} \sim 1/N^{1/2}$, so that for high intensities, i.e. large $\mathcal{E}_o$, α and N, then the variance becomes negligible. Thus for large intensities and therefore large photon numbers, equation (10) or (11) relating the average photon number N and the classical field amplitude $\mathcal{E}_o$ becomes very accurate. In fact N/V, the photon density is of the order of 10^{20} for intensities > 10^{12} W/cm^2. $|\alpha\rangle$ is related simply to the vacuum ground state $|0\rangle$ (zero photons) of the field by application of the unitary operator $\hat{T}$, [23], [27],

$$\hat{T}(\alpha) = \exp\left[\alpha\hat{a}^+ - \alpha^*\hat{a}\right] , \qquad (25)$$

$$|\alpha\rangle = \hat{T}|0\rangle , \qquad (26)$$

which displaces $\hat{a}$ and $\hat{a}^+$.

$$\hat{T}^{-1}\hat{a}\hat{T} = \hat{a} + \alpha \; ; \; \hat{T}^{-1}\hat{a}^+\hat{T} = \hat{a}^+ + \alpha^+ , \qquad (27)$$

where $\hat{T}^{-1} = \hat{T}^+$. This displacement of the vacuum state of the photons to create a nonvanishing field is illustrated in figure 1. Thus choosing the electric field $\mathcal{E}$ as a coordinate, the photon eigen functions $\langle n|\mathcal{E}\rangle$ are the harmonic oscillator wavefunctions in $\mathcal{E}$ space. A state of n photon has clearly a random distribution of $\mathcal{E}$ values, with an average of zero i.e. $\langle n|\mathcal{E}|n\rangle = 0$. A displaced photon state has always an average nonzero $\mathcal{E}$ field, i.e. $\langle\mathcal{E}\rangle = \alpha$, but now the photon number distribution obeys a Poisson distribution (24). Furthermore the field fluctuations are minimum, i.e., Gaussian, as in the vacuum state.

$$\{-\hbar^2 \frac{d^2}{d\mathcal{E}^2} + \frac{\omega^2}{2}\mathcal{E}^2\}\psi_n(\mathcal{E}) = (n + 1/2)\hbar\omega\psi_n(\mathcal{E})$$

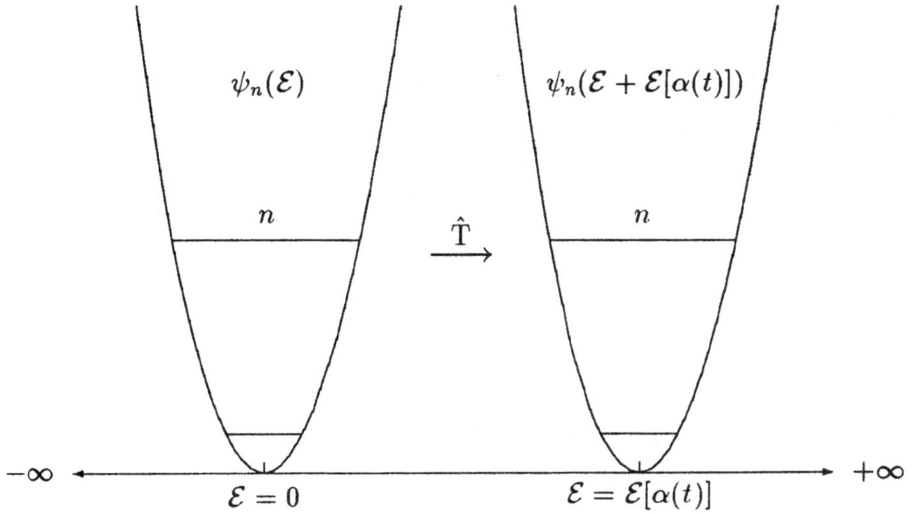

Figure 1. Photon wavefunctions $\psi_n(\mathcal{E})$ and effect of translation operator $\hat{T}$, equations (25),(35) in $\mathcal{E}$ space to give nonvanishing electric field $< \mathcal{E} >= \mathcal{E}[\alpha(t)]$, the classical value, equation (17).

3. Coupled Equations

We write the total time independent Hamiltonian of a molecule coupled to an electromagnetic field in the dipole approximation ($\vec{k} \cdot \vec{r} \approx 0$) in the electric field gauge as:

$$\hat{H} = \hat{H}_o - d \cdot \hat{\mathcal{E}} \quad , \quad \vec{d} = \sum_{\alpha} q_{\alpha} \vec{r}_{\alpha} \tag{28}$$

where $\hat{H}_o$ is the unperturbed total Hamiltonian and d is the total dipole moment for charges q_α. This can be derived (see next section) by a unitary transformation from the minimum coupling Coulomb ($\vec{A}\cdot\vec{p}$) gauge [2], [22-25]. The electric field gauge has the advantage in that one is working with measurable physical quantities such as the electric field $\mathcal{E}$ and is usually more rapidly convergent in computations [2]. Another possible gauge or representation is the Bloch-Nordsieck (BN) or Henneberger-Kramers (HK), sometimes also called space-translation representation is discussed in the next section. $\hat{H}_o$ is the sum of the two noninteracting systems, molecular $\hat{H}_m$ and field $\hat{H}_f$:

$$\hat{H}_o = \hat{H}_m + \hat{H}_f \ , \quad \hat{H}_f = \frac{1}{2}\hbar\omega\,(\hat{a}^+a + aa^+) \ . \tag{29}$$

We seek solutions of the usual time-dependent Schroedinger equation

$$i\hbar\,\frac{d|\Psi(t)\rangle}{dt} = \hat{H}|\Psi(t)\rangle \tag{30}$$

by choosing as eigenstates direct products of molecular states $|\psi\,(t)>$ and field states $|\mathcal{E}(t)>$ i.e., $|\psi\,(t)> = |\psi\,(t)>\ |\mathcal{E}(t)\rangle$ with the initial condition

$$|\Psi(t = -\infty)\rangle = |\psi_i\rangle\,|\alpha\rangle = |\psi_i\rangle\,\hat{T}|0\rangle \tag{31}$$

where $|\alpha\rangle$ is a coherent state of the initial radiation field, equation (26). Clearly, in $|\mathcal{E}(t)>$, the Poisson distribution (23) could be modified by absorptions and emissions in the presence of molecules. For low molecular densities and large field intensities or equivalently large N, molecule induced field modifications are negligible so that $|\mathcal{E}(t)> \approx \hat{T}\,|0>$ is taken as a valid approximation. We thus redefine a new basis

$$|\Psi(t)\rangle = \hat{T}|\psi(t)\rangle\,|0\rangle \ , \tag{32}$$

leading to a new Schroedinger equation

$$i\hbar\,\frac{d|\psi(t)\rangle}{dt} = \hat{H}'\,|\psi(t)\rangle \ , \tag{33}$$

with

$$\hat{H}'(t) = \hat{T}^{-1}\hat{H}\hat{T} - i\hbar\,\hat{T}^{-1}\,\frac{d\hat{T}}{dt} \ . \tag{34}$$

The zeroth order Hamiltonian, $\hat{H}'_0(t) = \hat{H}_0$ is invariant whereas from (26) where now one sets $\alpha(t) = \alpha e^{-i\omega t}$ (see (20-21)). One obtains then,

$$\hat{T}^{-1}\hat{\mathcal{E}}\hat{T} = \hat{\mathcal{E}} + \mathcal{E}(\alpha(t)) \quad . \tag{35}$$

Thus the quantum field operator $\hat{\mathcal{E}}$ becomes displaced by the classical field amplitude $\mathcal{E}(\alpha(t))$, the eigenvalue of the coherent state $|\alpha(t)\rangle$. Thus the new Hamiltonian $\hat{H}'$ with eigenstates $|\psi(t)\rangle$ $|0\rangle$ can now be written as:

$$\hat{H}'(t) = \hat{H}_0 - \vec{d}\cdot\left[\hat{\mathcal{E}} + \mathcal{E}(\alpha(t))\right] \quad . \tag{36}$$

The final expression shows that field induced effects, i.e. stimulated absorptions and emissions can be represented by the classical radiative interaction $\vec{d}\cdot\mathcal{E}(\alpha(t))$ where $\mathcal{E}(\alpha(t))$ is the classical electric field amplitude which is also an eigenvalue of the quantum coherent state $|\alpha(t)\rangle$, equation (17). This has no effect on the field vacuum state $|0\rangle$ since $\mathcal{E}(\alpha(t))$ is a scalar. The remaining quantum radiative interaction $\vec{d}\cdot\hat{\mathcal{E}}$ can change the vacuum state since $\hat{\mathcal{E}}$ is a quantum field operator. This gives rise to spontaneous emission accompanying all laser stimulated processes. Such spontaneous processes are proportional to ω^3 (the photon phase space) and can be neglected for frequencies smaller than UV frequencies. Alternatively, one solves for $\psi(t)$ exactly with the classic radiative interaction $d\cdot\mathcal{E}(t)$ and then evaluates the spontaneous effects by first order perturbation theory [11], [19].

3.1. TIME INDEPENDENT COUPLED EQUATIONS

We have seen in the previous section that the coherent state $|\alpha\rangle$ is a superposition of photon states with the most probable value of the photon number n being $\alpha = N^{1/2}$. We can therefore define stationary zeroth order stationary (time-independent) electron-field states,

$$|a, N\rangle = |a\rangle|N\rangle \quad , \tag{37}$$

as an appropriate expansion basis set. In particular, distinguishing the various time scales: electronic and nuclear, we define an electron-field zeroth order Hamiltonian

$$\hat{H}_0 = \hat{H}_{el}(\vec{r},\vec{R}) + \hat{H}_f \quad . \tag{38}$$

$$\hat{H}_{el}(\vec{r},\vec{R})|a\rangle = V_a(R)|a\rangle \quad . \tag{39}$$

$$\hat{H}_0|a,N\rangle = \left(V_a(R) + N\hbar\omega\right)|a,N\rangle \quad . \tag{40}$$

This leads to a dressed state representation of the dynamics of the molecule-field system.

Thus, specializing to a diatomic molecule with a single nuclear coordinate R, we next expand the total wavefunction in terms of the field-molecule states including the nuclear motion,

$$|\Psi(E)\rangle = \frac{1}{R} \sum_{a,N} F_{aN}(R)\,|a,N\rangle \ . \qquad (41)$$

The F's are nuclear functions for the coordinate R for propagation on the electronic-photon potential ($V_a(R)+N\hbar\omega$) of the electron-photon states $|a,N\rangle$. Thus by substituting equation (41) into the total time-independent Schroedinger equation $\hat{H}|\Psi(E)\rangle = E|\Psi(E)\rangle$, with $\hat{H}$ defined in (28), one obtains coupled second-order differential equations for the nuclear functions $F_{aN}(R)$,

$$\left\{\frac{d^2}{dR^2} + \frac{2M}{\hbar^2}\left[E - V_a(R) - N\hbar\omega\right]\right\} F_{aN}(R) = \frac{2M}{\hbar^2} \sum_{a',N'} V_{aN,a'N'}(R) F_{a'N'}(R), \qquad (42)$$

M is the reduced mass. Two types of potential terms occur: diagonal: $V_{aN} = (V_a(R) + N\hbar\omega)$, and nondiagonal $V_{aN\ ,a'N'}$. The latter describe the interstate couplings, both radiative and nonradiative. The nonradiative terms such as corrections to the Born-Oppenheimer approximation, spin-orbit couplings, etc., [28], are diagonal in photon number, i.e.,

$$V_{nr} = V_{aN,a'N}(R) \ . \qquad (43)$$

Neglecting spontaneous radiative processes, the field-induced radiative processes lead to the nondiagonal potential terms,

$$V_{aN,a'N'} = V_{aN,a'N\pm1} \ . \qquad (44)$$

Thus each potential $V_{aN}(R)$ is coupled to $V_{aN,a'N+1}$ by an emission and to $V_{aN,a'N-1}$ by absorption, as the electric field operator (14) is made up of a photon creation, $\hat{a}^+$, or annihilation $\hat{a}$ operator. For resonant transitions, only one coupling, i.e., emission or absorption is adequate, leading to what is usually called the rotating wave approximation, RWA. Neglect of the other nonresonant (virtual) transition, e.g. $V_{aN,a'N+1}$ versus $V_{aN,a'N-1}$ is not justified for a) nonresonant transitions where $\omega(\text{photon}) < \omega_{a,a'}$ or b) high intensities. This can readily be seen by the fact that the energy separation between the two dressed surfaces $V_{a',N+1}$ and $V_{a',N-1}$ is $2\hbar\omega(\text{photon})$. The radiative coupling is half the Rabi frequency, $\frac{\hbar\omega_R}{2} = \mu_{aa'}\,\mathcal{E}_o/2$ where $\mathcal{E}_o = 2\left(\frac{2\pi\hbar\omega N}{V}\right)^{1/2}$, equation (11). Thus RWA is valid for $\omega_R \ll \omega(\text{photon})$, so that for such conditions, resonant processes can be well described by two dressed surfaces only. This limit of neglect of virtual transitions leads to the most useful concept of laser-induced avoided crossings and laser-induced adiabatic states [2], [14-15], [29], as illustrated in figure 2.

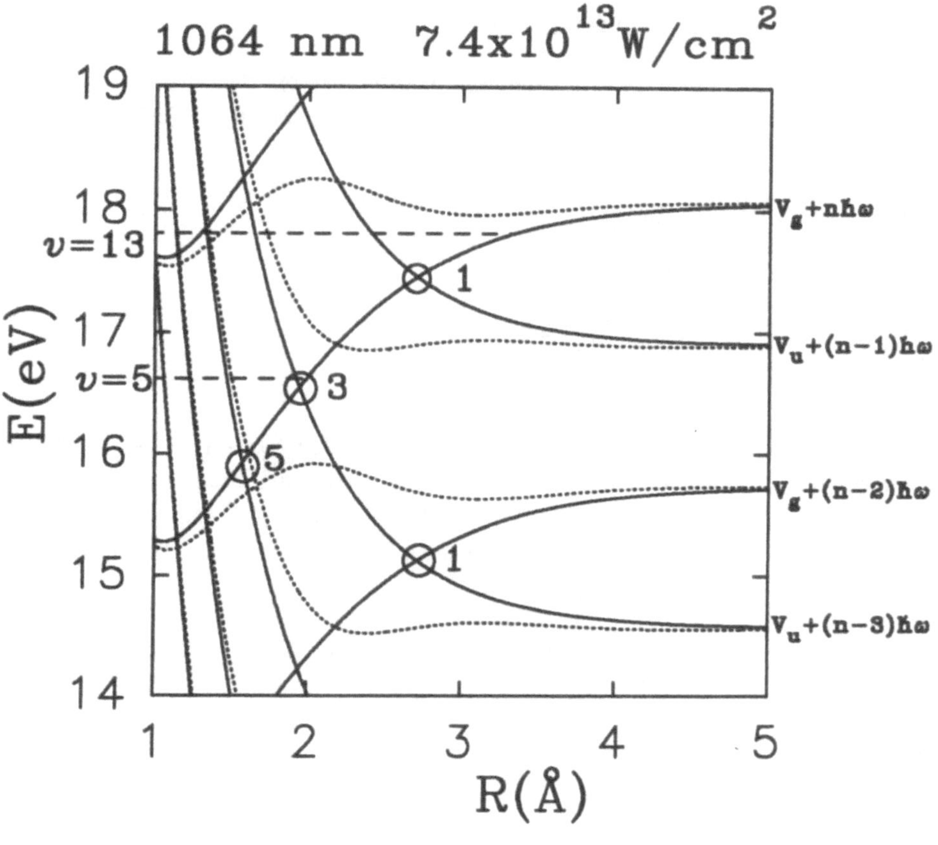

Figure 2. Dressed state representation of ${}^2\Sigma_g^+ \longrightarrow {}^2\Sigma_u^+$ photodissociation of $H_2{}^+$ at laser wavelength $\lambda = 1064$ nm and intensity $I = 7.4 \times 10^{13}$ W/cm^2. Solid lines are *diabatic* (unperturbed) electronic-field potentials ($V(R)+n\hbar\omega$) whereas dotted lines are *adiabatic* (field modified potentials – see ref. [19]). Circles indicate photon numbers absorbed-emitted at laser-induced avoided crossing. v = diabatic levels.

3.2. TIME DEPENDENT COUPLED EQUATIONS

The connection to the time-dependent approach is achieved by choosing an intermediate interaction picture, i.e., we remove only the radiation field Hamiltonian $\hat{H}_f$ (29) from the total Hamiltonian (28). We therefore define

$$
\begin{aligned}
|\Psi(t)\rangle &= \exp(-i\hat{H}_f t/\hbar)|\psi_I(t)\rangle \\
&= \sum_{a,n_a} e^{-in_a\omega t} P_{n_a}^{1/2}|a,n_a\rangle F_{an_a}(\vec{R},t)
\end{aligned}
\tag{45}
$$

The initial condition corresponds to $F_{an_a}(R,t = -\infty) = F_a(R)$. P_{n_a} is the coherent state Poisson distribution for photon numbers N_a associated with the initial channel $|a\rangle$ with a most probable photon number N_a. The time dependent nuclear amplitudes F_{an_a} will act as time varying coefficients modifying in principle the photon number distribution P_{n_a} in the initial coherent state during the molecule-field interaction.

The corresponding time dependent Hamiltonian for the new basis (45) is obtained via the transformation of the time independent Hamiltonian, (28),

$$
\begin{aligned}
\hat{H}_I(t) &= \exp\left(i\hat{H}_f t/\hbar\right)\hat{H}\exp(-i\hat{H}_f t/\hbar) \\
&= \hat{H}_m - d\cdot\hat{\mathcal{E}}(t)
\end{aligned}
\tag{46}
$$

$$
\hat{\mathcal{E}}(t) = i\left(\frac{2\pi\hbar\omega}{V}\right)^{1/2}\left[\hat{a}e^{-i\omega t} - \hat{a}^{+}e^{+i\omega t}\right] .
\tag{47}
$$

(note that we are using the dipole approximation, $\vec{k}\cdot\vec{r} = 0$ (see (14)).

Inserting the expansion (45) into the new time-dependent Schroedinger equation: $i\hbar\partial|\psi_I(t)\rangle/\partial t = \hat{H}_I(t)|\psi_I(t)\rangle$, $|\psi_I(t)\rangle = \sum_{a,n_a} P_{n_a}^{1/2}|a,n_a\rangle F_{a,n_a}(R,t)$,

leads to the coupled parabolic partial differential equations,

$$
i\hbar\frac{\partial F_a}{\partial t} = \langle a|\hat{H}_m|a\rangle F_a - \sum_b \langle a|d|b\rangle F_b\langle\alpha|\hat{\mathcal{E}}(t)|\alpha\rangle
\tag{48}
$$

In obtaining (48) we have made the following approximations: firstly, for large photon numbers, $P_{N_a} \approx P_{N_b}$, etc., with all coherent states peaking at some maximum value $N_a \approx N_b \approx N = |\alpha|^2$. Similarly the amplitudes $F_{a,n_a}(R,t)$ will be independent of n_a, i.e. this corresponds to the assumption that the coherent states are unperturbed by the molecules. Using equation (17) gives finally the completely time-dependent equations for the nuclear states F,

$$i\hbar\frac{\partial F}{\partial t}(\vec{R},t) = \frac{-\hbar^2}{2M}\left[F''(\vec{R},t)\right] + W(\vec{R},t)F(\vec{R},t) \tag{49}$$

where $\vec{F}$ is a vector with components F_a, $W_{aa} = V_a(R)$, $\langle a|\hat{H}_m|a\rangle = V_a(R)$,

$W_{a,a'}(t) = -d_{aa'}\,\mathcal{E}_0 \sin \omega t,\ \mathcal{E}_0 = 2\left(\frac{2\pi\hbar\omega N}{V}\right)^{1/2}$. The equation (49) is called the semiclassical time dependent coupled equations which we have derived above from the time-independent dressed state equations (42). We note equation (47) contains both absorptions ($\hat{a}$) and emissions ($\hat{a}^+$) so that equation (49) is exact for large photon numbers. Finally equation (47) corresponds to a single mode, plane wave electromagnetic field. Short pulses of course due to the uncertainty relation $\Delta\omega\tau \sim 1$ where τ is the pulse duration, have a frequency spread $\Delta\omega \sim 1/\tau$, so that equation (47) must then be generalized to multimodes [30].

Clearly the dressed state picture is useful for plane waves or for long pulses such that the field envelope f(t) for a pulse $\mathcal{E}(t) = \mathcal{E}_o f(t) \sin \omega t$ is slowly varying. Very short pulses at high intensities produce transients, i.e., high energy real and virtual transitions are produced by rapid turn-ons of pulses. The same is true of pulses with rapidly varying frequencies or phases. An illustrative example of the latter is the use of chirped pulses [4] in three level systems [31].

Figure 3 illustrates the effect of a linear chirp, i.e., linear frequency sweep, $\omega = \omega_o + \beta t$, on the complete inversion of a three level system as described in a time dependent dressed state representation in the RWA approximation. The transition is $|1,n\rangle \rightarrow |2n-1\rangle \rightarrow |3,n-2\rangle$. Avoided crossings occur on resonance between levels 1 and 2 at ω_{12} and 2 and 3 at ω_{23}. The lower path A corresponds to a positive frequency sweep, $\beta > 0$, and in the adiabatic limit, i.e., slow frequency sweep, $\beta << 1$, complete inversion is obtained with the intermediate state 2 being populated in the middle of the pulse. Another pathway B is the upper transition with a negative frequency sweep, $\beta < 0$. One notes in this case that the level 2 is never populated. The first lower pathway A corresponds to two successive resonant transitions at successive times τ_1 and τ_2. The upper pathway B is at all times nonresonant. The transition probabilities can be estimated from the Landau Zener transition probability $P = e^{-\pi\omega_R^2/\beta}$. In both cases, resonant (lower path A) and nonresonant (upper path B), a slow frequency sweep ($\beta << 1$) produces successive transitions between pairs of dressed states whereas rapid passage involves transitions between many dressed states. Examples of corrections to RWA in dressed state calculations of photodissociation have been discussed in [32] whereas transient effects due to short pulse or nonadiabatic field effects have been illustrated in [33].

In conclusion of this section, we have shown that multiphoton transitions can be treated numerically by appropriate coupled equations. Thus for CW laser excitations, the time-independent dressed state coupled equations (42) can be used to calculate accurately perturbative and nonperturbative, radiative and nonradiative transitions simultaneously. Bound and continuum states can be included in the coupled equations approach in a unified scattering or S-matrix formalism by introducing appropriate entrance and exit artificial continuum channels [29], [34]. In a recent application of this time-independent method we have treated the coherent control of predissociation of the Π_g states of Cl_2 by interfering simultaneous two and four photon transitions. Time-dependent problems as obtains in short pulse excitations are most conveniently treated by the time-dependent coupled equations (49). For this purpose we have developed highly efficient numerical procedures based on unitary split exponential operator methods in order to calculate the time-dependent propagator for coupled parabolic partial differential equations [21]. This has allowed us to explore the use of chirped pulses [4] in the multiphoton IR excitation of triatomics [5], the generation of high harmonics by molecules subject to intense laser fields [11], and the control of electrons in photodissociating molecules [35].

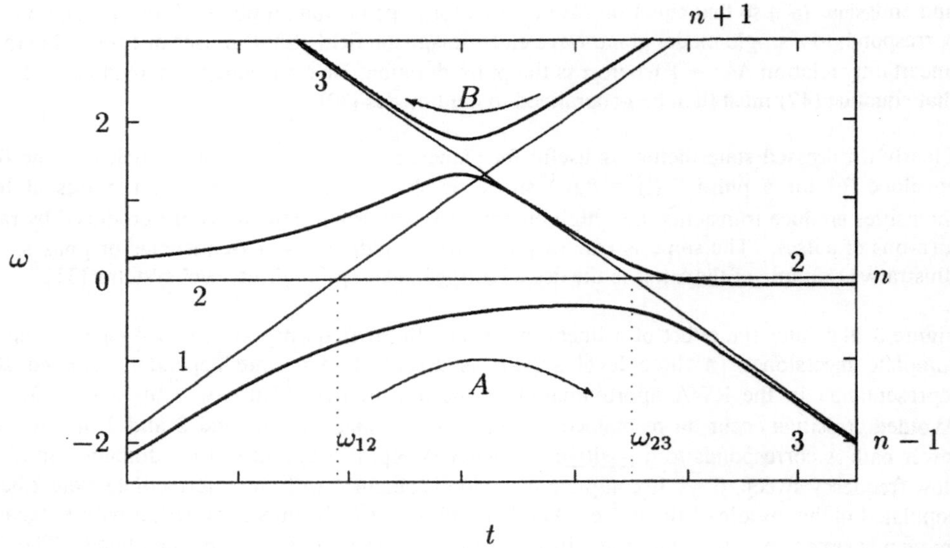

Figure 3. Adiabatic paths A and B for excitation of a three-level system by a linear chirped frequency pulse, $\omega(t) = \omega_0 + \beta t$ with resonances at ω_{12} and ω_{23} (see ref. [31]).

4. GAUGES, REPRESENTATIONS

Classically, a particle interacting with an electromagnetic field experiences a Lorentz force given by

$$F = e\left[\bar{\mathcal{E}} + \frac{\bar{v}}{c} \times \bar{B}\right] , \tag{50}$$

Reexpressing F in terms of the vector potential $\bar{A}$, equations (5-7) allows one to reexpress the charge-field dynamics in terms of an appropriate Lagrangean [2], [22-25] from which one can derive consistently generalized particle and field coordinater $\bar{r}$, $\bar{A}$ and the corresponding canonical momenta $\bar{p}$, $\bar{\mathcal{E}}$. One can by this method construct the appropriate Hamiltonian for which we just give the result for a single particle of mass m and charge e, in a static potential V(r),

$$H_p = \frac{1}{2}m\left(p - \frac{eA}{c}\right)^2 + V(r) . \tag{51}$$

One sees that the vector potential $\bar{A}$ acts not only as a potential but also changes the effective momentum of the charged particle. We emphasize immediately that in the above Hamiltonian, which is called the minimal coupling or Coulomb gauge Hamiltonian p is not the physical momentum $m\bar{v}$, where $\bar{v}$ is the instantaneous velocity, but rather $\bar{p}$ is the canonical momentum, i.e., $\bar{p} - e\bar{A}/c = m\bar{v}$, as can be verified by applying Hamilton's equations of motions to (51), $\dot{x} = \partial H/\partial p_x$, $\dot{p}_x = - \partial H/\partial x$. We also repeat that in the Coulomb gauge, the transversality condition, $\bar{\nabla}\cdot\bar{A} = 0$, leads to mutually perpendicular fields $\bar{\mathcal{E}}$, $\bar{B}$ and the propagation vector $\bar{k}$, (6-7). As a result of this transversality condition, one can separate the corresponding quantum Hamiltonian into a noninteracting part H_o and a radiative interaction part V_R,

$$H_o = -\frac{\hbar^2}{2m}\bar{\nabla}^2 + V(r) \; ; \; V_R = -\frac{e}{mc}\bar{A}\cdot\bar{p} + \frac{e}{mc^2}\bar{A}^2 . \tag{52}$$

(where we have used $(\bar{p}\cdot\bar{A})\varphi = \bar{A}\cdot\bar{p}\varphi$).

For any eigenstate of H_o, the average value $\langle\bar{p}\rangle = 0$ by symmetry so that the average value of V_R becomes using $A_o = c/\omega\ \mathcal{E}_o$,

$$\langle V_R \rangle = \frac{e}{4mc^2}A_o^2 = \frac{e^2}{4m\omega^2}\mathcal{E}_o^2 . \tag{53}$$

This is called the ponderomotive energy (PE) of a charged particle in an electromagnetic field [2]. This energy can be obtained also by solving the classical equations of motion of the free particle in the field $\mathcal{E}(t) = \mathcal{E}_o \cos \omega t$, i.e.

$$m\ddot{r} = e\mathcal{E}_o \cos \omega t \quad ; \quad \left\langle \frac{1}{2} m\dot{r}^2 \right\rangle = \frac{e\mathcal{E}_o^2}{4m\omega^2} \tag{54}$$

For an electron, this can be expressed in terms of laser wavelength λ as

$$PE(cm^{-1}) = 750 \times I(W/cm^2)\, \lambda^2(m) \tag{55}$$

Clearly, PE is the "wiggling" or oscillating energy an electron acquires in an electromagnetic field $\mathcal{E}(t)$ and will influence the electron dynamics at high intensities [2], [17]. As an example, at an intensity 100 TW/cm^2 (10^{14} W/cm^2) and $\lambda = 326$ nm, PE. = 8000 cm^{-1} = 1 eV.

In section (3), we showed how one could describe time-independent or time-dependent field-molecule interactions as coupled equations, which can therefore be solved numerically for both perturbative and nonperturbative regimes. The coupled equations (42) and (49) were written using the dipole-electric field interaction $\vec{d} \cdot \vec{\mathcal{E}}(t)$ (28) as the radiative interaction. In classical mechanics, this is obtained from the Coulomb gauge $(\vec{A} \cdot \vec{p})$ by a gauge or canonical transformation [2], [23-24]. In quantum mechanics this is obtained by a unitary transformation either on the time dependent Schroedinger equation (34) with a classical vector potential $\vec{A}(t)$ (we use here the dipole approximation $\vec{k} \cdot \vec{r} \approx 0$) in the Hamiltonian (51) or on the total time-independent coulomb gauge quantum particle-field Hamiltonian

$$\hat{H}_C = \hat{H}_p + \hat{H}_f \tag{56}$$

where $\hat{H}_p$ is the particle Hamiltonian (51) including the $\vec{A} \cdot \vec{p}$ radiative interaction, and H_f is the field Hamiltonian defined in (29). The electric field gauge Hamiltonian is obtained from the following $\vec{r}$ dependent unitary transformation,

$$\begin{aligned} H_E &= \exp(-ie\vec{r} \cdot \hat{A}/\hbar c)\, (\hat{H}_c)\, \exp(+ie\vec{r} \cdot \vec{A}/\hbar c) \\ &= \hat{p}^2/2m + V(r) - e\vec{r} \cdot \hat{\mathcal{E}} + \hat{H}_f \end{aligned} \tag{57}$$

where the field operators $\hat{\mathcal{E}}$ and $\hat{A}$ are related by the quantum canonical commutator (9). Corrections beyond the dipole approximation, i.e., using the exact field operators (13-14) can be obtained by more general unitary transformations, hence reproducing the classical expressions of electrostatics for fields and higher order electra and magnetic moments [36].

Another representation was used in the early days of quantum electrodynamics in order to eliminate infrared divergence problems and has found its use in intense field models. This we call the BN (Bloch-Nordsieck) representation [2], [7-9] and has been called by physicists the HK (Henneberger-Kramer) or space translation method [49]. This representation is obtained by a unitary transformation which has the advantage of emphasizing particle displacements induced by the field. The disadvantage is that it is not a gauge transformation as it introduces acceleration terms in the equations of motions [41-42]. The reason for this is that the appropriate unitary transformation for obtaining this representation is a p-dependent transformation.

$$\hat{H}_{BN} = \exp\left(ie\hat{p}\cdot\mathcal{E}/\hbar\omega^2\right)\left(\hat{H}_c\right)\exp\left(-ie\hat{p}\cdot\hat{\mathcal{E}}/\hbar m\omega^2\right) \quad . \tag{58}$$

The commutator $[\hat{\mathcal{E}}\ ,\ \hat{H}_f] = \frac{\hbar}{i}\frac{\hat{A}}{c}$, thus cancels the $\hat{A}\cdot\hat{p}$ term in H_p. It can be shown that all momenta remain invariant under this transformation but coordinates are displaced so that the new Hamiltonian becomes, [2], [42],

$$\hat{H}_{BN} = \frac{\hat{p}^2}{2m} + V\left(r + e\hat{\mathcal{E}}/m\omega^2\right) + \frac{e\hat{A}^2}{2mc^2} + \hat{H}_f \quad . \tag{59}$$

Comparing the two representations (57) and (59), one notes that in the electric field gauge, the radiative interaction is the simple classical expression, $\vec{d}\cdot\vec{\mathcal{E}}$, in the classical or large photon limit. The ponderomotive energy PE, equation (54) is a high order perturbative effect. In the BN representation, two physical effects are isolated immediately: c) the particle displacement $\Delta r(t) = e\hat{\mathcal{E}}/m\omega^2$ as it follows the field, and b) the ponderomotive term $e\hat{A}/2mc^2$. In the large photon number (high intensities) one can replace the electric field operator by its classical value $\mathcal{E}(t)$ (see section 2) so that particles undergo a maximum displacement r_o in the presence of a peak field amplitude $\mathcal{E}_o$,

$$r_o(a_o) = \frac{e\mathcal{E}_o}{m\omega^2} = 2.57 \times 10^{-12}\left[I(W/cm^2)\right]^{1/2}\lambda^2(nm) \quad . \tag{60}$$

Thus for an intensity $I = 10^{13}$ W/cm^2 and wavelength $\lambda = 1064$ nm (YAG laser), $r_o \approx 10a_o \approx 5$Å. Clearly one expects at this intensity electron orbits to be considerably distorted. This distortion increases with lower frequencies, leading in the infrared region to <u>tunnel ionization</u> [8-9]. Above 10^{14} W/cm^2, orbit deformations are at least one order of magnitude larger than molecular sizes, thus indicating rapid ionization, in agreement with recent exact calculations [11]. Even at lower intensities, these field induced particle displacements can modify considerably molecular nonadiabatic couplings [43].

The three gauges or representations, electric, Coulomb, and Bloch-Nordsieck (space translation) lead to three different Hamiltonians, $\hat{H}_E$ (equation (57)), $\hat{H}_C$ (equation (56)) and $\hat{H}_{BN}$ (equation (59)). The radiative couplings in the three representations lead to matrix elements between molecular states $|\alpha\rangle$ and $|\beta\rangle$ called [2], length: $\mathcal{E}_o r_{\alpha\beta}$, (61)

velocity: $A_o p_{\alpha\beta} / mc = i(\omega_{\beta\alpha} / \omega)\mathcal{E}_o r_{\alpha\beta}$, (62)

acceleration: $r_o(\bar{\nabla}V)_{\alpha\beta} = - \frac{(\omega_{\beta\alpha})^2}{\omega^2} \mathcal{E}_o r_{\alpha\beta}$. (63)

The last is obtained by expanding $V(r+r_o) = V(r) + r_o \bar{\nabla}V + ...$ where $r_o = e\mathcal{E}_o^2 / m\omega^2$ (equation (60)). One sees clearly from all three equations that all three gauges (representations) agree for resonant $\omega = \omega_{\alpha\beta}$ (on-shell) calculations only. For nonresonant (virtual) transitions, $\omega \neq \omega_{\alpha\beta}$, there will be considerable differences in perturbative calculations. Thus for low frequencies, high lying states make the matrix elements (62) and (63) much larger than (61), i.e., for low frequencies, the electric field gauge is preferable as it will give more convergent results. All three representations will of course give the same transition probabilities and energy shifts only for <u>exact</u> calculations, i.e., with complete states, bound and continuum. For high frequencies, (63) gives the smallest radiative couplings, implying that the BN (space translation) Hamiltonian (59) will be more appropriate. Indeed comparing the second order energy shifts of a bound state n gives the following results [44],

$$(\Delta E_n)^{EF} = \frac{\mathcal{E}_o^2}{2} \sum_m \frac{\omega_{nm}}{\omega_{nm}^2 - \omega^2} |r_{mn}|^2 \xrightarrow[\omega\to\infty]{} \frac{\mathcal{E}_o^2}{4m\omega^2} = \frac{eA_o^2}{4mc^2} , \quad (64)$$

$$(\Delta E_n)^{BN} = \frac{\mathcal{E}_o^2}{2\omega^2} \sum_m \frac{\omega_{nm}^3}{\omega_{nm}^2 - \omega^2} |r_{mn}|^2 \xrightarrow[\omega\to\infty]{} -\frac{\mathcal{E}_o^2}{4\omega^2}\left\langle \nabla^2 V \right\rangle_n = +\frac{\mathcal{E}_o^2}{4\omega^4}(4\pi\rho_n(o))$$

(65)

where $\rho_n(o)$ is the electronic density at the nucleus.

Both high frequency ($\omega\to\infty$) results are obtained by applying appropriate sum rules [45]. Thus in the EF gauge, the high frequency energy shift of any bound state is <u>positive</u> and is nothing else but the ponderomotive energy of a free electron, equation (53) and is already present in both Coulomb (equation (52)) and BN (equation (59)). Thus in the EF gauge, the ionization potential, IP, does not change in second order perturbation theory since (64) will cancel the ponderomotive energy (PE) of the free ionized energy. In the BN gauge, both bound and continuum states have the same PE, so that (65) is the net change in IP in that representation. Clearly, second order perturbation theory in the BN gauge gives high frequency results which vary as ω^{-4}, i.e. the perturbative convergence is now much better than the EF gauge. Our recent calculations of stabilization of H_2^+ against ionization shows that they can occur at high intensities, $I > 10^{16}$

W/cm^2 for short wavelength, $\lambda < 100$ nm, for which the BN representation gives a clear picture of field-induced charge redistribution, creating new stable molecules [46].

Acknowledgments. We thank the Natural Sciences and Engineering Research Council of Canada for supporting this work.

References

1. Warren, W.S., and Haner, M. (1988), in *Atomic and Molecular Processes with Short Intense Laser Pulses*, ed. A.D. Bandrauk, NATO ASI, vol. B171, Plenum Press, N.Y., pp. 1-11.
2. Bandrauk, A.D. (1993), *Molecules in Laser Fields*, Marcel Dekker, N.Y.
3. Bandrauk, A.D., Gauthier, and J.M., and McCann, J.F. (1993); *Chem. Phys. Lett.* **200**, 399; (1994); *J. Chem. Phys.* **100**, 340.
4. Chelkowski, S., Bandrauk, A.D., and Corkum, P.B. (1990), *Phys. Rev. Lett.* **65**, 2355.
5. Chelkowski, S., and Bandrauk, A.D. (1991), *Chem. Phys. Lett.* **186**, 264.
6. Allan, J. and Eberly, J.H. (1975), *Optical Resonances and Two-Level Atoms*, J. Wiley Press, N.Y.
7. Chelkowski, S., and Bandrauk, A.D. (1988), *J. Chem. Phys.* **89**, 3618.
8. Keldoysh, L.V. (1965), *Sov. Phys.* JETP **20**, 1307.
9. Corkum, P.B., Burnett, N.H. and Brunel, F. (1989), *Phys. Rev. Lett.* **62**, 1259.
10. Dietrich, P., and Corkum, P.B. (1992), *J. Chem. Phys.* **97**, 3187.
11. Zuo, T., Chelkowski, S., and Bandrauk, A.D. (1992), *Phys. Rev.* **A46**, 5342.
12. George, T.F., Zimmerman, I.H., Yuan, J.M., Laing, J.R., and Devries, P.L. (1977), *Acc. Chem. Res.* **10**, 449.
13. George, T.F. (1982), *J. Phys. Chem.* **86**, 10.
14. Bandrauk, A.D., and Sink, M.L. (1978), *Chem. Phys. Lett.* **57**, 569.
15. Bandrauk, A.D., and McCann, J.F. (1989), *Comments. Atom. Molec. Phys.* **22**, 325.
16. Verschmur, J., Noordham, L.D. and van Linden van den Heuvell, H.B., (1989), *Phys. Rev.* **A40**, 4383.
17. Zavriyev, A. and Bucksbaum, P.H. (1993), in *Molecules in Laser Fields*, ed. A.D. Bandrauk, Marcel Dekker, N.Y., Chap. 2.
18. Allendorf, S.W., and Szoke, A., (1991), *Phys. Rev.* **A44**, 518.
19. Aubanel, E.E., Zuo, T., and Bandrauk, A.D. (1994), *Phys. Rev.* **A49**, 3776.
20. Bandrauk, A.D. (1994), *Internatl. Rev. Phys. Chem.* **13**, 123.
21. Bandrauk, A.D., and Shen, H (1993), *J. Chem. Phys.* **176**, 428.
22. Lee, T.D. (1981), *Particle Physics and Field Theory*, Harwood Academic, N.Y.
23. Cohen-Tannoudji, C., Dupont-Roc, J., and Grynberg G. (1989), *Photons and Atoms*, J. Wiley Press, N.Y.
24. Mittleman, M.H. (1982), *Introduction to the Theory of Laser-Atom Interactions*, Plenum Press, N.Y.
25. Faisal, F.H.M. (1988), *Theory of Multiphoton Processes*, Plenum Press, N.Y.
26. Glauber, R.J. (1963), *Phys. Rev.* **130**, 2529; **131**, 2766.
27. Goldin, E. (1982), *Waves and Photons*, J. Wiley Publishers, N.Y., Chap. 7.
28. Lefebvre-Brion, H., and Field, R.F. (1986), *Perturbation in Spectra of Diatomic Molecules*, Academic Press, Orlando.
29. Bandrauk, A.D., and Atabek, O., (1987), *Adv. Chem. Phys.*, chap. 19, vol. 73.
30. Loudon, R. (1973), *Quantum Theory of Light*, Oxford Press, London.

31. Broers, B., van Linden van den Heuvell, H.B., and Noordham, L.D. (1992), *Phys. Rev. Lett.* **69**, 2062.
32. Aubanel, E., Gauthier, J.M., and Bandrauk, A.D. (1993), *Phys. Rev.* **A48**, 2145.
33. Aubanel, E., Bandrauk, A.D., and Rancourt, P. (1992), *Chem. Phys. Lett.* **197**, 419.
34. Miret-Artes, S., Atabek, O., and Bandrauk, A.D. (1992), *Phys. Rev.* **A45**, 8056.
36. Power, E.A., and Thirumamachandran, J. (1983), *Phys. Rev.* **A28**, 2649.
37. Bloch, F., and Nordsieck, A. (1937), *Phys. Rev.* **52**, 54.
38. Pauli, W., and Fierz, M. (1938), *Nuovo Ciments* **16**, 167.
39. Welton, T.A. (1963), *Phys. Rev.* **131**, 2766.
40. Henneberger, W.D. (1968), *Phys. Rev. Lett.* **21**, 838.
41. Chanmugan, G., and Schweber, S.S. (1969), *Phys. Rev.* **A1**, 1369.
42. Bandrauk, A.D., Kalman, O.F., and Nguyen-Dang, T.T. (1986), *J. Chem. Phys.* **84**, 6761.
43. Bandrauk, A.D., and Nguyen-Dang, T.T. (1985), *J. Chem. Phys.* **83**, 2840.
44. Ford, G.W., and O'Connell, R.F., (1976), *Phys. Rev.* **A13**, 1281.
45. Bethe, H.A.; and Salpeter, E.E., (1957), *Quantum Mechanics of 1 and 2-Electron Atoms*, Springer, Berlin.
46. Zuo, T.; and Bandrauk, A.D., to be submitted to *Phys. Rev. A*.

COHERENT AND INCOHERENT LASER CONTROL OF PHOTOCHEMICAL EVENTS

Moshe Shapiro
Dept. of Chemical Physics
The Weizmann Institute
Rehovot
Israel

and

Paul Brumer
Dept. of Chemistry
University of Toronto
Toronto, M5S1A1
Canada

1 Introduction

Selectivity is at the heart of Chemistry and the control of reactions using lasers has been a goal for decades. Recently, we[1]-[20] and other groups[21]-[30] have demonstrated theoretically that one can achieve this goal by using quantum interference phenomena. We showed that phases acquired by a quantum systems while excited by lasers enable one to control quantum interferences, and hence the outcome, of many dynamical processes. Initial experimental tests[31]-[36] of our approach, termed Coherent Control (CC), have confirmed many of the theoretical predictions and proven the viability of the method.

In this lecture we provide an introduction to the concepts[37] underlying CC and discuss its current status in both Chemistry and Physics. We start with an introduction to the basics of coherent control and follow it by a detailed discussion of two control scenarios. We then discuss the issue of control of a thermal ensemble. Finally, we discuss extensions of coherent control to the strong laser field domain.

1.1 Aspects of Scattering Theory and Reaction Dynamics

The processes we wish to control include branching "half" collisions,

$$ABC \quad \to A + BC \tag{1.1}$$

$$\to AB + C\,, \tag{1.2}$$

E. Yurtsever (ed.), Frontiers of Chemical Dynamics, 151–180.

and "full" collisions,

$$A+BC(m) \rightarrow A+BC(m') \quad (1.3)$$
$$\rightarrow AB(m'')+C\,. \quad (1.4)$$

In the above, A, B, C are either atoms, groups of atoms, electrons, or photons. m, m' denote the internal (vibrational, rotational, photon occupation) quantum numbers of the reactants or products.

Given $\Psi(t=0)$, the system wavefunction at an initial time, the evolution of the system is determined by the time dependent material Schrödinger Equation,

$$H_M\Psi(t) = i\hbar\partial\Psi(t)/\partial t. \quad (1.5)$$

where H_M is the system Hamiltonian. The wavefunction at long times, i.e. when the products are well separated, provides the probabilities of forming the products. The approach to be followed here consists of expressing the time evolution in terms of $|E_i\rangle$, the solutions of the time independent Schrödinger equation

$$H_M|E_i\rangle = E_i|E_i\rangle. \quad (1.6)$$

The long time behavior of $\Psi(t)$ is intimately connected with the nature of the time independent continuum energy eigenstates. For every continuum energy value E, each of the possible outcomes observed in the product region is represented by an independent wavefunction. The fact that such a set of degenerate wavefunctions of the separated products exists implies[38] the existence of a set of degenerate eigenfunctions of the total Hamiltonian, and a one-to-one correlation between the two sets. This "boundary" condition is expressed more precisely by denoting the different possible chemical products of the breakup of ABC in Eq. (1.2) by an index q (e.g. $q=1$ denotes the $A+BC$ products), and all additional identifying state labels by m. The set of continuum eigenfunctions of the material Hamiltonian,

$$H_M|E,m,q^-\rangle = E|E,m,q^-\rangle, \quad (1.7)$$

is now defined via the requirement that asymptotically every $|E,m,q^-\rangle$ state goes over to a state of the separated products, denoted $|E,m,q^0\rangle$, which is of energy E, chemical identity q and remaining quantum numbers m. The "minus" superscript serves to indicate this choice of boundary condition.

The description of the system in terms of $|E,m,q^-\rangle$ has an important advantage: Expressing the state of the system in the present in terms of these states, i.e., writing and initial continuum state as

$$\Psi(t=0) = \sum_{q,m}\int dE c_{q,m}(E)|E,m,q^-\rangle, \quad (1.8)$$

means that we know the fate of the system in the future. Since each of the $|E,m,q^-\rangle$ states correlates with a *single* product state, the probability of observing each $|E,m,q^0\rangle$ product state is simply given by $|c_{q,m}(E)|^2$ - the *preparation* probabilities. The probability of producing a chemical product q in the future is therefore given as

$$P_q = \sum_m \int dE|c_{q,m}(E)|^2. \quad (1.9)$$

The fact that the state of the system in the distant future is predetermined by the initially created state is, admittedly, intuitively obvious. However, consequences of this simple fact are often ignored. For example, arguments such as "intramolecular energy scrambling makes reaction control difficult", are misleading: A general wavepacket may show wondrously complicated temporal behaviour, yet Eq. (1.9) tells us that the probability of producing product q in the long time limit is merely the energy average of the preparation probabilities. Since the preparation coefficients are also determined by the energy eigenstates, we see that the long time limit is an average property of the states which make up a given wavepacket.

Another consequence of Eq. (1.9) is that pulse shaping which merely results in changing the phases of the preparation coefficients will have absolutely no effect on the q products yields.[13]. Likewise, shortening of a pulse, which results in broadening of the power spectrum of that pulse, will modify the $c_{q,m}(E)$ coefficients to *all* the q channels and will not necessarily "help beat out IVR".

Below we demonstrate that the key to laser control is to change one $c_{q,m}(E)$ coefficient relative to another $c_{q',m}(E)$ coefficient *at the same energy*. In order to understand how this can be done we discuss now the process of preparation.

1.2 Perturbation Theory, System Preparation and Coherence

Consider the effect of an electric field on an initially bound molecule. The molecule is assumed to be in an eigenstate $\mid E_g \rangle$ of the radiation-free Hamiltonian, H_M, before being subjected to a perturbing incident radiation field $\epsilon(t)$. The overall Hamiltonian is then given by:

$$H = H_M - \mathsf{d}[\bar{\epsilon}(t) + \bar{\epsilon}^*(t)] \tag{1.10}$$

where d is the component of the dipole moment along the electric field.

Consider now the case in which the impinging photon is energetic enough to dissociate the molecule. It is then necessary to expand $\mid \Psi(t) \rangle$ in the bound and scattering eigenstates of the radiation-free Hamiltonian,

$$\mid \Psi(t) \rangle = \sum_i c_i(t) \mid E_i \rangle \exp(-iE_it/\hbar) + \sum_{m,q} \int dE \, c_{E,m,q}(t) \mid E, m, q^- \rangle \exp(-iEt/\hbar). \tag{1.11}$$

Insertion of Eq. (1.11) into the time dependent Schrödinger Equation results in a set of first-order differential equations for the $c_\nu(t)$ coefficients, where ν represents either the bound (i) or scattering (E, m, q) indices.

For weak fields use of first order perturbation theory gives, for the post-pulse preparation coefficient,

$$c_{E,m,q}(t \gg \Gamma) = (\sqrt{2\pi}/i\hbar)\epsilon(\omega_{E,E_g})\langle E, m, q^- | \mathsf{d} | E_g \rangle \,, \tag{1.12}$$

where Γ is the pulse duration and

$$\epsilon(\omega) = (1/\sqrt{2\pi}) \int_{-\infty}^{\infty} \exp(i\omega t) \, \bar{\epsilon}(t) \, dt. \tag{1.13}$$

In this case $\omega = \omega_{E,E_g} = (E - E_g)/\hbar$.

The process described above amounts to the creation of a pure state (i.e. a state for which a phase may be defined) in the continuum by a well defined electric field. As long as there are no random collisions this state will remain pure (i.e., will retain its phase) - a feature of some importance to the discussion below.

It follows from Eq. (1.9) and Eq. (1.12) that the probability $P(E,q)$ of forming asymptotic product in arrangement q is,

$$P(E,q) = \sum_m |c_{E,m,q}(t \gg \Gamma)|^2 = (2\pi/\hbar^2) \sum_m |\epsilon(\omega_{E,E_g}) \langle E_g|\mathbf{d}|E,m,q^-\rangle|^2 \tag{1.14}$$

and that the branching ratio $R(1,2;E)$ between the $q=1$ products and the $q=2$ products at energy E is given as,

$$R(1,2;E) = \frac{\sum_m |< E_g|\mathbf{d}|\, E,m,1^- \rangle|^2}{\sum_m |< E_g|\mathbf{d}|\, E,m,2^- \rangle|^2} \tag{1.15}$$

1.3 Coherent Radiative Control of Chemical Reactions

We now address the issue of how to alter the above yield ratio $R(1,2;E)$ in a *systematic* fashion. Equation (1.15) makes clear that (at least in the weak field regime) this can not be achieved by altering the laser intensity, since the field strength cancels out in the expression for R. In fact, any other quantity which appears in a similar form in both the numerator and denominator, can not serve as a handle on yield control.

Quantum interference phenomena can, however, alter the numerator or denominator of R in an independent and controlled way. This can be achieved by accessing the final continuum state via two or more interfering pathways. One of the first examples which we studied [1], involves preparing a molecule in a superposition $c_1|\,\phi_1\,\rangle + c_2|\,\phi_2\,\rangle$ state and exciting the two components to the same final continuum energy E by using two CW sources [See Fig. 1]. The field employed is of the form,

$$\bar{\epsilon}(t) = \epsilon_1 e^{-i\omega_1 t + i\chi_1} + \epsilon_2 e^{-i\omega_2 t + i\chi_2} \tag{1.16}$$

where $\hbar\omega_i = E - E_i$. A straightforward computation[1] yields that,

$$R(1,2;E) = \frac{\sum_m |\langle \bar{\epsilon}_1 c_1 \phi_1 + \bar{\epsilon}_2 c_2 \phi_2 |\mathbf{d}|\, E,m,1^- \rangle|^2}{\sum_m |\langle \bar{\epsilon}_1 c_1 \phi_1 + \bar{\epsilon}_2 c_2 \phi_2 |\mathbf{d}|\, E,m,2^- \rangle|^2} \tag{1.17}$$

where $\bar{\epsilon}_i = \epsilon_i \exp(i\chi_i)$. Expanding the square gives:

$$R(1,2;E) =$$

$$\frac{\sum_m [|\bar{\epsilon}_1 c_1 \langle \phi_1 |\mathbf{d}|\, E,m,1^- \rangle|^2 + |\bar{\epsilon}_2 c_2 \langle \phi_2 |\mathbf{d}|\, E,m,1^- \rangle|^2 + 2Re[c_1 c_2^* \bar{\epsilon}_1 \bar{\epsilon}_2^* \langle \phi_1 |\mathbf{d}|\, E,m,1^- \rangle]}{\sum_m [|\bar{\epsilon}_1 c_1 \langle \phi_1 |\mathbf{d}|\, E,m,2^- \rangle|^2 + |\bar{\epsilon}_2 c_2 \langle \phi_2 |\mathbf{d}|\, E,m,2^- \rangle|^2 + 2Re[c_1 c_2^* \bar{\epsilon}_1 \bar{\epsilon}_2^* \langle \phi_1 |\mathbf{d}|\, E,m,2^- \rangle]} \tag{1.18}$$

The structure of the numerator and denominator of Eq. (1.18) is of the type desired, i.e. each has a term associated with the excitation of the $|\phi_1>$ state, a term associated with the excitation of the $|\phi_2>$ state, and a term corresponding to the interference between the two excitation routes. The interference term, which can either be constructive or destructive, is in general different for the two product channels. What makes Eq. (1.18) so important *in practice* is that the interference term has coefficients whose magnitude and sign depend upon *experimentally controllable* parameters. Thus the experimentalist can manipulate laboratory parameters and, in doing so, directly alter the reaction product yield by varying the magnitude of the interference term. In the case of Eq. (1.18) the experimental parameters which alter the yield[1] are contained in the complex quantity $A = \tilde{\epsilon}_2 c_2/\tilde{\epsilon}_1 c_1$. Both $x \equiv |A|$ and $\theta_1 - \theta_2 \equiv arg(A)$ can be controlled separately in the experiment.

Results of a specific computational example based upon Eq. (1.18) are shown in Fig. 2. Here we consider control over the relative probability of forming $I(^2P_{3/2})$ vs. $I(^2P_{1/2})$, denoted I and I*, in the dissociation of methyl iodide:

$$CH_3I \rightarrow CH_3 + I \quad (1.19)$$

$$\rightarrow CH_3 + I^*, \quad (1.20)$$

The computations were carried out with realistic potential surfaces[39, 40] within the framework of a fully quantum photodissociation theory[41, 40]. Fig. 2 shows a typical plot of the yield of I* as a function of $\theta_1 - \theta_2$ and $S \equiv x^2/(1+x^2)$. ($S = 0$ corresponds to $\tilde{\epsilon}_1 = 0$ and $S = 1$ corresponds to $\tilde{\epsilon}_2 = 0$). We see that our ability to control the process ("range of control") is almost complete: As we change S and $\theta_1 - \theta_2$, the yield varies from 30% I to 70%. Higher and lower ratios can also be achieved[42] with different choices of the initial pair of states $|\phi_1\rangle$ and $|\phi_2\rangle$.

These ideas can be naturally extended to the control of N products, using an initial superposition of N states[17]. Experimentally, the creation of an initial superposition of two (or more) states may be achieved by acting on a single ground state with a light pulse whose frequency width spans the levels of interest[9, 10, 11]. Alternatively, one can employ stimulated emission pumping through an intermediate electronic state[20]. The "real time" analogue of the above two CW frequencies scenario, in which the superposition state preparation is affected by a single broad-band pulse and the dissociation by a second pulse, is discussed in detail below.

2 Representative Control Scenarios

As mentioned above, the two step approach of Fig. 1 and 2 is but one particular implementation of coherent control; numerous other scenarios may be designed. They all rely upon the same "coherent control principle", that *in order to achieve control one must drive a state through multiple independent optical excitation routes to the same final state.*

It is helpful to think of coherent control as analogous to a double (multiple) slit experiment: The tuning in of a desired product ratio R, accomplished by varying the external laser parameters (e.g., A), is analogous to probing different regions of a screen on which the double slit interference patterns are imaged. Control arises because these interference patterns are different for different final channels (due to the different molecular phases).

It would seem that laser incoherence would lead to loss of control since incoherence implies that the phases of $\bar{\epsilon}_1$ and $\bar{\epsilon}_2$ in Eq. (1.18) are random. An ensemble average of these phases is expected to lead to the disappearance of the interference term. This is true, however, only in the fully chaotic limit. Control can persist in the presence of some laser incoherence[19] or when the initial state is described by a *mixed*, as distinct from *pure*, state[7]. Most surprising is the fact, described below, that by utilizing strong laser fields one can attain quantum interference control with completely *incoherent* sources[43].

We now describe in more detail two additional control scenarios.

2.1 Interference Between n-Photon and m-Photon Routes ("$n+m$" control)

So far, we exploited quantum interference phenomena by dissociating a superposition of several energy eigenstates with a single type (one photon absorption) process. It is possible instead to start with a *single* energy eigenstate and employ interference between optical routes of *different* types. Such is the interference between two multiphoton processes of different multiplicities. In order to satisfy the coherent control principle, which requires that we reach the same final energy E, we must use photons of commensurate frequencies, i.e. frequencies which satisfy an $m\omega_1 = n\omega_2$ relation, with integer m and n. Selection rules dictate the acceptable n, m pairs.

As the simplest example, we examine a one photon process interfering with a three photon process ("3+1" control). Let H_g and H_e be the nuclear Hamiltonians for a ground and excited electronic states. H_g is assumed to have a discrete spectrum and H_e to possess a continuous spectrum. The molecule, initially in an eigenstate $|E_i\rangle$ of H_g is subjected to two electric fields (see Fig. 3) given by

$$\epsilon(t) = \epsilon_1 \cos(\omega_1 t + \mathbf{k}_1 \cdot \mathbf{R} + \theta_1) + \epsilon_3 \cos(\omega_3 t + \mathbf{k}_3 \cdot \mathbf{R} + \theta_3)\,, \tag{2.1}$$

Here $\omega_3 = 3\omega_1$, $\epsilon_l = \epsilon_l \hat{\mathbf{e}}_l$, $l = 1, 3$; ϵ_l is the magnitude and $\hat{\mathbf{e}}_l$ is the polarization of the electric fields. The two fields are chosen parallel, with $\mathbf{k}_3 = 3\mathbf{k}_1$.

The probability $P(E, q; E_i)$ of producing product with energy E in arrangement q from a state $|E_i\rangle$ is given by

$$P(E,q;E_i) = P_3(E,q;E_i) + P_{13}(E,q;E_i) + P_1(E,q;E_i)\,. \tag{2.2}$$

where $P_1(E, q; E_i)$ and $P_3(E, q; E_i)$ are the probabilities of dissociation due to the ω_1 and ω_3 excitation, and $P_{13}(E, q; E_i)$ is the term due to interference between the two excitation routes.

In the weak field limit, $P_3(E, q; E_i)$ is given by

$$P_3(E,q;E_i) = \left(\frac{\pi}{\hbar}\right)^2 \epsilon_3^2 F_3^{(q)} \tag{2.3}$$

where

$$F_3^{(q)} = \sum_n |\langle E, n, q^- \,|\, (\hat{\mathbf{e}}_3 \cdot \mathbf{d})_{e,g} \,|\, E_i \rangle|^2\,. \tag{2.4}$$

$\mathbf{d}$ is the electric dipole operator, and

$$(\hat{\mathbf{e}}_3 \cdot \mathbf{d})_{e,g} = \langle e \,|\, \hat{\mathbf{e}}_3 \cdot \mathbf{d} \,|\, g \rangle\,, \tag{2.5}$$

with $|\,g\,\rangle$ and $|\,e\,\rangle$ denoting the ground and excited electronic states, respectively. $P_1(E,q;E_i)$ is given in third order perturbation theory by[6]

$$P_1(E,q;E_i) = (\frac{\pi}{\hbar})^2 \epsilon_1^6 F_1^{(q)}, \tag{2.6}$$

where,

$$F_1^{(q)} = \sum_n |\langle E,n,q^- \,|\, T \,|\, E_i \rangle|^2 , \tag{2.7}$$

with

$$T = (\hat{\epsilon}_1 \cdot \mathrm{d})_{e,g}(E_i - H_g + 2\hbar\omega_1)^{-1}(\hat{\epsilon}_1 \cdot \mathrm{d})_{g,e}(E_i - H_e + \hbar\omega_1)^{-1}(\hat{\epsilon}_1 \cdot \mathrm{d})_{e,g} \,. \tag{2.8}$$

We assumed that $E_i + 2\hbar\omega_1$ is below the dissociation threshold and that dissociation occurs from the excited electronic state only.

A similar derivation[6] gives the cross term in Eq. (2.2) as

$$P_{13}(E,q;E_i) = -2(\frac{\pi}{\hbar})^2 \epsilon_3 \epsilon_1^3 \cos(\theta_3 - 3\theta_1 + \delta_{13}^{(q)})|F_{13}^{(q)}| \tag{2.9}$$

with the amplitude $|F_{13}^{(q)}|$ and phase $\delta_{13}^{(q)}$ defined by

$$|F_{13}^{(q)}| \exp(i\delta_{13}^{(q)}) = \sum_n \langle E_i \,|\, T \,|\, E,n,q^- \rangle \langle E,n,q^- \,|\, (\hat{\epsilon}_3 \cdot \mathrm{d})_{e,g} \,|\, E_i \rangle \,. \tag{2.10}$$

The branching ratio $R_{qq'}$ between the q and q' products can then be written as

$$R_{qq'} = \frac{P(E,q;E_i)}{P(E,q';E_i)} = \frac{\epsilon_3^2 F_3^{(q)} - 2\epsilon_3\epsilon_1^3 \cos(\theta_3 - 3\theta_1 + \delta_{13}^{(q)})|F_{13}^{(q)}| + \epsilon_1^6 F_1^{(q)}}{\epsilon_3^2 F_3^{(q')} - 2\epsilon_3\epsilon_1^3 \cos(\theta_3 - 3\theta_1 + \delta_{13}^{(q')})|F_{13}^{(q')}| + \epsilon_1^6 F_1^{(q')}} \,. \tag{2.11}$$

Next, we rewrite Eq. (2.11) in a more convenient form. We define a dimensionless parameter $\bar{\epsilon}_i$ and a parameter x as follows:

$$\epsilon_l = \bar{\epsilon}_l \epsilon_0; \quad \text{for } l = 1,3 \qquad x = \bar{\epsilon}_1^3/\bar{\epsilon}_3. \tag{2.12}$$

The quantity ϵ_0 essentially carries the unit for the electric fields; variations of the magnitude of ϵ_0 can also be used to account for unknown transition dipole moments. Utilizing these parameters, Eq. (2.11) becomes

$$R_{qq'} = \frac{F_3^{(q)} - 2x\cos(\theta_3 - 3\theta_1 + \delta_{13}^{(q)})\epsilon_0^2|F_{13}^{(q)}| + x^2\epsilon_0^4 F_1^{(q)}}{F_3^{(q')} - 2x\cos(\theta_3 - 3\theta_1 + \delta_{13}^{(q')})\epsilon_0^2|F_{13}^{(q')}| + x^2\epsilon_0^4 F_1^{(q')}} \,. \tag{2.13}$$

The numerator and denominator of Eq. (2.13) contain contributions from two independent routes and an interference term. Since the interference term is controllable through variation of laboratory parameters, so too is the product ratio $R_{qq'}$. Thus the principle upon which this control scenario is based is the same as in the first example above, although the interference is introduced in an entirely different way.

Experimental control over $R_{qq'}$ is obtained by varying the difference $(\theta_3 - 3\theta_1)$ and the parameter x. The former is the phase difference between the ω_3 and the ω_1 laser fields and the latter, via Eq. (2.12), incorporates the ratio of the two lasers amplitudes.

Experimentally one envisions using "tripling" to produce ω_3 from ω_1, the subsequent variation of the phase of one of these beams provides a straightforward method of altering $\theta_3 - 3\theta_1$. Indeed, generating ω_3 from ω_1 allows for compensation of any phase jumps in the two laser sources. Thus the relative phase $\omega_3 - 3\omega$, is well defined.

With the qualitative principle of interfering pathways established, it remains to determine the quantitative extent to which coherent control alters the yield ratio in a realistic system. To this end we consider an application to one photon vs. three photon ("3+1") photodissociation of IBr. In particular, we focus on the energy regime where IBr dissociates to both $I(^2P_{3/2})$+$Br(^2P_{3/2})$ and $I(^2P_{3/2})$+$Br^*(^2P_{1/2})$. The IBr potential curves and coupling strengths used in the calculation, taken from the work of Child[44], are shown in Fig. 4.

A complete computation requires inclusion of angular momentum. A detailed discussion of the role of angular momentum is given below. Here we simply display the results of the quantum calculation which fully incorporates all the rotational states involved in the "3+1" CC of IBr. Two different cases were examined, those corresponding to fixed initial magnetic quantum numbers M_i and those corresponding to averaging over a random distribution of M_i, for fixed J_i. Results typical of those obtained are shown in Fig. 5 and 6, where we provide a contour plot of the yield of $Br^*(^2P_{1/2})$ for the case of excitation from $J_i = 1$, $M_i = 0$ and $J_i = 42$ with an average over M_i, as a function of laser control parameters (relative intensity and phase). The range of control in each case is vast with, remarkably, no loss of control with averaging over M_j.

As pointed out above, "3+1" is not necessarily the only viable control scenario in the "$n+m$" family. It has the advantage that one may generate one of the frequencies (the tripled photon) from the other. This is indeed the reason why the "3+1" route was the first control scenario to be implemented experimentally. (see discussion below).

As discussed below, control of *integral* (in contrast to *differential*) cross-sections requires that the $| E, n, q^- \rangle$ continuum states be made up of equal parity $| J, M \rangle$ angular momentum states. This means that in the "$m+n$" control scheme, the integer n must have the same parity as the integer m. Thus, studies of a "2+2" scheme for the control of the Na_2 photodissociation[18, 45] (discussed in detail in Section IV) and of a "2+4" scenario for the control of the Cl_2 photodissociation[46], have been published. In addition, studies of "3+1" control with strong fields, have also appeared[48, 49]. These studies and others[47] have verified that "$n+m$" control is viable even when strong fields are used, although the dependence on the x amplitude, and the $\theta_n - 3\theta_m$ phase factors is no longer as transparent as in the weak field case, discussed above.

The weak field "3+1" scenario has now been experimentally implemented in part in REMPI type experiments. The experiments demonstrated control of the total ionization rate, first in Hg[31], and then in HCl and CO[32] In the case of HCl[32], the molecule was excited to an intermediate $^3\Sigma^-(\Omega^+)$ vib-rotational resonance, using a combination of three ω_1 ($\lambda_1 = 336$ nm) photons and one ω_3 ($\lambda_3 = 112$ nm) photon. The ω_3 beam was generated from an ω_1 beam by tripling in a Kr gas cell. Ionization of the intermediate state takes place by absorption of one additional ω_1 photon.

The relative phase of the light fields was varied by passing the ω_1 and ω_2 beams through a second Ar or H_2 ("tuning") gas cell of variable pressure. The HCl REMPI experiments verified the prediction of a sinusoidal dependence of the ionization rates on the relative phase of the two exciting lasers of Eq. (2.13). The HCl experiment also verified the prediction of Eq. (2.13) of the dependence of the strength of the sinusoidal modulation of the ionization current on the x amplitude factor.

As discussed below, if one is content with controlling angular distributions, one can lift the equal-parity restriction. Absorption of two photons of perpendicular polarizations[5, 8], or of two photons interfering with their second-harmonic photon ("2+1" scenario)[8, 35, 36], result in states of different parities. Though such processes do not lead to control of integral quantities, they do allow for control of differential cross-sections. The "1+2" scenario (discussed in Section IV) has been implemented experimentally for the control of photo-current directionality in semiconductors, using no bias voltage[35].

2.2 The Pump-Dump Scheme

A useful extension of the scenario outlined in the above is a "pump-dump" scheme, in which an initial superposition of bound states is prepared with one laser pulse and subsequently dissociated with another. The scenario is shown qualitatively in Fig. 7. The pump and dump steps are assumed to be temporally separated by a time delay τ. The analysis below shows that under these circumstances the control parameters are the central frequency of the pump pulse, and the time delay between the two pulses.

Consider a molecule, initially ($t = 0$) in eigenstate $|\, E_g \rangle$ of Hamiltonian H_M, subjected to two transform limited light pulses. The field $\bar{\epsilon}(t)$ consists of two temporally separated pulses $\bar{\epsilon}(t) = \bar{\epsilon}_x(t) + \bar{\epsilon}_d(t)$, with the Fourier transform of $\bar{\epsilon}_x(t)$ denoted $\epsilon_x(\omega)$, etc. For convenience, we have chosen Gaussian pulses peaking at $t = t_x$ and t_d respectively. As discussed above, the $\bar{\epsilon}_x(t)$ pulse induces a transition to a linear combination of two excited bound electronic state with nuclear eigenfunctions $|\, E_1 \rangle$ and $|\, E_2 \rangle$, and the $\bar{\epsilon}_d(t)$ pulse dissociates the molecule by further exciting it to the continuous part of the spectrum. Both fields are chosen sufficiently weak for perturbation theory to be valid[50].

The superposition state prepared by the $\bar{\epsilon}_x(t)$ pulse, whose width is chosen to encompass just the two E_1 and E_2 levels, is given in first order perturbation theory as,

$$|\, \phi(t) \rangle = |\, E_g \rangle e^{-iE_g t/\hbar} + c_1 |\, E_1 \rangle e^{-iE_1 t/\hbar} + c_2 |\, E_2 \rangle e^{-iE_2 t/\hbar}, \tag{2.14}$$

where,

$$c_k = (\sqrt{2\pi}/i\hbar)\langle\, E_k\, |\mathrm{d}|\, E_g \rangle \epsilon_x(\omega_{kg}), \; k = 1, 2, \tag{2.15}$$

with $\omega_{kg} \equiv (E_k - E_g)/\hbar$.

After a delay time of $\tau \equiv t_d - t_x$ the system is subjected to the $\bar{\epsilon}_d(t)$ pulse. It follows from Eq. (2.14) that after this delay time each preparation coefficient has picked up an extra factor of $e^{-iE_k\tau/\hbar}$, $k = 1, 2$. Hence, the phase of c_1 relative to c_2 at that time increases by $[-(E_1 - E_2)\tau/\hbar = \omega_{2,1}\tau]$ Thus the natural two-state time evolution replaces the relative laser phase of the two-frequency control scenario above.

After the decay of the $\bar{\epsilon}_d(t)$ pulse the system wavefunction is given as,

$$|\, \psi(t) \rangle = |\, \phi(t) \rangle + \sum_{n,q} \int dE B(E, n, q|t) |\, E, n, q^- \rangle e^{-iEt/\hbar}. \tag{2.16}$$

The probability of observing the q fragments at total energy E in the remote future is therefore given as,

$$P(E, q) = \sum_n |B(E, n, q|t = \infty)|^2$$

$$= (2\pi/\hbar^2)\sum_n |\sum_{k=1,2} c_k \langle E, n, q^- |\mathbf{d}| E_k \rangle \epsilon_d(\omega_{EE_k})|^2 \tag{2.17}$$

where $\omega_{EE_k} = (E - E_k)/\hbar$, c_k is given by Eq. (2.15).

Expanding the square and using the Gaussian pulse shape gives:

$$P(E,q) = (2\pi/\hbar^2)[|c_1|^2 \mathsf{d}_{1,1}^{(q)}\epsilon_1^2 + |c_2|^2 \mathsf{d}_{2,2}^{(q)}\epsilon_2^2 + 2|c_1 c_2^* \epsilon_1 \epsilon_2 \mathsf{d}_{1,2}^{(q)}| \cos(\omega_{2,1}(t_d - t_x) + \alpha_{1,2}^{(q)}(E) + \phi)] \tag{2.18}$$

where $\epsilon_i = |\epsilon_d(\omega_{EE_i})|$, $\omega_{2,1} = (E_2 - E_1)/\hbar$ and the phases ϕ, $\alpha_{1,2}^{(q)}(E)$ are defined by

$$\langle E_1 |\mathbf{d}| E_g \rangle \langle E_g |\mathbf{d}| E_2 \rangle \equiv |\langle E_1 |\mathbf{d}| E_g \rangle \langle E_g |\mathbf{d}| E_2 \rangle| e^{i\phi}$$

$$\mathsf{d}_{i,k}^{(q)}(E) \equiv |\mathsf{d}_{i,k}^{(q)}(E)| e^{i\alpha_{i,k}^{(q)}(E)} = \sum_n \langle E, n, q^- |\mathbf{d}| E_i \rangle \langle E_k |\mathbf{d}| E, n, q^- \rangle \tag{2.19}$$

Integrating over E to encompass the full width of the second pulse, and forming the ratio, $Y = P(q)/[\sum_q P(q)]$, gives the ratio of products in each of the two arrangement channels, i.e. the quantity we wish to control. Once again it is the sum of two direct photodissociation contributions, plus an interference term.

Examination of Eq. (2.18) makes clear that the product ratio Y can be varied by changing the delay time $\tau = (t_d - t_x)$ or ratio $x = |c_1/c_2|$; the latter is most conveniently done by detuning the initial excitation pulse.

It is enlightening to consider this scenario as applied[9] to a model branching photodissociation reaction with masses of D and H, i.e.

$$\mathrm{H + HD \leftarrow DH_2 \rightarrow D + H_2}, \tag{2.20}$$

in which one uses the first pulse to excite a pair of states in a binding (Rydberg) electronic state and the second pulse to dissociate the system by de-exciting it back to the ground state. Typical results (see also Ref.[9]) for control are shown in Fig. 8 where the yield is seen to vary from 16% to 72% as the time delay and tuning of the initial excitation pulse are varied. This is an extreme range of control, especially in light of the fact that the two product channels differ only by mass factors.

It is highly instructive to examine the nature of the superposition state prepared in the initial excitation, [Eq. (2.14)] and its time evolution during the delay between pulses. An example of such a state is shown in Fig. 9 where we plot the wavefunction for a collinear model of the reaction of Eq. (2.20). Specifically, the coordinates are the reaction coordinate S and its orthogonal conjugate x. The wavefunction is shown evolving over 1/2 of its total possible period. Examination of Fig. 9 shows that de-exciting this superposition state during frame (b) would yield a substantially different product yield than de-exciting at the time of frame (e). However, there is clearly no particular preference of the wavefunction for large positive or large negative S at these particular times, which would be the case if the reaction control were a result of some spatial characteristics of the wavefunction. Rather, the essential control characteristics of the wavefunction are carried in the quantum amplitude and phase of the created superposition state.

A second example of pump-dump control[11] is provided by the example of IBr photodissociation. Specifically, we showed that it is possible to control the Br* vs. Br yield in this process, using two conveniently chosen picosecond pulses. The first pulse was chosen

to prepare a linear superposition of two bound states which arise from mixing of the X and A states. A subsequent pulse pumps this superposition to dissociation where the relative yields of Br and Br* are examined. Results typical of those obtained are shown in Fig. 10 where the relative yield is shown as a function of the delay between pulses and the detuning of the pump pulse from the energetic center of the two bound states in the initial superposition. The results show the vast range of control which is possible with this relatively simple experimental setup. Once again it is worth noting that both the potential energy surfaces and quantum photodissociation computations are "state-of-the-art", so that the results should be representative of results expected in the laboratory.

Theoretical work on similar pump dump scenarios for the control of the

$$\mathrm{D + OH \leftarrow HOD \rightarrow H + OD}$$

dissociation via the B-state[51] of HOD and the A-state[52] of HOD have recently been published. Experimental work on the control of this system is now in progress[53].

3 "2+2" Control of a Thermal Ensemble

In practice there are a number of sources of incoherence which tend to diminish control. Prominent amongst these are effects due to an initial thermal distribution of states and effects due to partial coherence of the laser source. Below we describe one approach, based upon a resonant "2+2" scenario, which deals effectively with both problems. An alternative method in which coherence is retained in the presence of collisions is discussed elsewhere[7].

The specific scheme we advocate is depicted, for the particular case of Na_2 photodissociation, in Fig. 11. Here the molecule is lifted from an initial bound state $|E_i, J_i, M_i\rangle$ to energy E via two independent two photon routes. To introduce notation, first consider a single such two photon route. Absorption of the first photon of frequency ω_1 lifts the system to a region close to an intermediate bound state $|E_m J_m M_m\rangle$, and a second photon of frequency ω_2 carries the system to the dissociating states $|E, \hat{\mathbf{k}}, q^-\rangle$, where the scattering angles are specified by $\hat{\mathbf{k}} = (\theta_k, \phi_k)$. Here the J's are the angular momentum, M's are their projection along the z-axis, and the values of energy, E_i and E_m, include specification of the vibrational quantum numbers. Specifically, if we denote the phases of the coherent states by ϕ_1 and ϕ_2, the wavevectors by $\mathbf{k}_1$ and $\mathbf{k}_2$ with overall phases $\theta_i = \mathbf{k}_i \cdot \mathbf{R} + \phi_i$ $(i = 1, 2)$ and the electric field amplitudes by ϵ_1 and ϵ_2, then the probability amplitude for resonant two photon $(\omega_1 + \omega_2)$ photodissociation is given[18, 45] by

$$\begin{aligned}
&T_{\hat{\mathbf{k}}q,i}(E, E_iJ_iM_i, \omega_2, \omega_1) = \\
&= \sum_{E_m,J_m} \frac{\langle E, \hat{\mathbf{k}}, q^-|\mathbf{d}_2\epsilon_2|E_mJ_mM_i\rangle\langle E_mJ_mM_i|\mathbf{d}_1\epsilon_1|E_iJ_iM_i\rangle}{\omega_1 - (E_m + \delta_m - E_i) + i\Gamma_m} \exp[i(\theta_1 + \theta_2)] \\
&= \frac{\sqrt{2\mu k_q}}{h} \sum_{J,p,\lambda\geq 0} \sum_{E_m,J_m} \begin{pmatrix} J & 1 & J_m \\ -M_i & 0 & M_i \end{pmatrix} \begin{pmatrix} J_m & 1 & J_i \\ -M_i & 0 & M_i \end{pmatrix} \\
&\times \sqrt{2J+1}\, D^{Jp}_{\lambda,M_i}(\theta_k, \phi_k, 0) t(E, E_iJ_i, \omega_2, \omega_1, q|Jp\lambda, E_mJ_m) \exp[i(\theta_1 + \theta_2)] \quad (3.1)
\end{aligned}$$

Here $\mathbf{d}_i$ is the component of the dipole moment along the electric-field vector of the ith laser mode, $E = E_i + (\omega_1 + \omega_2)$, δ_m and Γ_m are respectively the radiative shift and width of the intermediate state, μ the reduced mass, and k_q is the relative momentum of the dissociated product in q-channel. The D^{Jp}_{λ,M_i} is the parity adapted rotation matrix[55] with λ the magnitude of the projection on the internuclear axis of the electronic angular momentum and $(-1)^J p$ the parity of the rotation matrix. We have set $\hbar \equiv 1$, and assumed for simplicity lasers which are linearly-polarized and with parallel electric-field vectors. Note that the T-matrix element in Eq. (3.1) is a complex quantity, whose phase is the sum of the laser phase $\theta_1 + \theta_2$ and the molecular phase, i.e. the phase of $\mathbf{t}$.

The probability of producing the fragments in q-channel is obtained by integrating the square of Eq. (3.1) over the scattering angles $\hat{\mathbf{k}}$, with the result:

$$
\begin{aligned}
P^{(q)}(E, E_iJ_iM_i, \omega_2, \omega_1) &= \int d\hat{\mathbf{k}} |T_{\hat{\mathbf{k}}q,i}(E, E_iJ_iM_i, \omega_2, \omega_1)|^2 \\
&= \frac{8\pi\mu k_q}{h^2} \sum_{J,p,\lambda\geq 0} \left| \sum_{E_m,J_m} \begin{pmatrix} J & 1 & J_m \\ -M_i & 0 & M_i \end{pmatrix} \begin{pmatrix} J_m & 1 & J_i \\ -M_i & 0 & M_i \end{pmatrix} \right. \\
&\quad \left. \times t(E, E_iJ_i, \omega_2, \omega_1, q|Jp\lambda, E_mJ_m) \right|^2, \qquad (3.2)
\end{aligned}
$$

Because the t-matrix element contains a factor of $[\omega_1 - (E_m + \delta_m - E_i) + i\Gamma_m]^{-1}$ the probability is greatly enhanced by the approximate inverse square of the detuning $\Delta = \omega_1 - (E_m + \delta_m - E_i)$ as long as the line width Γ_m is less than Δ. Hence only the levels closest to the resonance $\Delta = 0$ contribute significantly to the dissociation probability. *This allows us to selectively photodissociate molecules from a thermal bath, reestablishing coherence necessary for quantum interference based control and overcoming dephasing effects due to collisions.*

Consider then the following coherent control scenario. A molecule is irradiated with three interrelated frequencies, $\omega_0, \omega_+, \omega_-$ where photodissociation occurs at $E = E_i + 2\omega_0 = E_i + (\omega_+ + \omega_-)$ and where ω_0 and ω_+ are chosen resonant with intermediate bound state levels. The probability of photodissociation at energy E into arrangement channel q is then given by the square of the sum of the T matrix elements from pathway "a" $(\omega_0 + \omega_0)$ and pathway "b" $(\omega_+ + \omega_-)$. That is, the probability into channel q

$$
\begin{aligned}
P_q(E, E_iJ_iM_i; \omega_0, \omega_+, \omega_-) &\equiv \int d\hat{\mathbf{k}} \left| T_{\hat{\mathbf{k}}q,i}(E, E_iJ_iM_i, \omega_0, \omega_0) + T_{\hat{\mathbf{k}}q,i}(E, E_iJ_iM_i, \omega_+, \omega_-) \right|^2 \\
&\equiv P^{(q)}(a) + P^{(q)}(b) + P^{(q)}(ab) \qquad (3.3)
\end{aligned}
$$

Here $P^{(q)}(a)$ and $P^{(q)}(b)$ are the independent photodissociation probabilities associated with routes a and b respectively and $P^{(q)}(ab)$ is the interference term between them, discussed below. Note that the two T matrix elements in Eq. (3.3) are associated with different lasers and as such contain different laser phases. Specifically, the overall phase of the three laser fields are $\theta_0 = \mathbf{k}_0 \cdot \mathbf{R} + \phi_0, \theta_+ = \mathbf{k}_+ \cdot \mathbf{R} + \phi_+$ and $\theta_- = \mathbf{k}_- \cdot \mathbf{R} + \phi_-$, where ϕ_0, ϕ_+ and ϕ_- are the photon phases, and $\mathbf{k}_0$, $\mathbf{k}_+$, and $\mathbf{k}_-$ are the wavevectors of the laser modes ω_0, ω_+ and ω_-, whose electric field strengths are $\epsilon_0, \epsilon_+, \epsilon_-$ and intensities I_0, I_+, I_-.

The optical path-path interference term $P^{(q)}(ab)$ is given by

$$
P^{(q)}(ab) = 2|F^{(q)}(ab)| \cos(\alpha_a^q - \alpha_b^q) \qquad (3.4)
$$

with relative phase

$$\alpha_a^q - \alpha_b^q = (\delta_a^q - \delta_b^q) + (2\theta_0 - \theta_+ - \theta_-). \tag{3.5}$$

where the amplitude $|F^{(q)}(ab)|$ and the molecular phase difference $(\delta_a^q - \delta_b^q)$ are defined by

$$\begin{aligned} &|F^{(q)}(ab)| \exp[i(\delta_a^q - \delta_b^q)] \\ &= \frac{8\pi\mu k_q}{h^2} \sum_{J,p,\lambda\geq 0} \sum_{E_m,J_m} \sum_{E'_m,J'_m} \begin{pmatrix} J & 1 & J_m \\ -M_i & 0 & M_i \end{pmatrix} \begin{pmatrix} J_m & 1 & J_i \\ -M_i & 0 & M_i \end{pmatrix} \begin{pmatrix} J & 1 & J'_m \\ -M_i & 0 & M_i \end{pmatrix} \\ &\times \begin{pmatrix} J'_m & 1 & J_i \\ -M_i & 0 & M_i \end{pmatrix} t(E, E_iJ_i, \omega_0, \omega_0, q|Jp\lambda, E_mJ_m) t^*(E, E_iJ_i, \omega_-, \omega_+, q|Jp\lambda, E'_mJ'_m). \end{aligned} \tag{3.6}$$

Consider now the quantity of interest $R_{qq'}$, the branching ratio of the product in q-channel to that in q'-channel. Noting that in the weak field case $P^{(q)}(a)$ is proportional to ϵ_0^4, $P^{(q)}(b)$ to $\epsilon_+^2\epsilon_-^2$, and $P^{(q)}(ab)$ to $\epsilon_0^2\epsilon_+\epsilon_-$ we can write

$$R_{qq'} = \frac{\mu_{aa}^{(q)} + x^2\mu_{bb}^{(q)} + 2x|\mu_{ab}^{(q)}|\cos(\alpha_a^q - \alpha_b^q) + (B^{(q)}/\epsilon_0^4)}{\mu_{aa}^{(q')} + x^2\mu_{bb}^{(q')} + 2x|\mu_{ab}^{(q')}|\cos(\alpha_a^{q'} - \alpha_b^{q'}) + (B^{(q')}/\epsilon_0^4)} \tag{3.7}$$

where $\mu_{aa}^{(q)} = P^{(q)}(a)/\epsilon_0^4$, $\mu_{bb}^{(q)} = P^{(q)}(b)/(\epsilon_+^2\epsilon_-^2)$ and $|\mu_{ab}^{(q)}| = |F^{(q)}(ab)|/(\epsilon_0^2\epsilon_+\epsilon_-)$ and $x = \epsilon_+\epsilon_-/\epsilon_0^2 = \sqrt{I_+I_-}/I_0$. The terms with $B^{(q)}, B^{(q')}$, described below, correspond to resonant photodissociation routes to energies other than $E = E_i + 2\hbar\omega_0$ and hence[4] to terms which do not coherently interfere with the a and b pathways. Minimization of these terms, due to absorption of $(\omega_0 + \omega_-)$, $(\omega_0 + \omega_+)$, $(\omega_+ + \omega_0)$ or $(\omega_+ + \omega_+)$, is discussed elsewhere[18, 45]. Here we just emphasize that the product ratio in Eq. (3.7) depends upon both the laser intensities and relative laser phase. Hence manipulating these laboratory parameters allows for control over the relative cross section between channels.

The proposed scenario, embodied in Eq. (3.7), also provides a means by which control can be improved by eliminating effects due to laser jitter. Specifically, the term $2\phi_0 - \phi_+ - \phi_-$ contained in the relative phase $\alpha_a^q - \alpha_b^q$ can be subject to the phase fluctuations arising from laser instabilities. If such fluctuations are sufficiently large then the interference term in Eq. (3.7), and hence control, disappears[19]. The following experimentally desirable implementation of the above 2-photon plus 2-photon scenario readily compensates for this problem. Specifically, consider generating $\omega_+ = \omega_0 + \delta$ and $\omega_- = \omega_0 - \delta$ in a parametric process by passing a beam of frequency $2\omega_0$ through a nonlinear crystal. This latter beam is assumed generated by second harmonic generation from the laser ω_0 with the phase ϕ_0. Then the quantity $2\phi_0 - \phi_+ - \phi_-$ in the phase difference between the $(\omega_0 + \omega_0)$ and $(\omega_+ + \omega_-)$ routes is a constant. That is, in this particular scenario, fluctuations in ϕ_0 cancel and have no effect on the relative phase $\alpha_a^q - \alpha_b^q$. *Thus th 2-photon plus 2-photon scenario is insensitive to the laser jitter of the incident laser fields.*

To examine the range of control afforded by this scheme consider the photodissociation of Na_2 in the regime below the Na(3d) threshold where dissociation is to two product channels Na(3s) + Na(3p) and Na(3s) + Na(4s). Two photon dissociation of Na_2 from a bound state of the $^1\Sigma_g^+$ state occurs[18, 45] in this region by initial excitation to an excited intermediate bound state $|E_mJ_mM_m\rangle$. The latter is a superposition of states of the $A^1\Sigma_u^+$ and $b^3\Pi_u$ electronic curves, a consequence of spin-orbit coupling. That is, the two photon photodissociation can be viewed[45] as occurring via intersystem crossing subsequent to absorption of the first photon. The continuum states reached in the excitation can be either of singlet or triplet character but, despite the multitude of electronic states involved in the computation, the predominant contributions to the products Na(3p) and Na(4s) are found to come from the $^3\Pi_g$ and $^3\Sigma_g^+$ states, respectively. Methods for computing the required photodissociation amplitude, which involves eleven electronic states are discussed elsewhere[45]. Since the resonant character of the two photon excitation allows us to select a single initial state from a thermal ensemble we consider here the specific case of $v_i = J_i = 0$ without loss of generality, where v_i, J_i denote the vibrational and rotational quantum numbers of the initial state.

The ratio $R_{qq'}$ depends on a number of laboratory control parameters including the relative laser intensities x, relative laser phase, and the ratio of ϵ_+ and ϵ_- via η. In addition, the relative cross sections can be altered by modifying the detuning. Typical control results are shown in Fig. 12 which provides contour plots of the Na(3p) yield (i.e., the ratio of the probability of observing Na(3p) to the sum of the probabilities to form Na(3p) plus Na(4s)). The figure axes are the ratio of the laser amplitudes x, and the relative laser phase $\delta\theta = 2\theta_0 - \theta_+ - \theta_-$. Here $\omega_0 = 631.899$, $\omega_+ = 562.833$ and $\omega_- = 720.284$ nm and control is seem to be large, ranging from 30% Na(3p) to 90% as $\delta\theta$ and x are varied.

Note that the proposed approach is not limited to the specific frequency scheme discussed above. Essentially all that is required is that the two resonant photodissociation routes lead to interference and that the cumulative laser phases of the two routes be independent of laser jitter. As one sample extension, consider the case where paths a and b are composed of totally different photons, $\omega_+^{(a)}$ and $\omega_-^{(a)}$ and $\omega_+^{(b)}$ and $\omega_-^{(b)}$, with $\omega_+^{(a)} + \omega_-^{(a)} = \omega_+^{(b)} + \omega_-^{(b)}$. Both these sets of frequencies can be generated, for example, by passing $2\omega_0$ light through nonlinear crystals, hence yielding two pathways whose relative phase is independent of laser jitter in the initial 2 ω_0 source. Given these four frequencies we now have an additional degree of freedom in order to optimize control, although the experiment is considerably more complicated than in the three frequency case. Typical results for Na_2 are provided elsewhere[18, 45]. Note also that the control is not limited to two-product channels, such as those discussed above. Recent computations[45] on higher energy Na_2 photodissociation, where more product arrangement channels are available, show equally large ranges of control for the three channel case.

4 Control with Intense Laser Fields

We now discuss some extensions of CC to strong laser fields. Parallel work involving other strong field scenarios has been done by Bandrauk et al.[46], Corkum et al.[47], Bardsley et al.[48] Lambropoulos et al.[58] and Guisti-Suzor et al.[49]. Here we concentrate on a strong field control scenario in which the dependence on the relative phase between the two laser beams, hence on laser coherence, disappears. As a result, coherence plays no role in this scenario (save for being intimately linked with the existence of the narrow-band laser sources needed for its execution). Although the unimportance of coherence means that we lose phase control, the effect still depends on quantum interference phenomena.

The scenario is therefore called Interference Control.

To illustrate interference control we look at the control of the electronic states of Na atoms generated by the photodissociation of Na_2, a process treated in the context of weak field CC above. We envision a scenario, depicted in Fig. 13, in which we employ two laser sources: One laser, (not necessarily intense) with center frequency ω_1 is used to excite a molecule from an initially populated bound state $|E_i\rangle$ to a dissociative state $|E,m,q^-\rangle$. A second laser, with frequency ω_2, is used to couple ("dress") the continuum with some (initially unpopulated) bound states $|E_j\rangle$. With both lasers on, dissociation to $|E,m,q^-\rangle$ occurs via one direct, $|E_i\rangle \to |E,m,q^-\rangle$, and a multitude of indirect, e.g., $|E_i\rangle \to |E,m,q^-\rangle \to |E_j\rangle \to |E,m,q^-\rangle$, pathways. The interference between these pathways to form a given channel q at product energy E can be either constructive or destructive. As we show below, varying the frequencies and intensities of the two excitation lasers strongly affects this interference term, providing a means of controlling the photodissociation line shape and the branching ratio into different products.

With this scenario in mind we now briefly discuss the methodology of dealing with strong laser fields and the extension of CC ideas to this domain. We consider the photodissociation of a molecule with Hamiltonian H_M in the presence of a radiation field with Hamiltonian H_R, whose eigenstates are the Fock states $|n_k\rangle$ with energy $n_k\hbar\omega_k$. (In the case of several frequencies the repeated index in $n_k\omega_k$ implies the sum over the modes.)

Strong field dynamics is completely embodied[59] in the fully interacting eigenstates of the total Hamiltonian H, $H = H_M + H_R + V$, where V is the light-matter interaction, denoted $|(E,m,q^-),n_k^-\rangle$

$$H|(E,m,q^-),n_k^-\rangle = (E + n_k\hbar\omega_k)|(E,m,q^-),n_k^-\rangle \,. \tag{4.1}$$

The minus superscript on n_k is used in exactly the same way as in the weak field domain: it is a reminder that each $|(E,m,q^-),n_k^-\rangle$ state correlates to a non-interacting $|(E,m,q^-),n_k\rangle \equiv |E,m,q^-\rangle|n_k\rangle$ state when the light-matter interaction V is switched off.

If the system is initially in the $|E_i,n_i\rangle \equiv |E_i\rangle|n_i\rangle$ state and we suddenly switch on V, the photodissociation amplitude to form in the future the product state $|E,m,q^-\rangle|n_k\rangle$ is simply given[59] as the overlap between the initial and fully interacting state $\langle(E,m,q^-),n_k^-|E_i,n_i\rangle$. This overlap assumes the convenient form

$$\langle(E,m,q^-),n_k^-|E_i,n_i\rangle = \langle(E,m,q^-),n_k|VG(E^+ + n_k\hbar\omega_k)|E_i,n_i\rangle, \tag{4.2}$$

by using the Lippmann-Schwinger equation

$$\langle(E,m,q^-),n_k^-| = \langle(E,m,q^-),n_k| + \langle(E,m,q^-),n_k|VG(E^+ + n_k\hbar\omega_k). \tag{4.3}$$

Here $G(\mathcal{E}) = 1/(\mathcal{E} - H)$ and $E^+ = E + i\delta$, with $\delta \to 0^+$ at the end of the computation. Equation (4.2) is exact and provides a connection between the photodissociation amplitude and the VG matrix element. It is the latter which we compute exactly using a high field extension of the artificial channel method[60, 61].

Two quantities are of interest: the channel specific line shape,

$$A(E,q,n_k|E_i,n_i) = \int d\hat{\mathbf{k}}\, |\langle(E,\hat{\mathbf{k}},q^-),n_k^-|E_i,n_i\rangle|^2, \tag{4.4}$$

and the total dissociation probability to channel q

$$P(q) = \sum_{n_k} \int dE\, A(E, q, n_k | E_i, n_i). \tag{4.5}$$

In Eq. (4.5) the sum is over photons that excite the molecule above the dissociation threshold. In writing Eq. (4.4) diatomic dissociation is assumed, so that $m = \hat{\mathbf{k}}$.

Consider for example the photodissociation of Na_2 from the $| E_i \rangle = | v = 19, {}^3\Pi_u \rangle$ initial state, where v denotes the vibrational quantum number in the ${}^3\Pi_u$ electronic potential[62] (see Fig. 13). $| E_i \rangle$ is assumed to have been prepared by previous excitation from the ground electronic state. Excitations from $| E_i \rangle$ by ω_1 and mixing of the initially unpopulated $| E_j \rangle$ by ω_2 to the dissociating continua produce Na(3s)+Na(3p) and Na(3s)+Na(4s). Computations were done with ω_1 chosen within the range 15,430 $cm^{-1} < \omega_1 < 15{,}700\ cm^{-1}$ with intensity $I_1 \sim 10^{10}$ W/cm^2, which is sufficiently energetic to dissociate levels of the ${}^3\Pi_u$ state with $v \geq 19$ to both Na(3s) + Na(3p) and Na(3s) + Na(4s). The second laser has fixed frequency $\omega_2 = 13{,}964$ cm^{-1} and intensity $I_2 = 3.2 \times 10^{11}$ W/cm^2 and can dissociate levels with $v \geq 26$ to both products. Under these circumstances the contribution of above threshold dissociation is found to be negligible. However cognizance must be taken of the possibility of dissociation of $| E_i \rangle$ by ω_2 and of $| E_j \rangle$ by ω_1. These processes do not interfere and can not be controlled. Hence we must find the range of parameters that minimizes them.

Figure 14 shows computed line shapes $A(E, q, n_k | E_i, n_i)$ (on a logarithmic scale) as a function of the product translational energy E, with $\omega_2 = 13{,}964$ cm^{-1}, $\omega_1 = 15{,}456$ cm^{-1}, $I_1 = 5.5 \times 10^9$ W/cm^2 and $I_2 = 3.51 \times 10^{10}$ W/cm^2. Results for both the Na(3p) +Na(3s) and Na(4s) + Na(3s) product channels are shown. Figure 15 contains similar results, but with $\omega_1 = 15{,}511$. In addition, the line shape for excitation with the laser of frequency $\omega_1 = 15{,}456$ cm^{-1} only (ω_2 laser off) is shown in Fig. 14 for the Na(3s)+Na(4s) product; the Na(3p) + Na(4s) result is similar.

Consider first $A(E, q, n_k | E_i, n_i)$ associated with excitation by a single laser [Fig. 14, curve (c)]. The line shape is comprised of a series of non-Lorentzian peaks and dips corresponding to resonance contributions from the dressed $v = 19, 20, 21$ vibrational states. The predominant contribution is the direct $v_i = 19$ excitation, with smaller $v = 20, 21$ contributions arising from stimulated emission and absorption from and to the continuum. Further, the overall shape between the peaks shows Fano-type interference between the photodissociation pathways arising from the pairs of adjacent vibrational states. Since significant dissociation is observed from states other than the initially populated $v_i = 19$ it is clear that the power broadening is on the same order of magnitude as the vibrational level spacing.

With both ω_1 and ω_2 lasers on, each peak splits into two peaks in a manner which is dependent both upon asymptotic channel [compare curves (a) and (b) within each of Figs. 14 and 15] and frequency ω_1 [compare Fig. 14 with Fig. 15]. An analysis of this structure is provided later below. Here we note the significant implication that by varying ω_1 we can control the channel specific line shapes $A(E, n_k | \epsilon_i, n_i)$. For example, comparing Figs. 14 and 15 shows that the increase in ω_1 results in a shift of the dominant peaks to higher E. Products at $E \approx 9025$ cm^{-1} are strongly enhanced relative to the case in Fig. 14 and products at $E \approx 8980$ cm^{-1} are suppressed,etc. Tuning ω_2 or changing the laser intensities also changes the line shapes, as discussed elsewhere[43].

Integrating $A(E, q, n_k|\epsilon_i, n_i)$ over E [Eq. (4.5)] for various ω_1 values gives $P(q)$ as a function of ω_1. The result of these computations are shown in Fig. 16 for both Na(3s)+Na(3p) [curve (a)] and Na(3s)+Na(4s) [curve (b)] channels, with $I_1 = 8.7 \times 10^9$ W/cm^2, $I_2 = 3.51 \times 10^{10}$ W/cm^2 and ω_2=13,964 cm^{-1}. The probability $P(q)$ is seen to oscillate strongly as a function of ω_1, with the distance between the peaks (or dips) being the vibrational spacing between $v = 31$ and 32. The oscillations for the two product channels are out of phase. Hence, for example, the probabilities of producing Na(3s)+Na(4s) and Na(3s) + Na(3p) at $\omega_1 = 15,494$ cm^{-1}are 0.198 and 0.730, respectively. The reverse situation occurs at ω_1= 15,573 cm^{-1}where the total dissociation probability remains 0.93 but where 68% of product is Na(3s) + Na(4s). Thus varying ω_1 provides a straightforward method to control the branching ratio into final product channels. Furthermore, and significantly, computations show that arbitrarily changing the relative phase between the ω_1 and ω_2 does not alter Figs. 14-16, indicating that the control process is independent of the relative laser phase. This is consistent with the model discussed below.

Reducing the laser power in these computations[43] narrows the frequency range over which the product probability oscillates, clearly indicating that this range is at least partially determined by the power broadening.

The qualitative behavior seen in Figs. 14-16 can be readily understood in terms of a simple model which assumes excitation of the initial state $|\,1\,\rangle \equiv |\,E_i, n_1, n_2\,\rangle$ with laser ω_1 to the continuum $|\,E_q\,\rangle \equiv |\,(E, q^-), n_1 - 1, n_2\,\rangle$, which is coupled to state $|\,2\,\rangle \equiv |\,E_j, n_1 - 1, n_2 + 1\,\rangle$ with laser ω_2. If these are the only contributing states then the photodissociation amplitude is given by

$$\langle\, E_q\,|VG(\mathcal{E})|\,1\,\rangle = \langle\, E_q\,|V|\,1\,\rangle\langle\,1\,|G(\mathcal{E})|\,1\,\rangle + \langle\, E_q\,|V|\,2\,\rangle\langle\,2\,|G(\mathcal{E})|\,1\,\rangle, \tag{4.6}$$

where $\mathcal{E} = E^+ + (n_1 - 1)\hbar\omega_1 + n_2\hbar\omega_2$. Using $(\mathcal{E} - H_0 - V)G(\mathcal{E}) = 1$, we obtain coupled equations for the matrix elements of G which can be solved. Substituting the result into Eq. (4.6), gives

$$\langle\, E_q\,|VG(\mathcal{E})|\,1\,\rangle = \frac{(\mathcal{E} - E_2 - \pi_{2,2}(E))\langle\, E_q\,|V|\,1\,\rangle + \langle\, E_q\,|V|\,2\,\rangle\pi_{2,1}(E)}{[E - E_1 - \pi_{1,1}(E)][E - E_2 - \pi_{2,2}(E)] - \pi_{1,2}(E)\pi_{2,1}(E)}, \tag{4.7}$$

where $E_1 = E_i + n_1\hbar\omega_1 + n_2\hbar\omega_2$, $E_2 = E_j + (n_1 - 1)\hbar\omega_1 + (n_2 + 1)\hbar\omega_2$, and $\pi_{a,b}$ $(a, b = 1, 2)$ is given by

$$\pi_{a,b}(E) = \sum_q \int dE' \frac{\langle\, a\,|V|\,E'_q\,\rangle\langle\, E'_q\,|V|\,b\,\rangle}{E - E' - (n_1 - 1)\hbar\omega_1 - n_2\hbar\omega_2}. \tag{4.8}$$

The E dependence of Eq. (4.7) can be exposed by substituting $\mathcal{E} = E^+ + (n_1 - 1)\hbar\omega_1 + n_2\hbar\omega_2$ into Eq. (4.7). Denoting $\pi_{a,b}(E)$ at this energy by $\pi_{a,b}$, we have

$$|\langle\, E_q\,|VG(\mathcal{E})|\,1\,\rangle|^2 = \left|\frac{(\mathcal{E} - E_j - \hbar\omega_2 - \pi_{2,2})\langle\, E_q\,|V|\,1\,\rangle + \langle\, E_q\,|V|\,2\,\rangle\pi_{2,1}}{(\mathcal{E} - \chi_+)(\mathcal{E} - \chi_-)}\right|^2 \tag{4.9}$$

where

$$2\chi_\pm = (E_i + \pi_{1,1} + \hbar\omega_1) + (E_j + \pi_{2,2} + \hbar\omega_2) \pm \sqrt{[(E_i + \pi_{1,1} + \hbar\omega_1) - (E_j + \pi_{2,2} + \hbar\omega_2)]^2 + 4\pi_{1,2}\pi_{2,1}} \tag{4.10}$$

$\chi_{\pm}$ are the eigenvalues associated with the diagonalization of the matrix coupling the two dressed states, of energy $(E_i + \pi_{1,1} + \hbar\omega_1)$ and $(E_j + \pi_{2,2} + \hbar\omega_2)$ via the continuum. The real and imaginary parts of $\pi_{a,a}$ $(a = 1,2)$ give the shifts and broadenings of the two levels.

Equation (4.9) shows photodissociation occurring via two pathways, $|\,E_i, n_1, n_2\,\rangle \rightarrow |\,(E, q^-), n_1 - 1, n_2\,\rangle$ and $|\,E_i, n_1, n_2\,\rangle \rightarrow |\,E_j, n_1 - 1, n_2 + 1\,\rangle \rightarrow |\,(E, q^-), n_1 - 1, n_2\,\rangle$; interference between them can be constructive or destructive, depending on the relative sign of the two terms. This interference can be manipulated by varying the laser frequencies. The double peak structure seen in Figs. 14 and 15 is consistent with the form of Eq. (4.9) wherein two peaks are predicted as the function of $\mathcal{E}$, at the two roots of the equations $(\mathcal{E} - \chi_{\pm}) = 0$. For example, in the case of Fig. 14, the first double peak arises from the interaction between the dressed $v = 19$ and $v = 31$ levels of the $^3\Pi_u$ state whereas the decrease in photodissociation (compared to the curve (c)) in the middle of the double peak results from the destructive interference between them. A similar explanation applies to the second and third double peaks, which result mainly from the combined excitations of v=20 and 32, and of v=21 and 33, respectively. Note that the locations of the peaks are channel-independent but that the ratio of the heights of the peaks, given by the ratio of $|\langle\, E_q\,|V|\,1\,\rangle\,[\mathcal{E} - E_j - \hbar\omega_2 - \pi_{2,2}] + \langle\, E_q\,|V|\,2\,\rangle\pi 2, 1|^2$ evaluated at $\mathcal{E} = \chi^+$ and χ^-, respectively, depends strongly on the laser frequencies, intensities and the channel index q. Thus Eq. (4.10) encompasses the channel dependence of the interference and hence the control over product probabilities.

Note also that Eq. (4.9) is consistent with a photodissociation amplitude wherein control of the line shape and product probabilities is independent of the relative phase of the two routes. That is, if ϕ_1 and ϕ_2 are the phases of the two lasers (including the spatial phases $\mathbf{K_1} \cdot \mathbf{r}, \mathbf{K_2} \cdot \mathbf{r}$) then absorption of an ω_1 or ω_2 photon contributes a phase factor $\exp[i\phi_1]$ or $\exp[i\phi_2]$ to the matrix elements of V. Similarly, stimulated emission of one photon of ω_1 or ω_2 contributes a phase factor $\exp[-i\phi_1]$ or $\exp[-i\phi_2]$ to the matrix elements of V. Therefore, the second term in the nominator of Eq.(4.9) carries an overall phase factor $\exp[i\phi_2] \times \exp[i\phi_1 - i\phi_2] = \exp[i\phi_1]$, which is the same as the phase factor in the first term. The *relative* phase of the two routes, which enters the interference term, is therefore independent of both ϕ_1 and ϕ_2.

This model fails, however, to include excitation and dissociation of neighboring vibrational states of $|\,E_i\,\rangle$ and $|\,E_j\,\rangle$. Nonetheless the computations in Figs. 14-16, which incorporate all vibrational states, clearly demonstrate the desired control. Further computations[45], which include rotations have also been performed. Inclusion of these rotational states leads to a series of multiple peak-and-dip structure in the line shape corresponding to the resonance contributions of multiple rotational states. Because of this, the dependence of the channel specific dissociation yield on ω_1 changes, but control over line shapes and product yields is still strong.

5 Conclusions

Our discussion makes clear that the characteristic features which we invoke in order to control chemical reactions are purely quantum in nature. There is, for example, little classical about the time dependent picture where the ultimate outcome of the deexcitation, i.e.

product H + HD or H_2 + D depends entirely upon the phase and amplitude characteristics of the wavefunction. Indeed, as repeatedly emphasized above, if, e.g. collisional effects are sufficiently strong so as to randomize the phases then reaction control is lost. Hence reaction dynamics is intimately linked to the wavefunction phases which are controllable through coherent optical phase excitation.

These results must be viewed in light of the history of molecular reaction dynamics over the past two decades. Possibly the most useful result of the reaction dynamics research effort has been the recognition that the vast majority of qualitatively important phenomena in reaction dynamics are well described by classical mechanics. Quantum and semiclassical mechanics were viewed as necessary only insofar as they correct quantitative failures of classical mechanics for unusual circumstances and/or for the dynamics of very light particles. Considering reaction dynamics in traditional chemistry to be essentially classical in character therefore appeared to be essentially correct for the vast majority of naturally occurring molecular processes. Coherence played no role. The approach which we have introduced above makes clear, however, that coherence phenomena have great potential for application. The quantum phase is always present and can be used to our advantage, even though it is irrelevant to traditional chemistry. By calling attention to the extreme importance of coherence phenomena to controlled chemistry we herald the introduction of a new focus in atomic and molecular science, i.e. introducing coherence in controlled environments to modify molecular processes, thus defining the area of coherence chemistry.

References

[1] P.Brumer and M. Shapiro, Chem. Phys. Lett. **126**, 541 (1986).

[2] M. Shapiro and P. Brumer, J. Chem. Phys. **84**, 4103 (1986).

[3] P. Brumer and M. Shapiro, Faraday Disc. Chem. Soc. **82**, 177 (1986).

[4] M. Shapiro and P. Brumer, J. Chem. Phys. **84**, 4103 (1986).

[5] C. Asaro, P. Brumer and M. Shapiro, Phys. Rev. Lett. **60**, 1634 (1988).

[6] M. Shapiro, J. Hepburn and P. Brumer, Chem. Phys. Lett. **149**, 451 (1988).

[7] P. Brumer and M. Shapiro, J. Chem. Phys. **90**, 6179 (1989).

[8] G. Kurizki, M. Shapiro and P. Brumer, Phys. Rev. B **39**, 3435 (1989).

[9] T. Seideman, M. Shapiro and P. Brumer, J. Chem. Phys. **90**, 7136 (1989).

[10] J. Krause, M. Shapiro and P. Brumer, J. Chem. Phys. **92**, 1126 (1990).

[11] I. Levy, M. Shapiro and P. Brumer J. Chem. Phys. **93**, 2493 (1990).

[12] P. Brumer and M. Shapiro, Accounts Chem. Res. **22**, 407 (1989).

[13] P. Brumer and M. Shapiro, Chem. Phys. **139**, 221 (1989).

[14] C.K. Chan , P. Brumer and M. Shapiro, J. Chem. Phys. **94**, 2688 (1991).

[15] M. Shapiro and P. Brumer, J. Chem. Phys. **95**, 8658 (1991).

[16] P. Brumer and M. Shapiro, Ann. Rev. Phys. Chem., **43**, 257 (1992).

[17] M. Shapiro and P. Brumer, J. Chem. Phys. **97**, 6259 (1992).

[18] Z. Chen, P. Brumer and M. Shapiro, Chem. Phys. Lett. **198**, 498 (1992).

[19] X-P. Jiang, P. Brumer and M. Shapiro "Partially Coherent Laser Pulses in Coherent Control of Chemical Reactions" J. Chem. Phys. , (to be submitted).

[20] J. Dods, P. Brumer and M. Shapiro, "Two Color Coherent Control with SEP Preparation: Electronic Branching in the Na_2 Photodissociation" Can. J. Chem., submitted.

[21] D. J. Tannor and S. A. Rice, J. Chem. Phys. **83**, 5013 (1985); D. J. Tannor, R. Kosloff and S. A. Rice, J. Chem. Phys. **85**, 5805 (1986).

[22] S.A. Rice, D.J. Tannor, and R. Kosloff, J. Chem. Soc. Faraday Trans. 2, **82**, 2423 (1986);

[23] D.J. Tannor, and S.A. Rice, Adv. Chem. Phys. **70**, 441 (1988).

[24] R. Kosloff, S.A. Rice, P. Gaspard, S. Tersigni, and D.J. Tannor , Chem. Phys. **139**, 201 (1989).

[25] S. Tersigni , P. Gaspard and S.A. Rice, J. Chem. Phys. **93**, 1670 (1990).

[26] S. Shi, A. Woody, and H. Rabitz, J. Chem. Phys. **88**, 6870 (1988); S. Shi and H. Rabitz, Chem. Phys. **139**, 185 (1989).

[27] A.P. Peirce, M. Dahleh, and H. Rabitz, Phys. Rev. A **37**, 4950 (1988).

[28] S. Shi, and H. Rabitz, J. Chem. Phys. **92**, 364 (1990).

[29] J.L. Krause, R.M. Whitnell, K.R. Wilson, Y. Yan and S. Mukamel, J. Chem. Phys. **99**, 6562 (1993)

[30] W. Jakubetz, B. Just, J. Manz, and H.-J. Schreier, J. Phys. Chem. **94**, 2294 (1990).

[31] C. Chen, Y-Y. Yin, and D.S. Elliott, Phys. Rev. Lett. **64**, 507 (1990); *ibid*, **65**, 1737 (1990).

[32] S.M. Park, S-P. Lu, and R.J. Gordon, J. Chem. Phys. **94**, 8622 (1991); S-P. Lu, S.M. Park, Y. Xie, and R.J. Gordon, J. Chem. Phys. **96**, 6613 (1992).

[33] N.F. Scherer, A.J. Ruggiero, M. Du, G.R. Fleming, J. Chem. Phys. **93**, 856 (1990).

[34] K.J. Boller, A. Imamoglu and S.E. Harris, Phys. Rev. Lett. **66**, 2593 (1991).

[35] B.A. Baranova, A.N. Chudinov, and B. Ya Zel'dovitch, Opt. Comm., **79**, 116 (1990).

[36] Y-Y. Yin, C. Chen, D.S. Elliott, and A.V. Smith, Phys. Rev. Lett. **69**, 2353 (1992)

[37] For a discussion of the basic principles of coherence, quantum interference, time dependence, which are fundamental to coherent control, see, e.g., J.D. Macomber, "The Dynamics of Spectroscopic Transitions", (Wiley, N.Y., 1976).

[38] This the asymptotic condition of scattering theory [see, J.R. Taylor, "Scattering Theory", (J. Wiley, N.Y., 1972)].

[39] M. Shapiro and R. Bersohn, J. Chem. Phys. **73**, 3810 (1980).

[40] M. Shapiro, J.Phys. Chem, **90**, 3644 (1986).

[41] M. Shapiro, J. Chem. Phys. **56** 2582 (1972);

[42] M. Shapiro and P. Brumer, in "Methods of Laser Spectroscopy", ed. A. Prior, A. Ben-Reuven and M. Rosenbluh (Plenum, N.Y., 1986).

[43] Z. Chen, M. Shapiro and P. Brumer, "Control of Photodissociation Branching Ratios via Two Color Frequency Tuning of Intense Laser Fields" Phys. Rev. Lett. - submitted.

[44] M.S. Child, Mol. Phys. **32**, 495 (1976).

[45] Z. Chen, P. Brumer, and M. Shapiro, J. Chem. Phys. **98**, 6843 (1993).

[46] S. Chelkowski and A.D. Bandrauk, Chem. Phys. Lett. **186**, 284 (1991); A.D. Bandrauk, J.M. Gauthier, J.F. McCann, Chem. Phys. Lett. **200**, 399 (1992)

[47] S. Chelkowski, A.D. Bandrauk, and P.B. Corkum, Phys. Rev. Lett. **65**, 2355 (1990).

[48] A. Szöke, K.C. Kulander, and J.N. Bardsley, J. Phys. B **24**, 3165 (1991); R.M. Potvliege and P.H.G. Smith, J. Phys. B **25**, 2501 (1992)

[49] E. Charron, A. Guisti-Suzor and F.H. Mies, Phys. Rev. Lett. **71**, 692 (1993).

[50] Contrary to popular expectation, perturbation theory does not imply a small total photodissociation yield. Computational results (P. Brumer and M. Shapiro - to be published) indicate that perturbation theory is quantitatively correct for dissociation probabilities as large as 0.2.

[51] M. Shapiro and P. Brumer, J. Chem. Phys. **98**, 201 (1993).

[52] N.E. Henriksen and B. Amstrup, Chem. Phys. Lett. **213**, 65 (1993); J. Chem. Phys. **97**, 8285 (1993)

[53] K.R. Wilson, private communication.

[54] A.R. Edmonds, "Angular Momentum in Quantum Mechanics" (Princeton University Press, Princeton 2nd edition, 1960).

[55] I. Levy and M. Shapiro, J. Chem. Phys. **89**, 2900 (1988).

[56] R. Bavli and H. Metiu, Phys. Rev. Lett. , ().

[57] M. Yu Ivanov, P.B. Corkum, and P. Dietrich, Laser Physics, **3**, 375 (1993).

[58] T. Nakajima and P. Lambropoulos, Phys. Rev. Lett. **70**, 1081 (1993).

[59] P.Brumer and M.Shapiro, *Adv. Chem. Phys.* **60** 371, K. P. Lawley, Ed., (Wiley-Interscience Pub., 1986).

[60] M. Shapiro and H. Bony, *J. Chem. Phys.* **83**, 1588 (1985); G. G. Balint-Kurti and M.Shapiro, Adv. Chem. Phys. **60** 403, K. P. Lawley, Ed., (Wiley-Interscience Pub., 1986).

[61] A. D. Bandrauk and O. Atabek, *Adv. Chem. Phys.* **73**, 823 (1989)

[62] The potential curves and the relevant electronic dipole moments are taken from: I. Schmidt, Ph.D. Thesis, Kaiserslautern University, 1987.

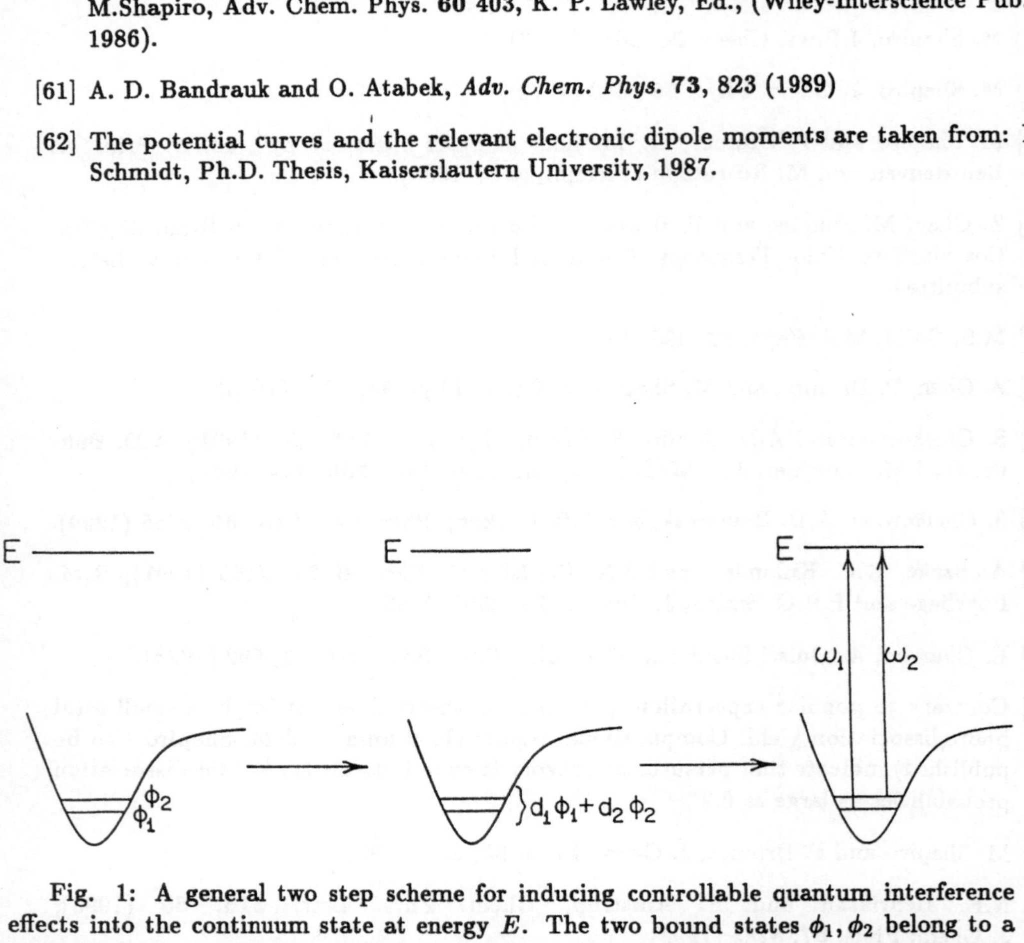

Fig. 1: A general two step scheme for inducing controllable quantum interference effects into the continuum state at energy E. The two bound states ϕ_1, ϕ_2 belong to a lower electronic state whereas the level at energy E is that of an excited electronic state. Coherence introduced in the first step is carried into the continuum. (From Ref. [12]).

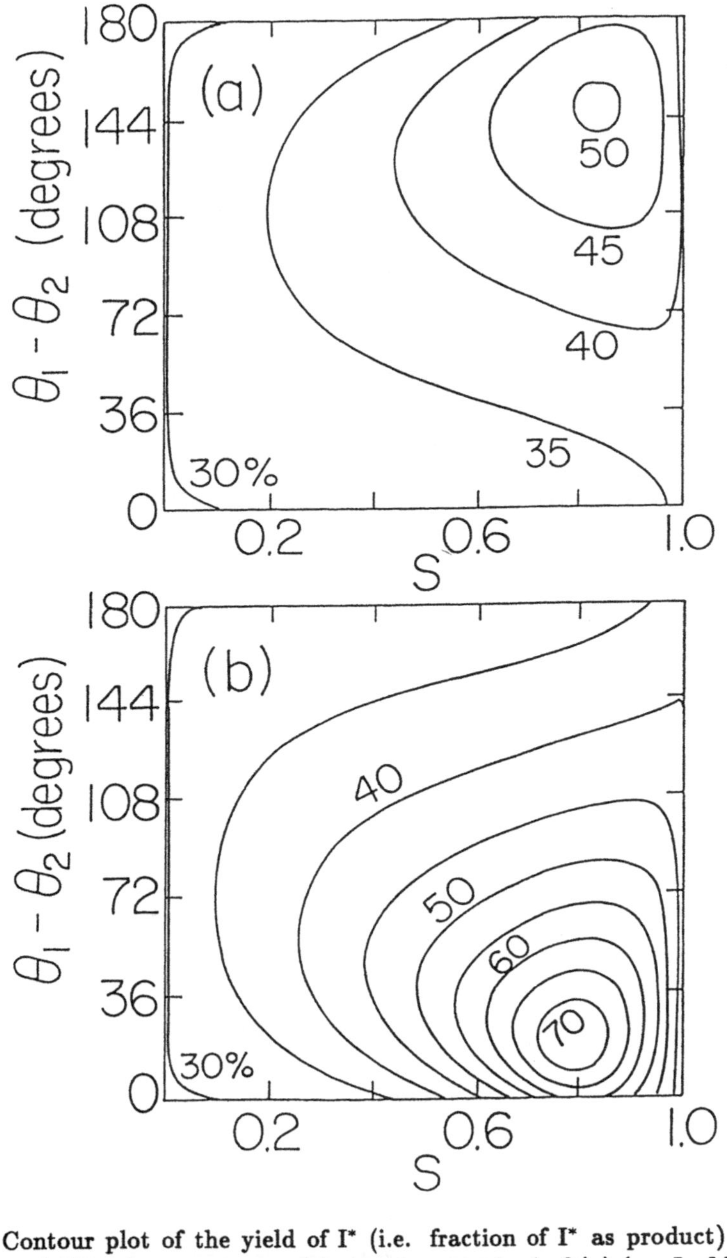

Fig. 2: Contour plot of the yield of I* (i.e. fraction of I* as product) in the photodissociation of CH_3I from a superposition state comprised of (a) $(v_1, J_1, M_1) = (0,0,0)$ + $(v_2, J_2, M_2) = (0,1,0)$ and (b) $(0,0,0) + (0,2,0)$. Here v_i, J_i, M_i are the vibrational, rotational and rotational projection quantum numbers of the ith bound state. (From Ref. [1]).

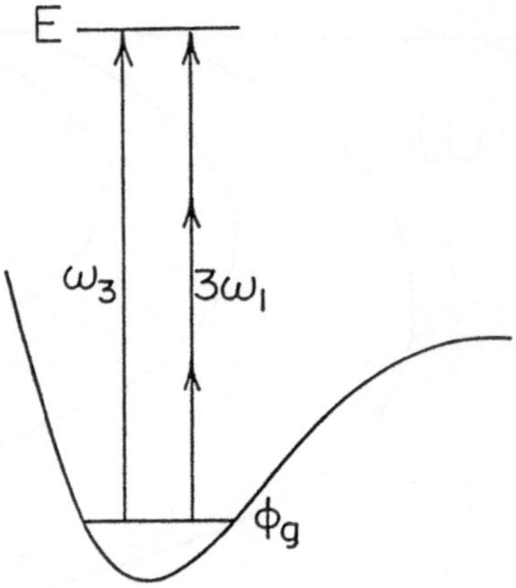

Fig. 3: A multiple optical-route scheme to inducing controllable quantum interference effects into the continuum state at energy E. Here the level ϕ_g is a bound state of a lower electronic state and that at E is a continuum state of the excited electronic state. Simultaneous application of frequencies ω_1 and $\omega_3 = 3\omega_1$ leads to interference in the continuum state. (From Ref. [12]).

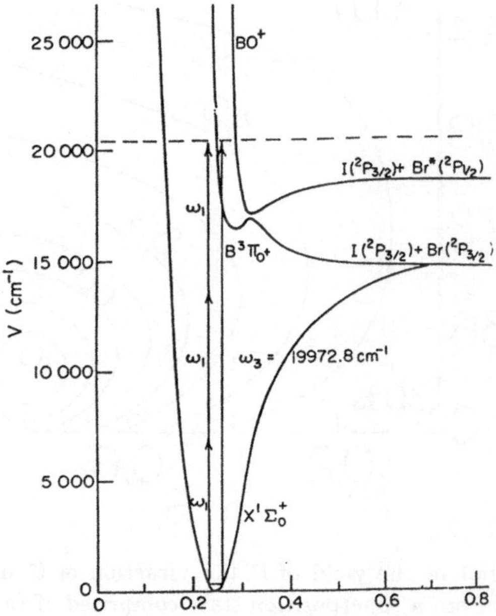

Fig. 4: IBr potential curves relevant in the one-plus-three photon induced dissociation. (From Ref. [14]).

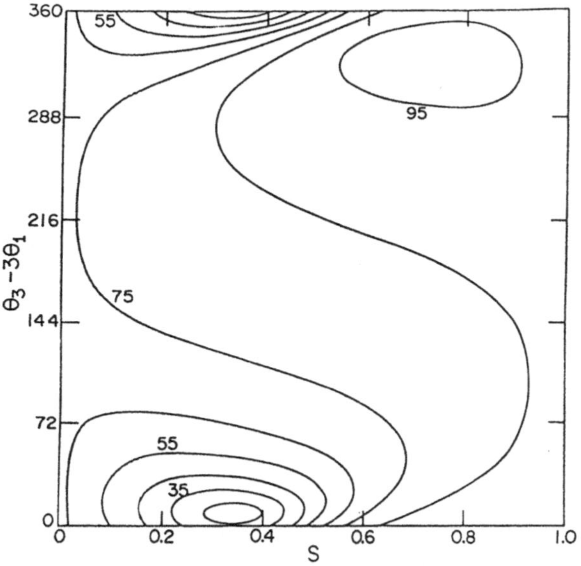

Fig. 5: Contour plot of the yield of Br*($^2P_{1/2}$) (percentage of Br* as product) in the photodissociation of IBr from an initial bound state in $X^1\sum_0^+$ with $v = 0$, $J_i = 1$, $M_i = 0$. Results arise from simultaneous (ω_1, ω_3) excitation ($\omega_3 = 3\omega_1$), with $\omega_1 = 6657.5\ \mathrm{cm}^{-1}$ (From Ref. [14]).

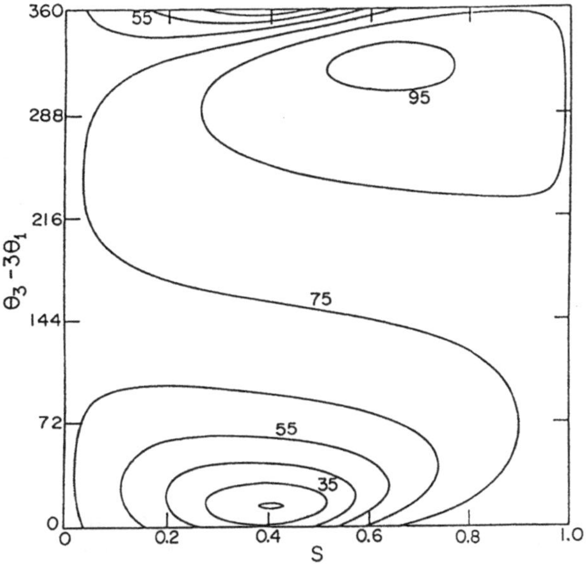

Fig. 6: As in Fig. 5 but for $v = 0$, $J_i = 42$, $\omega_1 = 6635.0\ \mathrm{cm}^{-1}$. and M-averaged ($\epsilon_0 = 1/8$). (From Ref. [14]).

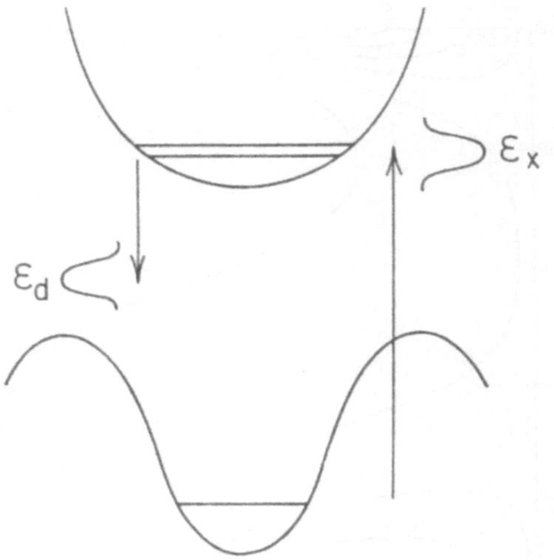

Fig. 7: Coherent radiative control via a picosecond pulse scheme. In this case a single level is excited with a laser pulse to produce a superposition of two bound states in an excited electronic state. Subsequent deexcitation of this state to the continuum of the ground state allows control over the reaction on the ground state surface. (From Ref. [12]).

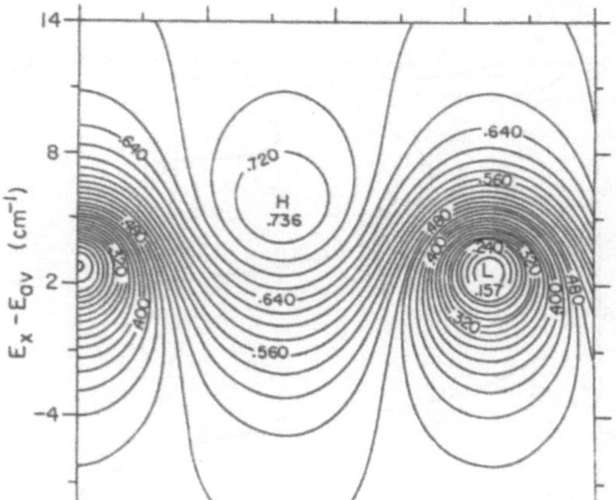

Fig. 8: Contour plot of the DH yield in the reaction D + H_2 → DH + H. The control parameters are the difference in energy between the excitation pulse center E_x and the average of the energy of the two excited levels E_{av} and the time between the pulses τ. Although the abscissa begins at zero and spans approximately one period, the results are periodic in the delay time. (From Ref. [9]).

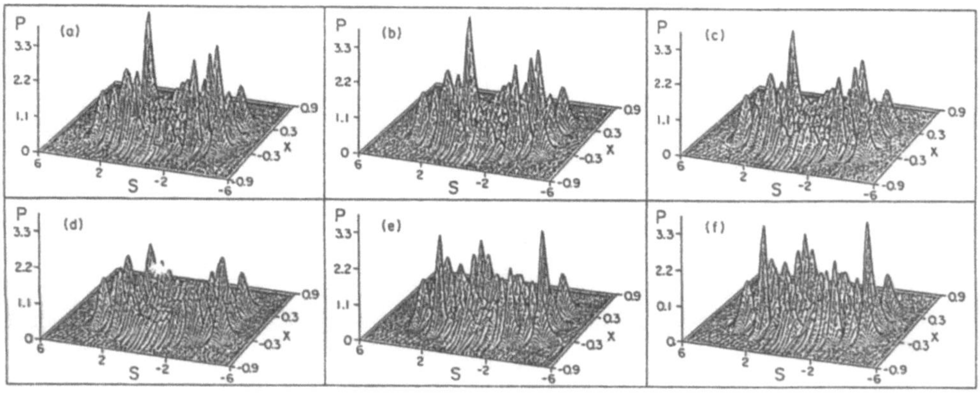

Fig. 9: Time evolution of the square of the wavefunction for a superposition state comprised of levels 56 and 57 of the G1 surface of H_3. The probability is shown as a function of the reaction coordinate S and orthogonal distance x at times (a) 0, (b) 0.0825 psec, (c) 0.165 psec, (d) 0.33 psec, (e) 0.495 psec, (f) 0.66 psec, which correspond to equal fraction of one half the period $2\pi/\omega_{2,1}$. (From Ref. [9]).

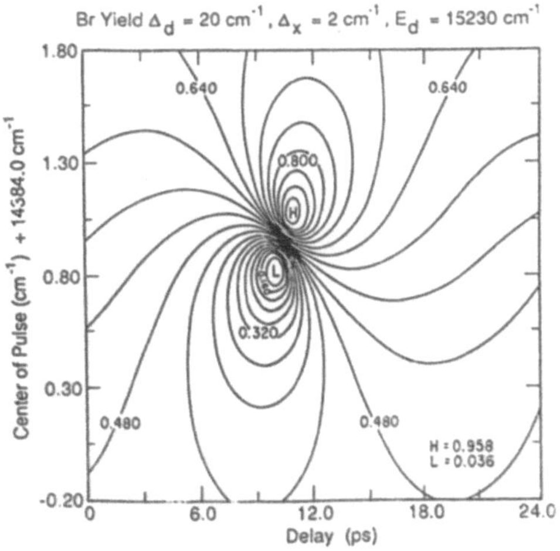

Fig. 10: Computed control over the Br yield as a function of E_x - the excitation pulse detuning, and τ - the time delay between the pulses. Parameters are of Fig. 6, Ref. [11]. (From Ref. [11]).

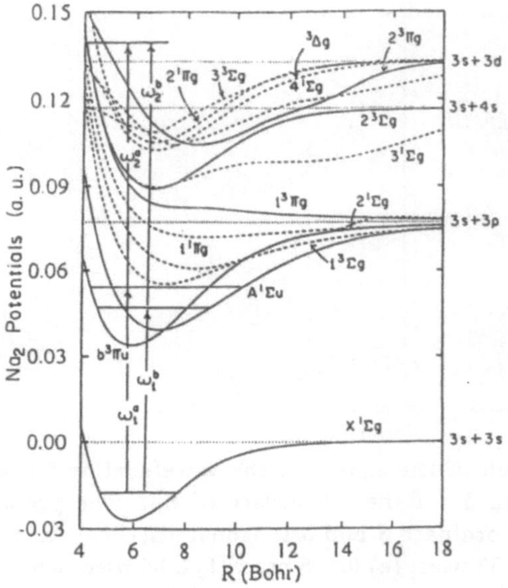

Fig. 11: Two resonant two-photon paths in the photodissociation of Na_2. (From Ref. [18]).

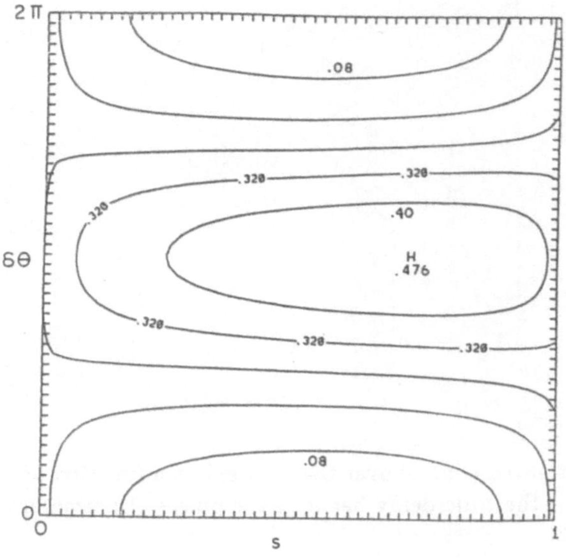

Fig. 12: Contours of equal Na(3*p*) yield. Ordinate is the relative laser phase and the abscissa is $S = x^2(1 + x^2)$ where x is the field intensity ratio. Here $\omega_0 = 627.584$, $\omega_+ = 611.207$, $\omega_- = 644.863$ nm and $\eta = 0.5$. See Ref. [18] for a discussion of η which can be used to minimize background contributions. (From Ref. [45]).

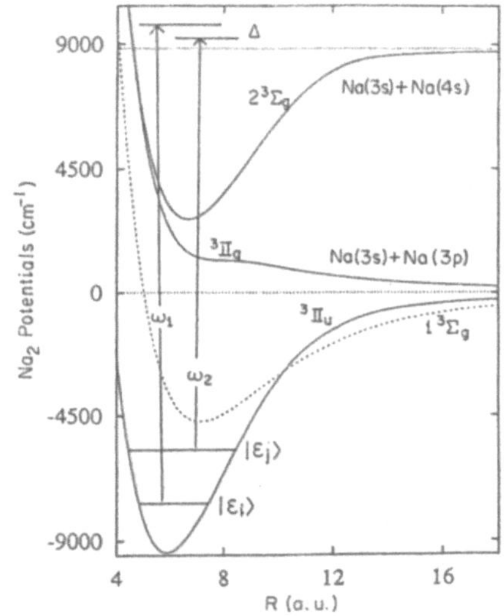

Fig. 13: Control scenario applied to the photodissociation of the $^3\Pi_u$ state of Na_2. For the case considered in this paper $|\,E_i\,\rangle$ corresponds to $v = 19$ with $E_{v=19} = -6512.8$ $cm^{-1}$and $|\,E_j\,\rangle$ is $v = 31$ with $E_{v=31} = -4966.04$ cm^{-1}.

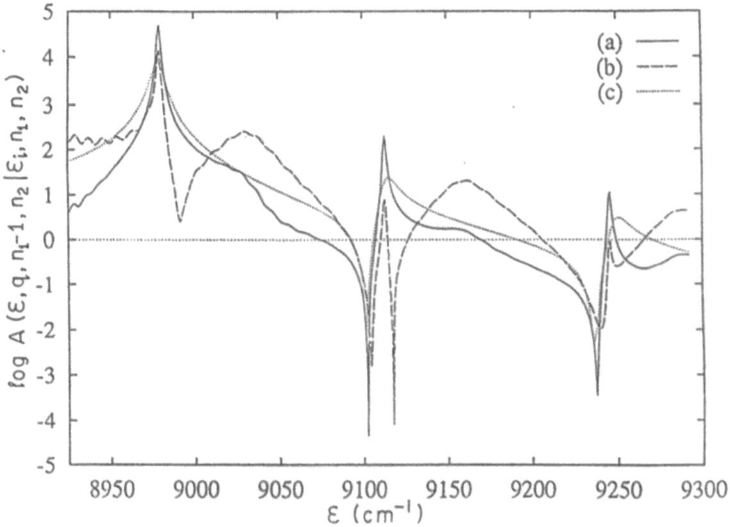

Fig. 14: log $A(E, q, n_k|\epsilon_i, n_i)$ as a function of E (where the Na(3p) + Na(3s) asymptote defines the zero energy). (a) Na(3s)+Na(3p) product and (b) Na(3s)+Na(4s) product, with both lasers on; (c) for the Na(3s)+Na(4s) product with only one laser (ω_1) on. Here $\omega_1 = 15,456$ cm^{-1}, $\omega_2 = 13,964$ cm^{-1}, $I_1 = 5.5 \times 10^9$ W/cm^{-1} and $I_2 = 3.51 \times 10^{10}$ W/cm^{-1}.

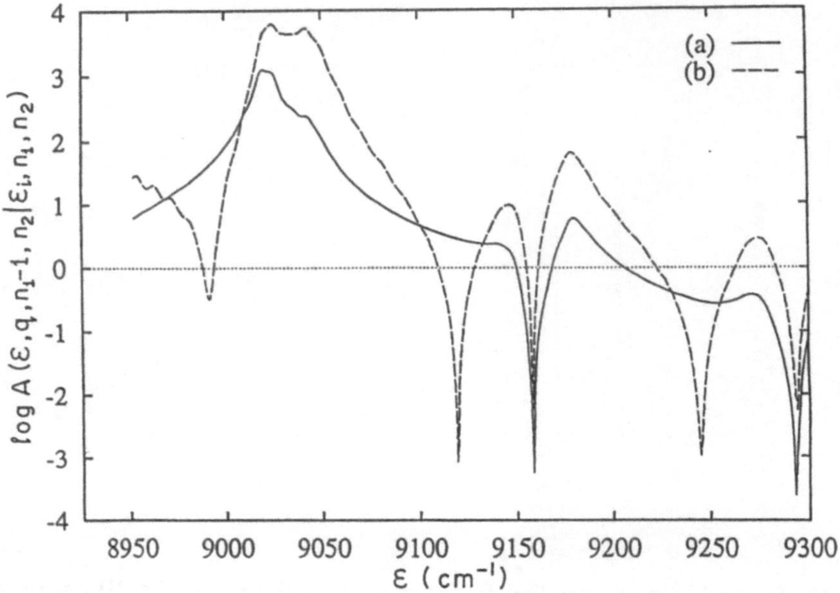

Fig. 15: (a) and (b): As in Fig. 15 but with $\omega_1 = 15,511$ cm^{-1}.

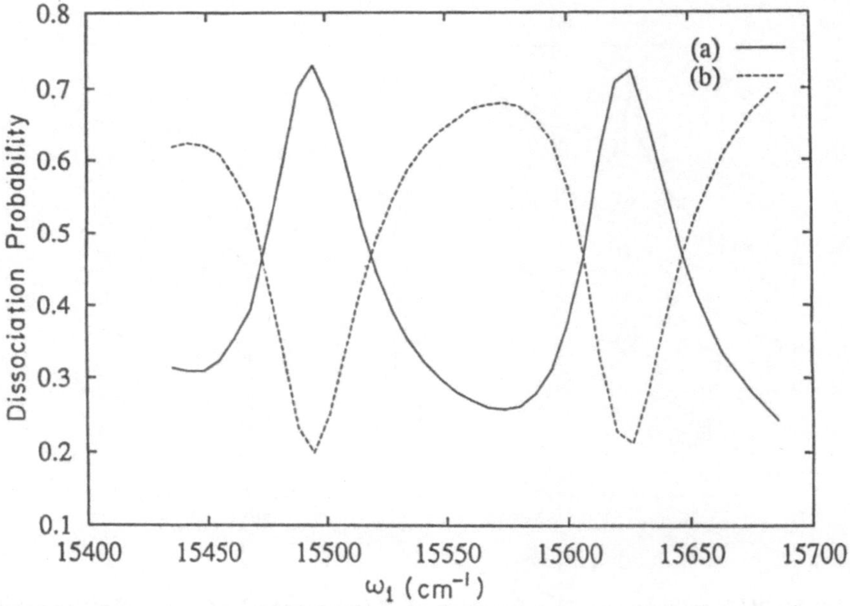

Fig. 16: Probability of forming (a) Na(3s)+Na(3p) and (b) Na(3s)+Na(4s) as a function of ω_1, with $\omega_2 = 13,964$ cm^{-1}, $I_1 = 8.7 \times 10^9$ W/cm^{-1} and $I_2 = 3.51 \times 10^{10}$ W/cm^{-1}.

ADAPTIVE FEEDBACK CONTROL OF MOLECULAR MOTION

H. RABITZ
Princeton University
Department of Chemistry
Princeton, New Jersey, U.S.A., 08544

Abstract

Control of molecular-scale events, including chemical reactions, has been a long sought-after goal. A central problem is to design control fields such that particular molecular objectives are achieved while suppressing undesirable processes. The techniques of optimal control theory within quantum mechanics provides the framework for carrying out the designs. By replacing the model of the molecule in the computer with the actual molecule in the laboratory, various design problems can be overcome. In this fashion, the molecule acts as an analog computer, to solve its own dynamical equations in appropriate pump-probe experiments performed iteratively, and guided by a learning algorithm to ultimately achieve the desired molecular control objective. The practicality of this approach and some future directions of the field will also be discussed.

1. Introduction

Although using lasers to control events at the molecular scale has been a long sought-after dream, until recently, little has come from this effort, except considerable frustration in the laboratory. Much of the prior theoretical research was largely directed towards understanding intramolecular dynamics. A rebirth of the molecular control field is now under way, due to two developments. First, the general theoretical and conceptual basis for designing control laser fields is now understood[1,2], and second, in the laboratory, the necessary laser tools (principally, optical pulse shaping) are now becoming available.[3] The subject is wide open for study again, and the accelerating activity suggests that many exciting discoveries are ahead.

Historically, the goal in this area has been one of controlling molecular motion. One of the difficulties is that we do not often know the input Hamiltonian reliably well, although the tools of adaptive feedback control[4] are capable of circumventing this problem. The determination of Hamiltonian features (i.e., the potential surface) is an additional goal of considerable importance for a basic understanding of molecular structure, dynamics, and spectroscopy. Recent research strongly suggests that adaptation of some of the same tools for controlling molecular motion can be turned around and used for molecular inversion purposes to extract potential surfaces.

E. Yurtsever (ed.), Frontiers of Chemical Dynamics, 181–193.

The remainder of this paper will present the basic theoretical foundations for molecular control, with the aim of arguing the necessity of employing adaptive feedback for this purpose. The essential aspects of the emerging laboratory techniques for molecular control will be discussed at a level of detail appropriate for the theoretical considerations. Some indications of the future directions of the field will also be suggested.

2. Foundations of Molecular Control

Control of molecular motion may be viewed as active intervention into quantum dynamics phenomena. Thus, given the Hamiltonian $H = H_0 + U(\underline{x}, t)$, the question is whether we may find an external control interaction $U(\underline{x}, t)$ to steer the quantum mechanical system from an initial state $|\psi(0)\rangle$ to a desired final state $|\psi(T)\rangle$ at the target time T. Here, H_0 refers to the Hamiltonian of the molecule without control. Although the control $U(\underline{x}, t)$ is written as a fully general function of position $\underline{x}$ and time t variables, in the case of optical fields, it typically has the form of an electric dipole interaction $U(\underline{x}, t) \rightarrow \mu(\underline{x})\cdot\varepsilon(t)$, where $\mu(\underline{x})$ is the dipole moment operator and $\varepsilon(t)$ is the control optical electric field. The electric dipole interaction forms the foundation of molecular spectroscopy, and the only distinction here is that the field $\varepsilon(t)$ is being applied to actively intervene into the molecular dynamics, rather than just determine the resonance frequencies and intensities.

Other forms of intervention in quantum mechanics can also be envisioned, including the case where $U(\underline{x}, t)$ is just a spatially-dependent function $V(\underline{x})$. Such a circumstance of control at the molecular scale seems far-fetched since the time-independent Hamiltonian is fixed by nature. In the case of electron transport in semiconductors, one may envision control of this type by growing or otherwise modifying the solid state material.[5] The potential $V(\underline{x})$ may be created by design to manipulate the electron motion. Ultimately, it may be possible to combine such solid state stationary control with the imposition of optical fields, to actually realize the general control framework $U(\underline{x}, t)$. This paper will primarily concern itself with the circumstance of optical control, as this corresponds to the presently most active aspect of the subject.

Quantum mechanics is inherently a wave phenomena and therefore, involves interference processes. We may schematically understand the process of control as one of steering the time-dependent wavepacket

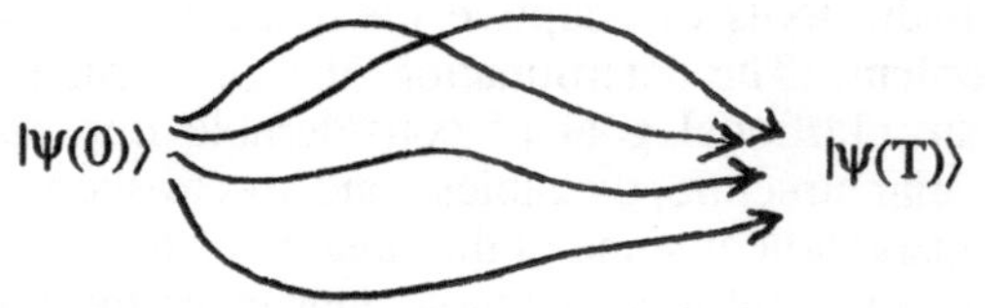

over many, if not an infinite number of, paths from the initial state to the final state. These paths, indicated by the multiple arrows above, set up an interference pattern. The fundamental principle of molecular control is that of manipulating the constructive and destructive interferences such that the wavefunction has high amplitude in a particular desired state, with minimal amplitude elsewhere, perhaps including along the entire evolutionary history of the wavepacket. Thus, the problem of control reduces to determining the field ε(t) that can effectively manipulate these interferences.[1,2]

Various physical regimes can be envisioned with the above control paradigm in mind. In one limit, only a few molecular states are involved and the optical fields are weak. In this regime, perturbation theory, and possibly, intuitive arguments, may suggest field structure for practical control. On the other hand, many molecules have high densities of states. Secondly, stronger fields open up the prospect for more active control, and especially for creating high yields of desired final states.[6] In the latter regime, manipulation of interferences, again, forms the mechanism of control, but the strong mixing induced by the field would likely make intuitive arguments unreliable. Regardless of the regime, there is a single comprehensive theoretical framework of quantum mechanical molecular control, which covers all circumstances. One goal of this section will be the establishment of this framework, and a number of other basic physical issues.

2.1 CONTROLLABILITY

Controllability addresses the general question of whether any conceivable control field ε(t) exists to steer a quantum mechanical system from initial state $|\psi(0)\rangle$ to an arbitrary final state $|\psi(T)\rangle$. Various other restricted forms of the controllability question can be raised, such as whether it is possible to steer an initial state to an expectation value $\langle\psi(T)|O|\psi(T)\rangle$ of the observable operator O. An algorithm for assessing the controllability might not provide an explicit field. Thus far, within quantum mechanics, controllability has been addressed within the former framework of controlling the wavefunction. Presently, there is no constructive algorithm available for assessing controllability for arbitrary quantum mechanical problems described by the partial differential equation (i.e., an infinite dimensional problem)

$$i\hbar\frac{\partial}{\partial t}|\psi(t)\rangle = [H_0 + \mu\cdot\varepsilon(t)]|\psi(t)\rangle \quad (1)$$

However, for two restrictive cases, information is available. First, in the regime where the control field is weak, a perturbation approximation to eq. (1) may be generated, based on the expansion $|\psi(t)\rangle = |\psi_0(t)\rangle + |\psi_1(t)\rangle + \ldots$, resulting in

$$i\hbar\frac{\partial}{\partial t}|\psi_1(t)\rangle = H_0|\psi_0(t)\rangle + \mu\cdot\varepsilon(t)|\psi_0(t)\rangle \quad (2)$$

where $|\psi_0(t)\rangle$ satisfies the unperturbed time-dependent Schrödinger equation. In the case of eq. (2), an explicit algorithm exists[7] to test for controllability, with the assumption that the initial state is an eigenstate of H_0. The physical conclusion from the analysis is that within first order perturbation theory, a quantum mechanical system will be controllable, provided the final states are non-degenerate and are dipole-coupled to the initial state. This result is intuitively grounded in simple physical arguments. The second case, where a controllability algorithm exists, corresponds to a finite dimensional analog of eq. (1) (i.e., a projection of the equation onto a finite basis set). In this case, the Lie algebra generated by the operators H_0

and $\mu(t)$ can be used to establish a controllability test. Only very limited applications have been made with this more general algorithm, at this point.

Knowledge of the controllability of a given quantum mechanical system is of fundamental importance, as it dictates whether a laboratory experiment has, in principle, the possibility of exactly achieving a desired target state. However, even if a system is not totally controllable, that does not imply all is lost. Arriving in the general neighborhood of a target state will often be quite acceptable. For example, if the goal is to steer a chemical reaction product out channel 1 *versus* that of channel 2, it is not necessary to exactly achieve a 100% yield out of channel 1. Thinking along these lines naturally leads to considering optimization algorithms for designing control fields which can achieve the final state as best as possible. This matter will be returned to again below.

2.2 TRACKING OF AN EVOLVING QUANTUM MECHANICAL STATE

Before consideration of the optimization process referred to above, a very simple perspective on wavepacket control occurs if the quantum mechanical system is exactly controllable. Assuming the latter situation will be the case, the basic premise of tracking can be stated as follows: Given a prescribed trajectory $\langle\psi(t)|O|\psi(t)\rangle = f(t)$, $O \leq t \leq T$, find the control field $\varepsilon(t)$ that exactly meets the molecular wavepacket trajectory f(t). One may view tracking as a means for design of the control field $\varepsilon(t)$, with the given path f(t) prescribed. In turn, this process may also be viewed as a test of one's intuition for the choice of f(t). The choice would be deemed as good if it leads to a physically attractive laboratory realizable control $\varepsilon(t)$.

One of the main attractions of tracking is the resultant simple algorithm associated with the inversion $f(t) \rightarrow \varepsilon(t)$. The algorithm[8] starts by utilizing the Heisenberg equation of motion

$$i\hbar\frac{\partial}{\partial t}\langle O(t)\rangle = \langle[H_0, O]\rangle + \varepsilon(t)\langle[\mu, O]\rangle \tag{3}$$

where, for simplicity, we have assumed that the operator O is not explicitly a function of time and the electric field is polarized along a single fixed direction. Further assuming that $\langle[\mu, O]\rangle$ is not identically zero, we may then solve eq. (3) for the desired field

$$\varepsilon(t) = \left[i\hbar\frac{\partial}{\partial t}f(t) - \langle[H_0, O]\rangle\right] \Big/ \langle[\mu, O]\rangle \tag{4}$$

where we have substituted in the demand of exact tracking $\langle O(t)\rangle = f(t)$. Equation (4) is not a closed solution, as the expectation values on the right-hand side depend on the unknown wavefunction. However, we know that the wavefunction must satisfy Schrödinger's equation (1), and we may substitute the field in eq. (4) into this equation to obtain

$$i\hbar\frac{\partial}{\partial t}|\psi_1(t)\rangle = [H_0 + \mu\cdot\varepsilon(t,\psi)]|\psi(t)\rangle \tag{5}$$

The field in eq. (5) contains the wavefunction in its argument to emphasize that the field in eq. (4) has an explicit dependence on the wavefunction. The final step of the algorithm consists of solving eq. (5) from the initial condition $|\psi(0)\rangle$, and substituting the result into eq. (4). By construction, this algorithm will produce a control field that exactly meets the tracking demand. However, there is no way of *a priori* telling if the resultant control field will be physically acceptable, without actually numerically executing the algorithm.

There are two points to note in this tracking approach to molecular control. First, molecular tracking is a highly nonlinear problem. That is, although we are dealing with Schrödinger's equation, when we pose the problem in the inverse sense, given the objective $\langle\psi(t)|O|\psi(t)\rangle = f(t)$, $0 \leq t \leq T$ and find the control $\varepsilon(t)$, then the process is not linear. This point is evident in eq. (5). Superficially, it is Schrödinger's equation, but in fact, it is highly nonlinear due to wavefunction dependence in the field. This point of nonlinearity of molecular control will arise again in the different format of molecular optimal control below. Second, the tracking algorithm calls for no iteration of the dynamical equations. One single solution of the nonlinear eq. (5) suffices to give the control field meeting the objective in eq. (4).

Thus far, experience with utilizing tracking has shown that there is an enormous demand placed on the quality of one's physical intuition on the input path f(t) in order for the control field $\varepsilon(t)$ to have an acceptable physical form. For example, it may happen that a control path f(t) gives a field with undesirable characteristics, while only a slight shift of path could alter that circumstance. Once again, this suggests that an optimal control approach would be attractive. Another general difficulty with tracking arises when there are multiple objectives of interest $\langle O_i(t)\rangle = f_i(t)$, $i = 1, 2, \ldots$. Since we have only one control field $\varepsilon(t)$, it is evident that generally, it will not be possible to simultaneously specify multiple objective paths and expect to get a single consistent control field. However, it is possible to slightly modify the algorithm to ask for a single field that does as best as possible collectively treating the competing paths. Notwithstanding these various objections, tracking has an enormous appeal due to its simplicity, and it is anticipated that it will receive further attention.

2.3 OPTIMAL CONTROL OF QUANTUM DYNAMICS: ACHIEVING THE BEST POSSIBLE SOLUTION

In realistic molecular control problems, there inevitably will be competitive demands placed on the molecular motion, and possibly also including laboratory laser constraints. For example, while we desire to reach the objective $\langle\psi(T)|O|\psi(T)\rangle$ at the target time T, we may also desire that the expectation value $\langle\psi(t)|O'|\psi(t)\rangle$ of another operator O' be as small as possible over the entire control interval $0 \leq t \leq T$. An example of this situation might arise in a triatomic molecule, where the objective operator O is one bond length, and the penalty operator O' is the other bond length. Many other situations can also be envisioned, including hard penalties of some expectation values being constrained during the dynamics. The full generality of molecular optimal control will not be expressed here, but a typical form will be treated, for illustration.

Given that molecular control is often a problem of balancing competitive factors, the natural approach is to introduce a variational principle through a cost functional J

$$J = (\langle\psi(T)|O|\psi(T)\rangle - O^*)^2 + \int_0^T dt\, |\langle\psi(t)|O'|\psi(t)\rangle|^2 + \int_0^T \varepsilon^2(t)dt$$

$$+ \text{Im}\int_0^T \langle\lambda(t)|i\hbar\frac{\partial}{\partial t} - H_0 - \mu\cdot\varepsilon(t)|\psi(t)\rangle dt \qquad (6)$$

where O^* is the desired target value. By construction, $J \geq 0$, and the goal is to minimize the cost functional. The first term in J corresponds to the desired physical objective, while the second and third terms are penalties respectively associated with an additional operator O', and the optical field fluence. The last integral in eq. (6) introduces the Lagrange multiplier state function $|\lambda(t)\rangle$ for the purpose of assuring that Schrödinger's equation is satisfied as a constraint. The cost functional depends on the three unknowns $|\lambda(t)\rangle$, $|\psi(t)\rangle$, and $\varepsilon(t)$. Setting to zero the first variations of J, with respect to these functions, produces three design equations

$$i\hbar\frac{\partial}{\partial t}|\psi(t)\rangle = [H_0 + \mu\cdot\varepsilon(t)]|\psi(t)\rangle \quad , \quad |\psi(0)\rangle \qquad (7)$$

$$i\hbar\frac{\partial}{\partial t}|\lambda(t)\rangle = [H_0 + \mu\cdot\varepsilon(t)]|\lambda(t)\rangle - \langle\psi(t)|O'|\psi(t)\rangle O|\psi(t)\rangle \qquad (8a)$$

$$|\lambda(T)\rangle = 2(\langle\psi(T)|O|\psi(T)\rangle - O^*)O\,|\psi(t)\rangle \qquad (8b)$$

$$\varepsilon(t) = \text{Im}\langle\lambda(t)|\mu|\psi(t)\rangle \qquad (9)$$

Some general comments are warranted about these control field design equations. First, these equations are nonlinear, including the Schrödinger equation in eq. (7). This follows since the field in eq. (9) itself depends on the wavefunction and the Lagrange multiplier function. Second, the Lagrange multiplier function contains an additional nonlinearity due to the term with the penalty operator O'. These design equations have the structure of two coupled generalized nonlinear Schrödinger equations. Equations of this overall type have been examined in other areas of physics and applied mathematics,[9] with the curious point that they can have both highly regular and irregular solutions. Both cases may also arise in the present context. There is also the tantalizing prospect that perhaps by including the generalized control interaction $U(\underline{x},t)$, solitonic or non-dispersive wavefunction character may arise. Another curious point regarding these equations is that eq. (7) is an inital value problem, while eq. (8a) is a final value problem, through the condition in eq. (8b). Thus, the two sets of coupled nonlinear Schrödinger equations generate a boundary value problem in the time variable. This boundary value character in time leads to the possibility of there being multiple solutions corresponding to the generalized eigenvalue nature of the equations. The existence of multiple solutions has been proven under rather mild assumptions. From a simple physical perspective, it is not surprising that multiple solutions can exist, as basically, we start at an initial state $|\psi(0)\rangle$, and strive to meet a final objective $\langle\psi(T)|O|\psi(T)\rangle$ without any path specified in between. Each of the multiple solutions correspond to a different path from the starting point to the final point.

Each solution of the design equations will produce a control field in eq. (9) corresponding to a local minimum of the cost functional in eq. (6). In practice, it is not necessary to find the absolute minimum of the cost functional, but rather a physically viable solution. The primary difficulty of achieving a local solution is not due to the nonlinearity of the design equations. Rather, their boundary value nature in time necessitates iteration or some type of shooting method. This point can simply be seen, since we cannot integrate eq. (7) forward in time without knowledge of the control field, which is initially not known. Numerous iterative schemes can be employed for solving the design equations, and it remains to be seen what approach is best for this purpose. Without awaiting such an analysis, numerous examples have been treated as illustrations in the literature.

3. General Conclusions from Molecular Control Design Studies

Many design studies of molecular control have been conducted in the past few years by several researchers.[6,10] In particular, case studies have considered control of electronic, rotational, and vibrational degrees of freedom, and in some cases, all three simultaneously. Objectives of chemical dissociation, as well as coherent state preparation have been examined. The most elaborate studies have naturally employed simple molecules. However, in the case of model Hamiltonians, particularly that of harmonic oscillators, polyatomic molecules with ten or more atoms have been subject to control designs. Naturally, each molecule and each objective within a given molecule produces field-driven dynamical evolution of a particular nature. However, some general points are beginning to emerge from these studies, and it is these points that are perhaps most important at this juncture in the development of the subject. A list of some key observations is given below:

(1) To successfully achieve control, the field must work cooperatively with the dynamical capabilities of the molecule.

(2) Successful control at high yields of transfer from the initial state calls for the development of significant action, such that $\mu\varepsilon T/\hbar \gtrsim \pi$, where ε is the characteristic mean amplitude of the control field. This relation also shows that the field amplitude and the target time are inversely related. Control over short periods of time will call for intense fields, while the opposite may be true for long control intervals.

(3) Practical trade-offs can exist. Driving the target to its objective in a short period of time may avoid deleterious side effects, but the resultant intense fields may produce their own complexities for creation in the laboratory. Second, quantum mechanical designs with intense fields may be difficult to carry out due to possible uncontrolled electronic polarizations. Weak fields

may have a strong resonant character, with a modest number of frequency components. Although this may be desirable for weak fields, coherence needs to be maintained for a longer period of time, and it may also be difficult to fight against certain competing molecular dynamical processes over long periods of control.

(4) Sub-optimal or local solutions to the control cost functional may be more desirable than finding the absolute minimum. This comment is based on the fact that to achieve perfect control may call for undue complexity in the control field, while backing off even a modest amount may significantly diminish the structure of the field. The introduction of frequency filtering or other constraints on the form of the field can aid in finding sub-optimal solutions.

(5) Robustness to Hamiltonian uncertainties or field errors is not guaranteed, unless such requests are included in the design process. Including robustness can dramatically improve this important property, by guiding the solution to have desirable characteristics.

(6) The computations involved in creating designs can be quite intense, as they include solving Schrödinger's equation, as well as its adjoint, repeatedly in an iterative mode until convergence is achieved. Nevertheless, it is reasonable to state that if a problem may be reliably modelled, then it may be subject to control design efforts.

(7) Perhaps the most important conclusion is that numerous test studies have produced viable control solutions yielding high quality results. Due caution is called for, since the systems are often simple and are modelled with less complexity than will actually occur in the laboratory. Nevertheless, given that there are multiple solutions and likely, many poor ones amongst them, it is very encouraging that quality answers have been found without an undue degree of searching.

These general points may be drawn together to form an overall positive conclusion for the future of molecular control. In particular, for at least simple systems where the Hamiltonian is known well (e.g., rotational motion), the design equations may be successfully solved. In such cases where the resultant fields may be generated in the laboratory, it is reasonable to expect success. However, a number of serious issues still remain to be treated. In general, (a) we do not know the Hamiltonian well for virtually any molecule, (b) solving the dynamical equations can be quite difficult, and (c) errors will be created in the laboratory-generated fields. It is these issues that are addressed by adaptive feedback molecular control.

4. Adaptive Feedback Molecular Control: A Necessity

The type of control introduced above is referred to as open loop, in the sense that a design is achieved from computational modelling. The resultant field is generated in the laboratory, and applied to the molecular sample for its action, with an observation finally performed to assure that the desired state is achieved. There is no feedback in the process, and the effort will either be successful or not. Given the overall complexities of design, including points (5) and (6) raised above, there are relatively few molecular systems that likely will be successfully controlled in an open loop fashion. Indeed, traditional engineering control applications are rarely performed open loop for the same reason. Rather, some sort of feedback is introduced in order to stabilize the problem, or generally, to deal with the numerous uncertainties involved. Feedback traditionally suggests observing the system during its control evolution, and feeding corrections back to the control laser, to account for deviations that may arise. Such a literal achievement of feedback at the molecular scale seems highly unlikely, given the ultrafast nature of molecular events. That is, although ultrafast pump (control) and observations (probe) could be performed, actually, digesting the results and feeding them back to correct the control laser does not seem to be feasible. On the other hand, an *adaptive* variation on this theme is quite feasible, and will generally be necessary for molecular control to generally become a reality.

The key observations leading to an adaptive molecular control feedback algorithm are as follows:[4]

(1) Although we may not know the Hamiltonian of any given molecule to a high degree of precision, the actual molecule itself knows its Hamiltonian quite precisely.

(2) Although we may have difficulty solving Schrödinger's equation, the actual target molecule solves its own dynamical equations precisely and in real time, when a control field is applied.

(3) Control field generation is becoming a more routinely available laboratory tool, and most importantly, it may be computerized. The latter point is important since it then becomes possible to literally dial up one pulse shape after another at a very high duty cycle. Many, if not thousands or millions of, distinct pulse shapes may be generated per second.

(4) A canonical control experiment (at least as they have historically been posed) will consist of applying a control field, followed by a probe pulse which can detect if the desired state, say (A), is created *versus* the undesirable state, say (B). Typically such a simple distinction does not call for sophisticated off-line analysis of the data to conclude whether the desired state is achieved.

(5) Fast learning algorithms of various types exist. A learning algorithm is software that observes the action of the control pulse, and ideally, compares that information with all previous control experiments, to then suggest a new experiment to give improved results.

All of these facts may be integrated together, to suggest a closed loop adaptive feedback algorithm for achieving molecular control, as depicted in Figure 1. The distinction between adaptive molecular feedback and traditional feedback[11] lies in the fact that, in the molecular case, a sequence of experiments replaces the real time feedback events.

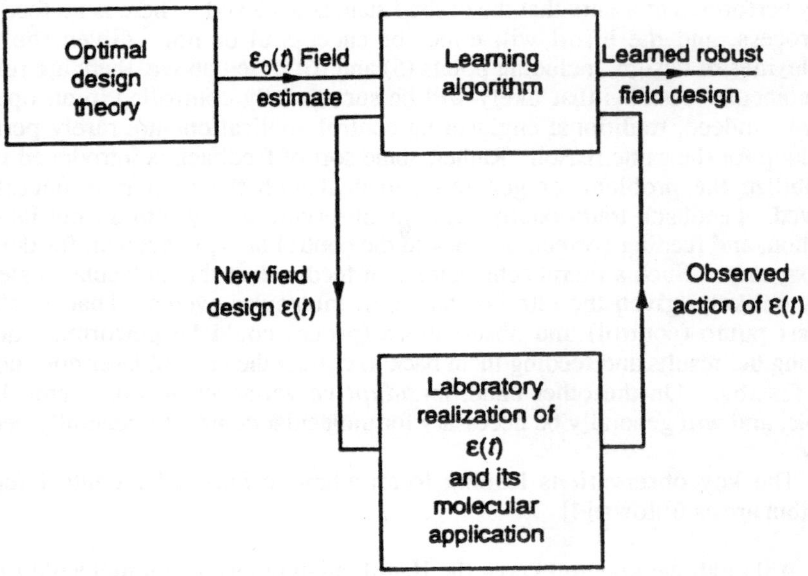

Figure 1: A schematic of an adaptive learning algorithm approach for teach lasers to control molecules. The algorithm is initiated by an optimal control estimate $\varepsilon_0(t)$ of the control field, followed by its laboratory refinement in a computer-controlled sequence of experiments, coupled to a pattern-recognizing learning algorithm.

A number of simulations have been carried out on the molecular adaptive feedback algorithm above, and the results are highly encouraging. Perhaps most intriguing is the case[4] where an optimal control design estimate $\varepsilon_0(t)$ was not performed to simulate an example of "black box" control without any knowledge of the molecular Hamiltonian. This particular illustration involved controlling rotational motion, and employed a genetic algorithm, with the results being highly successful. In general, a number of factors need to be considered when carrying out adaptive molecular feedback control. First, it is essential that the probe actually detect a signal! This latter circumstance may not happen, especially in complex situations, where, initially, the pumped molecule has essentially no component of amplitude in the desired final state. Thus, it may be necessary to introduce intermediate targets, or at least probe intermediate evolution, and accordingly, give guidance to a new sequence of experiments. The nature of the learning algorithm can also be important. Gradient-based methods, genetic algorithms, and simulated annealing have all been tried with varying degrees of success, and it remains to be seen which approach is best. The role of the initial field design $\varepsilon_0(t)$ may be increasingly important for more complex molecules, where the molecular motion might let the wavepacket literally wander off. The point above about performing intermediate probe measurements may also be helpful in this regard. Finally, it is important to consider the effects random and systematic errors in the laboratory. A study of these points has recently shown that a high degree of systematic errors in the control field may be tolerated, while random errors could become serious. On the other hand, noise in the detector may be tolerated by performing repeated measurements, to improve the signal-to-noise ratio, while systematic errors in the probe could steer the results in the wrong direction. These are standard points in any feedback process. In summary, it is argued that adaptive feedback will be an integral component of virtually all future molecular control experiments. This approach draws on the best capabilities of theory and laboratory tools.

5. Molecular Control as a Precision Tool for Learning About Molecules

The notion of molecular control suggests the creation of a particular state, or making some event happen. As we have argued, from a design perspective, a key limiting factor is lack of full knowledge about molecular Hamiltonians. Fortunately, the adaptive feedback route is a way of circumventing this problem. Nevertheless, knowledge of Hamiltonians, particularly including potential energy surfaces and optical coupling coefficients (e.g., transition dipole moments) is valuable for fundamentally understanding molecules and their chemical bonding properties. Thus, a major portion of laboratory spectroscopic and other types of measurements are often performed for the purpose of inversion to extract such information. It is not sufficient to perform the experiments alone, as an algorithm is required to invert the laboratory data in a stable and reliable fashion. This latter algorithm provides the critical link between the laboratory measurements and the sought-after Hamiltonian information. Unfortunately, except for diatomic molecules, until now, no such algorithm has existed.

Although the prospects of using molecular control ideas for data inversion purposes are tantalizing, once again, the issue of having an algorithm is an essential point. A potentially very attractive approach to this end can be built on arguments analogous to those involved in tracking control. In particular, returning to eq. (3), rather than taking the observable $\langle O(t)\rangle = f(t)$ as prescribed *a priori*, we may now consider $\langle O(t)\rangle$ as actually measured in the laboratory, due to the imposition of a field $\varepsilon(t)$. We may consider the field as known and focus in on the goal of determining the potential function $V(x)$ portion of the reference Hamiltonian H_0, or perhaps the dipole function $\mu(x)$. In either case, eq. (3) may be formally inverted, as it is explicitly linear, in either $V(x)$ or $\mu(x)$. This inversion may be effected based on the observation that eq. (3) is a Fredholm interval equation for either of these quantities. However, this inverted solution is only a formal one, since the unknown wavefunctions appear in the inverted expression, along with the data. Once again, following logic similar to tracking control, we may substitute either the formally inverted potential $V(x)$ or the dipole function $\mu(x)$ into eq. (1), and integrate it forward in time, assuming that the initial condition is known. Finally, the resultant wavepacket may be substituted back into the formal inverted expressions for $V(x)$ and $\mu(x)$ to finally yield a solution. This algorithm may be referred to as direct, since no iteration is called for, and it perhaps may stand unique amongst molecular inversion algorithms. Note also that it is highly nonlinear, consistent with the fact that data and Hamiltonians are generally not linearly related. Only preliminary simulations of this algorithm have been performed, but the results are very encouraging. The algorithm has the capability of forming a viable means for extracting valuable Hamiltonian information from dynamical and spectroscopic measurements.

At this point, the algorithm makes no direct use of formal control concepts, except through the notion that the field $\varepsilon(t)$ appears in driving the dynamical system. At a second level, optimal control will enter for designing a field $\varepsilon(t)$ for the purpose of inverting the resultant laboratory data. Little is known of the theory behind this process at this point, and the subject of optimally designed pump-probe measurements should be a rich area for future study.

6. Conclusion

This paper has attempted to present an overview of the current status of molecular control theory, and also indicate the connection between the theoretical developments and emerging laboratory studies. It is anticipated that these laboratory studies will demonstrate the basic elements of molecular control in the near future. Observations of the manipulation of molecular interferences have already been done.[12] The historically-posed challenges of coherently controlling chemical reactions will remain as an important goal in the field. However, new objectives may supplant the old goals, including the general prospect of using control concepts to learn about molecules and materials by optimally designing the experiments and providing inversion algorithms with this goal in mind. Although many of the problems being addressed are old, the field is now young given the fact that the laboratory and theoretical tools for properly addressing molecular control are only now becoming available. Many exciting developments and surprises lie ahead.

7. Acknowledgments

The author acknowledges support for this work from the Office of Naval Research and the Army Research Office.

8. References

1. Brumer, P. and Shapiro, M. (1989) Coherence Chemistry: Controlling Chemical Reactions with Lasers *Acc. Chem. Res.* **22**, 407 - 413.
2. Neuheuser, D. and Rabitz, H. (1993) Paradigms and Algorithms for Controlling Molecular Motion *Acc. Chem. Res.* **26**, 496 - 501.
3. Weiner, A. and Heritage, H. (1987) Picosecond and Femtosecond Fourier Pulse Shape Synthesis *Rev. Phys. Appl.* **22**, 1619 - 1628; Haner, M. and Warren, W. (1988) Synthesis of crafted optical pulses by time dominion modulation in a fiber-grating compressor *Appl. Phys. Lett.* **52**, 1458 - 1460
4. Judson, R. and Rabitz, H. (1992) Teaching Lasers to Control Molecules *Phys. Rev. Lett.* **68**, 1500 - 1503.
5. Gross, P., Ramakrishna, V., Vilallonga, E., Rabitz, H., Littman, M., Lyon, S., and Shayegan, M. (1994) Optimally designed potentials for control of electron-wave scattering in semiconductor nanodevices *Phys. Rev. B* **49**, 11100 - 11110.

6. Warren, W., Rabitz, H., and Dahleh, M. (1993) Coherent Control of Quantum Dynamics: The Dream is Alive *Science* **259**, 1581 - 1589.
7. Shen, L, Shi, S., and Rabitz, H. (1993) Control of Coherent Wave Functions: A Linearized Molecular Dynamics View *J. Phys. Chem.* **97**, 8874 - 8880.
8. Gross, P., Singh, H., Rabitz, H., Mease, K., and Huang, G.M. (1993) Inverse Quantum-Mechanical Control: A Means for Design and a Test of Intuition *Phys. Rev. A* **47**, 4593 - 4604.
9. Drazin, P. and Johnson, R. (1989) *Solitons: An Introduction*, Cambridge University Press, New York.
10. Krause, J., Whitnell, R., Wilson, K., Yan, Y., and Mukamel, S. (1993) Optical control of molecular dynamics: Molecular cannons, reflectrons, and wave-packet focusers *J. Chem. Phys.* **99**, 6562 -.6578.
11. Luenberger, D. (1979) *Introduction to Dynamic Systems: Theory, Models and Applications*, Wiley , New York.
12. Park, S., Lu, S., and Gordon, R. (1991) Coherent laser control of the resonance-enhanced multiphoton ionization of HCl *J. Chem. Phys.* **94**, 8622 - 8624.

INFORMATION THEORY APPROACH IN CHEMICAL DYNAMICS

A Tutorial Introduction

R. D. LEVINE
The Fritz Haber Research Center for Molecular Dynamics
The Hebrew University
Jerusalem 91904, Israel

The thesis that 'chemical reactions proceed in the most statistical way, subject to constraints' is discussed with examples drawn mainly from direct reactions. Attention is centered on the technical aspects and their motivation, with special reference to the notions of 'entropy', 'the prior distribution' and that of the constraints. Surprisal analysis is emphasized and the Lagrangian form, which is particularly useful when several constraints are imposed, is discussed. More advanced topics such as surprisal synthesis are reviewed.

1. Introduction

The idea that the products of a chemical reaction are 'as statistical as possible' is an old one [1-3]. Two, somewhat different points of view, have predisposed us towards this conclusion. The first derives from transition state theory (TST) [1-4]. When the reactants are in a thermal equilibrium and if one can identify a 'configuration of no return' which any reactive trajectory crosses once and only once then, at that configuration, the ensemble of trajectories is in thermal equilibrium. The other approach does not require the reactants to be in equilibrium. Rather, it argues that due to the strong coupling during the reactivve event, the system tends to erase any details. In pictorial terms, the system 'forgets' where it came from. The distribution is thus 'statistical' or, in more technical terms, the available phase space is uniformly populated.

The experimental observations with much support from computational studies show that often the products can be far from statistical. This is true both on the detailed level of the distribution of quantum states of the products and on the more averaged level of the branching

E. Yurtsever (ed.), Frontiers of Chemical Dynamics, 195–216.

into chemically distinct product channels. Any recent volume of the Annual Reviews of Physical Chemistry will contain one or more chapters providing evidence on this point. The purpose of this chapter is to discuss what can be done when the available phase space is not uniformly populated. It addresses the technical question of how do we rephrase the statement 'the available phase space is not uniformly populated' in a constructive way. In other words, we seek a description that does apply rather than a statement about what is observed not to be the case.

We started with the naive expectation that the available phase space will be uniformly populated. The first step is to rephrase this statement by an equivalent one, but one that can be generalized. This is done using the notion of *entropy*. When the available phase space is uniformly populated, the entropy is at its maximum. The generalization that is discussed below is that the entropy is always maximal [5]. What distinguishes the different types of behavior is not this property of the entropy but rather the conditions imposed (to be called the *constraints*) on the set of possibilities over which the maximum is searched. If the serach is unrestricted, except for the ever present conditions such that the total energy is conserved, the result is the familiar statistical limit in which the available phase space is uniformly populated. Imposing constraints leads to other types of distributions and even very few (one, two) simple constraints suffice to recover the essence of observed rather non statistical ('population inversion') products state distributions.

The technical discussion begins with the definition of the physical entropy with special reference to experiments in which there is an incomplete resolution of the final quantum states. This is most conveniently discussed using a simple property of the entropy known in the mathematical literature as the 'grouping property'. Long before entropy became a respectable mathematical object [6], this property was well known to chemists as the 'entropy of mixing'. From this follows the notion of the *prior distribution* [7,8].

The second technical aspect is how to impose constraints. We discuss it in detail for two reasons. The first is conceptual. Our derivation will explicitly show a rather important observation: the functional form of the distribution in phase space is explicilty known as soon as the physical nature of the constraints is specified. In other words, we do not require quantitative input in order to know the form of the resulting distribution. This is what *surprisal analysis* is about: A functional form of the distribution of products states in a form that contains one or more ('undetrmined') parameters is fitted to the data. The fit determines the value of the parameters. We shall show that the values of the parameters as determined from the fit is excatly equal to their theoretical value (of course, up to the noise level of the experimental data [9]).

The empirical utility of surprisal analysis follows from this property that one knows the functional form of the ditribution without having first to provide a quantitative input. The other advantage of the approach using the method of Lagrange undetermined parameters is that it provides for a very practical numerical method [10] of fitting either the experimental data or, equivalently, the constraints. The equivalence is also a direct implication of the method.

The points made above follow from the stationary property of the entropy, i.e., that its first variation about the maximum vanishes. Currently we are very interested in physical implications of the second variation of the entropy (the one whose sign distinguishes a maximum from a minimum).

2. Entropy

A mathematical object, known as *entropy*, can be associated with a probability distribution $\{p_i\}$, over n events, $i = 1,...n$. by the definition

$$S = -\sum_{i=1}^{n} p_i \ln p_i \tag{2.1}$$

We take it from now on that the n events are mutually exclusive and collectively exhaustive so that the distribution is normalized

$$\sum_{i=1}^{n} p_i = 1 . \tag{2.2}$$

Physics begins when we adopt the following understanding: physical entropy is given by equattion (2.1) only when the index i specifies a single quantum state of the system. In other words, one cannot compute the physical entropy of any given distribution simply by plugging into the mathematical definition (2.1). Yet experimental distributions often fail to fully reslove quantum states. How do we proceed?

2.1 THE GROUPING PROPERTY

An experimental distribution can be very coarse grained, say if it is a distribution over chemically distinct products or it can be more resolved, say if it is the products vibrational state distribution, etc. An experimental distribution is thus a distribution over groups of quantum states and a particular eigenstate is taken to belong to only one such group of states*. The more coarese grained the distribution is, the more quantum states belong to each group. The technical question is the form of the entropy under such circumstances, when what is given are the probabilities p_α of the different gropus α., α = 1,..,A.

The probability of any particular final quantu state i can be written in terms of the probability p_α of the group to which it belongs times the fraction, $P(i \,|\alpha\,)$, of quantum states in α that are i :

$$p_i = P(i|\alpha)p_\alpha \quad , \quad \sum_{i\subset\alpha} P(i|\alpha) = 1 \quad , \alpha = 1,....,A. \tag{2.3}$$

The normalization in equation (2.3) is the condition that summing over all quantum states that belong to the group α should yield the probability,p_α, of the group. Using (2.3), equation (2.1) for the entropy leads to

$$\begin{aligned} S &= -\sum_i p_i \ln p_i = -\sum_\alpha p_\alpha \sum_{i\subset\alpha} P(i|\alpha)(\ln p_\alpha + \ln P(i|\alpha)) \\ &= -\sum_{a=1}^{A} p_\alpha \ln p_\alpha + \sum_{\alpha=1}^{A} p_\alpha \left(-\sum_{i\subset\alpha} P(i|\alpha)\ln P(i|\alpha) \right) \\ &\equiv -\sum_{a=1}^{A} p_\alpha \ln p_\alpha + \sum_{\alpha=1}^{A} p_\alpha S_\alpha \equiv -\sum_{a=1}^{A} p_\alpha \ln(p_\alpha / g_\alpha). \end{aligned} \tag{2.4}$$

Equation (2.4) expresses the grouping property of the entropy: when quantum states are grouped, the entropy is the entropy of the distribution over the groups ('the mixing term') plus the weighted sum of the entropy, S_α,

$$S_\alpha \equiv -\sum_{i\subset\alpha} P(i|\alpha)\ln P(i|\alpha) \tag{2.5}$$

of each group. The entropy S_α can also be written as an effective number of states S_α = ln(g_α), as in the last expression in (2.4). Below we shall return to the interpretation of g_α as a number of states.

---

* The distinct quantum states of a system are mutually exclusive . However, one can consider a non stationary distribution for which our discussion needs qualifications.

The physical entropy is, in general, of the form (2.4) and not (2.1). Of course, the two forms are mathematically equivalent provided that one adds the physical statement that the definition (2.1) of the entropy applies only to the distribution of quantum states.

3. Maximum Entropy

When is a not fully resolved distribution of final states statistical?. We need to maximize the entropy (2.4) and no constraints of dynamical origin are to be imposed. Yet, in the process of seeking the maximum of the entropy we cannot make arbitrary variations in the probabilities because (i) the probabilities must be non negative and (ii) the variations msut be such that the normalization condition is maitained. The procedure of seeking a maximum subject to constraints is based on the introduction of a Lagrangian whose unconstrained maximum is the desired solution. For seeking the maximum of (2.1) subject to the normalization constraint, the Lagrangian L is given by

$$L = S - (\lambda_0 - 1)\sum_{i=1}^{n} p_i \,. \tag{3.1}$$

Here λ_0 is the Lagarange multiplier whose numerical value is, so far, undeterined. (We write the multiplier, for convenience as $\lambda_0 - 1$ since its value has not yet been detrmined). The unconstrained varaition of L leads to

$$\begin{aligned} \delta L = \delta S - (\lambda_0 - 1)\sum_i \delta p_i &= -\sum_i \delta p_i(\ln p_i + 1) - (\lambda_0 - 1)\sum_i \delta p_i \\ &= -\sum_i \delta p_i(\ln p_i + \lambda_0) \,. \end{aligned} \tag{3.2}$$

Since the variations δp_i are now arbitrary, the distribution for which the Lagrangian is stationary is, from (3.2),

$$\ln p_i = -\lambda_0 \qquad \text{or} \quad p_i = \exp(-\lambda_0) \tag{3.3}$$

The value of the Lagarange multiplier λ_0 is still undetrmined but it is already clear what the solution is: At the maximum of the entropy, (subject only to the normalization of the distribution) all the quantum states i have the same probability. If one wants one can determine the value of λ_0 from the normalization condition

$$\sum_{i=1}^{n} p_i = \sum_{i=1}^{n} \exp(-\lambda_0) = 1 \quad \text{or} \quad p_i = \exp(-\lambda_0) = 1/n\,. \tag{3.4}$$

The physical conclusion that is is the uniform distribution that is the solution is however already obvious without computing the numerical value of the Lagrange multiplier.

Two further aspect of the solution deserve comment. The first is to show that the stationary solution that we got is a maximum and not a minimum of the entropy. That is easy, by considering the second variation of the entropy

$$\delta^2 S = -\sum_i (\delta p_i)^2 / p_i \tag{3.5}$$

It follows that as long the probabilities are physically realistic (i.e., non negative), the stationary point of the entropy is always a maximum. The mathematical statement is that the entropy is a *convex function* of the probabilities, figure 1. We return to this point and provide an alternative proof in equation (3.7) of section 3.1.

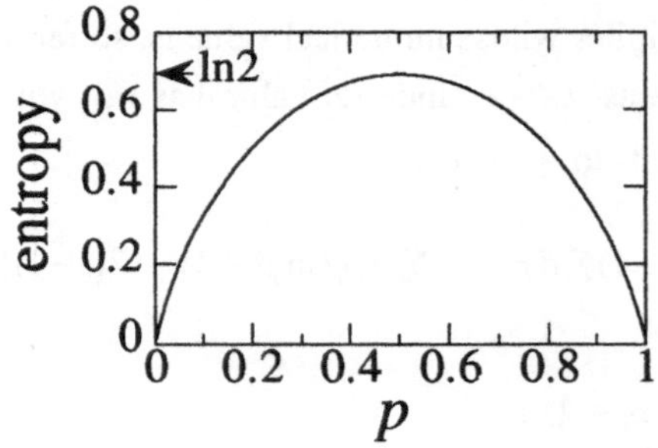

Figure 1. The entropy, equation (2.1) plotted for the two state case, vs. the probability

What is seemingly unexpected is that the probabilities in (3.4) came out to be non negative without our having imposed this constraint in (3.2). This is our first example of a constraint that is inherently satisfied (sometimes called a *non informative constraint.*) In this case it is possible to understand why the constraint does not restrict the searach for the maximal entropy. To see this one needs to think how, in principle, one would search for a maximum. Choose some initial trial values, all positive, for the probabilities and compute the entropy for this set of trial values. Next, compute the entropy for adjacent values of the probabilities. If the entropy is higher, move to the new values as a new set of trial values, etc., untill a maximum is located. A constraint is some limitation on the new trial values. But in the vicinity of a set of positive probabilities, all possible new sets are also sets of probabilities which are all positive. The inequality constraint $p_i \geq 0$ does not limit the possible variations of the trial probabilities as long as the p_i's are positive. If one includes a Lagrange multiplier for a non informative constraint, the numerical value of the parameter will be zero and we will prove this in section 4.4 below.

3.1 THE BOLTZMANN, GIBBS,.... INEQUALITY

An inequality, which goes under an unusually great variety of names, provides a simple comparison of the entropy of two distributions which satisfy the same set of constraints. While the inequality in essence is simply the statement that the entropy is a convex function, the inequality is such a convenient tool that it is worth stating explicitly. We consider two normalized distributions, the old p_i's and a new set, the q_i's. Then

$$\sum_{i=1}^{n} p_i \ln(p_i / q_i) \geq 0 \tag{3.6}$$

with equality <u>if and only if</u> $p_i = q_i$ for all i's. The 'only if' part is important because it is the condition that insures that the distribution whose entropy is maximal, is unique.

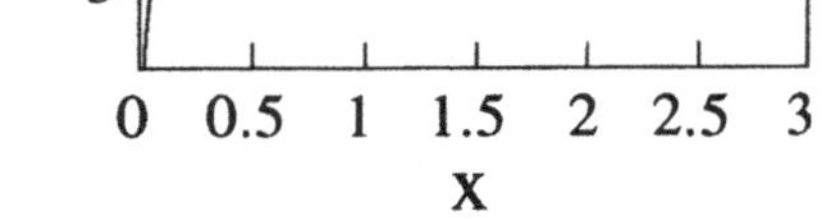

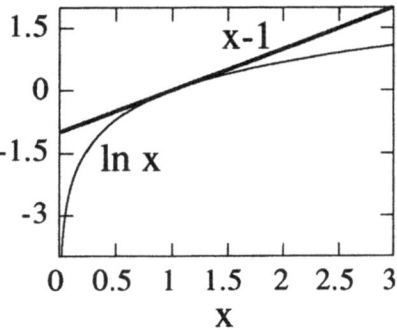

Figure 2. The logarithmic inequality

The proof of (3.6) is based on the observation that $\ln x \leq x - 1$ with equlity if and only if $x = 1.$, cf. figure 2 above. Now write $\ln(p_i / q_i) = -\ln(q_i / p_i)$ in (3.6) and put $x = (q_i / p_i)$. Then $-\sum p_i \ln(q_i / p_i) \geq \sum p_i(1 - (q_i / p_i)) = \sum p_i - \sum q_i = 0$,Q.E.D. An immedate application is to take $q_i = 1/n$ namely the uniform distribution and p_i to be some other normalized distribution:

$$0 \le \sum_{i=1}^{n} p_i \ln\left(p_i / \frac{1}{n}\right) = \ln n - \left(-\sum_{i=1}^{n} p_i \ln p_i\right) \tag{3.7}$$

thereby proving that amongst all normalized distributions over n states, the uniform distribution is the unique one, (recall 'if and only if'), of maximal entropy. At the unique maximum, the value of the entropy is $\ln n$. For physical probabilities, the minimal value, zero, of the entropy is not unique. At the minimum, one event occurs with certainty.

3.2 THE PRIOR DISTRIBUTION

We return to the grouping of quantum states. Say there are n_α , $\Sigma_\alpha n_\alpha = n$, states in the group α.. For a uniform distribution it follows that the maximum of the entropy is not when all groups of states are equally probable. Rather, the maximum is when the occupation of the group is proportional to the number of states in it

$$q_\alpha = \sum_{i=1}^{n_\alpha} \frac{1}{n} = \frac{n_\alpha}{n} \tag{3.8}$$

It is instructive to examine this result from the point of view of the grouping property, section 2.1. For a uniform distribution $Q(i|\alpha) \equiv q_i / q_\alpha = (1/n)/(n_\alpha/n) = 1/n_\alpha$ therefore $S_\alpha = \ln n_\alpha$ or $g_\alpha \equiv \exp(S_\alpha) = n_\alpha$. It follows from (2.4) that when all states within each group are equally probable, the entropy can be written as

$$\begin{aligned} S &= -\sum_\alpha p_\alpha \ln(p_\alpha / n_\alpha) \\ &= \ln n - \sum_\alpha p_\alpha \ln\left(p_\alpha \Big/ \frac{n_\alpha}{n}\right) \end{aligned} \tag{3.9}$$

Since $\ln n$ is the maximal value of the entropy, it follows from (3.6) that the entropy is maximal if and only if the probability of the group is $p_\alpha = n_\alpha / n$. This result is just what one would expect from (3.8). What (3.9) tells in addition is that any distribution over groups of quantum states that deviates from this has an entropy which is below its global maximum. In other words, a distribution which deviates from (3.8) signifies the presence of additional constraints* beyond normalization.

The distribution for which the probability of a group of states is proportional to the number of staates in the group (or equals the fraction of all quantum stats that belong to the group, cf. (3.8)) is the *prior distribution.*

The prior distribution is the immediate generalization of the concept of the uniform distribution. It is needed bcause the entropy of a physical system is defined in terms of the distribution over quantum states. Hence, when the observation is an incompletely resolved distributiion, the entropy is given by (2.4) rather than by (2.1).

3.3 MODELS

The need to introduce a prior distribution arises because experiments often fail to fully resolve the final stats. What is measured is therefore the probabilities p_α of the different groups α of final states. On the other hand, computing the physical entropy requires the fully resolved distribution of final states. The problem is to determine the p_i's given the p_α's . In principle this can be done by the method of next section but a short cut is to proceed as follows. The entropy of the coarse grained distribution is given by (2.4), with unknown entropies S_α for the distribution of states within each group. With n_α as the number of states within the group a, we have already shown that $S_\alpha \leq \ln n_\alpha$ with equality if and only if all states within the group are equally probable or, in symbols, iff $P(i|\alpha) = 1/n_\alpha$. It follows from (2.4) that for a given coarse grained distribution, p_α , the entropy is maximal if all states within the group are equally probable, $p_i = p_\alpha / n_\alpha$, for $i \subset \alpha$. The entropy is then given by equation (3.9).

The paragraph above redefined the *prior distribution.* Because of its central role, we will rederive it once more in section 4.1 below. It is sometimes the case that one has definite ideas about the distribution of states even when the experiment fails to resolve them. This is most often the case when it is the distribution of final relative kinetic energy that is being measured. Within a narrow range of final kinetic energies there can be many distinct combinations of vibrational and rotational states of the products which are consistent with the conservation of

---

* Adding a constraint cannot increase the value of the entropy. The reason is that one seeks the maximal value of the entropy subject to this constraint. The constraint restricts the range of possible distributions over which the maximum is searched. In general, this will lower the possible value of the maximum. It may be that the constraint is uninformative, in which case the value of the maximum is unchanged. What cannot be is that the value of the maximum will increase.

total energy. Particularly if the products are polyatomic and/or the internal energy is high, one may consider that not all possible states are equally probable. Strictly speaking, one should impose such a bias by the method of the next section. A short cut is to introduce a model for the distribution of states within each group. From the assumed distribution one computes the entropy of states, S_α, equation (2.5), within the group so that the entropy is given by (2.4). Equivalently, one can introduce a *model distribution* q_α which we write as $q_\alpha = g_\alpha / \sum_\alpha g_\alpha$ where this defines g_α. For the model distribution, the entropy is given by

$$\begin{aligned} S &= -\sum_\alpha p_\alpha \ln(p_\alpha / g_\alpha) \\ &= (\ln(\sum_\alpha g_\alpha)) - \sum_\alpha p_\alpha \ln(p_\alpha / q_\alpha) \end{aligned} \tag{3.10}$$

The prior distribution, which leads to equation (3.9), is cleaely a special case of (3.10) which corresponds to $g_\alpha = n_\alpha$ so that $\sum_\alpha g_\alpha = n$. Note also that, in the general case $S_\alpha \leq \ln n_\alpha$ so that $g_\alpha \equiv \exp(S_\alpha) \leq n_\alpha$ i.e., for any model other than the prior, the entropy is lower.

An example of a use of a model is when the total energy is not precisely known. The 'model' is an assumed value of this unknown energy. For each choice of value for the total energy one can compute a prior distribution. Each such distribution is a different model. For each model one uses the given observed distribution to compute the entropy. The 'best' value for the unknown energy is the one for which the computed entropy is maximal.

4. The Lagrangian

This section discusses a general approach for seeking the distribution whose entropy is maximal. The need for such a method arises because, in general, the measured distribution will not agree with either the prior or even with a model. By our thesis, this implies the prsence of constraints which lower the maximal possible value of the entropy. We here show how to find that distribution which satisfies the constraints and whose entropy is maximal. One needs such a method because the number of constraints is, by assumption, fewer (or even far fewer) than the number, n , of unknown probabilities. There will therefore be more than one distribution that satisfies the constraints. The task is to find that one whose entropy is maximal.

We will consider a simple example, that of a vibrational state distributiion p_v of the products of an exoergic reaction [1,2,8]. It is a coarse grained distribution becuse to each vibrational state there correspond several different rotational states with the balance of the enregy being in translatlation. The constraint we shall use is on the mean vibrational energy

$$\langle E_{vib} \rangle = \sum_{v} E_v p_v \tag{4.1}$$

For future reference we note that one can just as well write the constraint as a condition on the distribution of final quantum states

$$\langle E_{vib} \rangle = \sum_{i=1}^{n} E_v(i) p_i \tag{4.2}$$

where $E_v(i)$ is the vibrational energy of the i'th final quantum state. If several products' vibrational states can be accessed, there will be many (in fact, a continous range of) distributions which satisfy the constraint.

4.1 THE UNDETERMINED MULTIPLIERS

Amongst the distributions that satisfy the constraint we specify the one of maximal entropy in two steps. The first one selects the functional form of the distribution. The technique is, as in section 3, to replace seeking the constrained maxium of the entropy by seeking the unconstrained maximum of a Lagrangian. There are three constraints: (i) the probabilities need to be non negative, (ii) the probabilities need to sum up to unity, known as the normalization constraint and (iii) equation (4.2). The first two constraints are the ones we already used and as before we shall, for the moment, disregard the first one because, with hindsight, we know that it is non informative. The Lagrangian includes two *Lagrange multipliers*, one for each constraint that is used. The Lagrange multiplier for the normalization will be denoted, as before, by λ_0 . The Lagrange multiplier for the mean vibrational energy constraint will be denoted by λ_v . Note however that λ_v really means λ_{vib}. λ_v is a number and not a function of the vibrational quantum number v.

At this point, and throughout the first stage, the numerical value of the two Lagrange multipliers is undetrmined. It will be assigned a unique value in the second stage.

Writing the unknown distribution as q_iThe Lagrangian is

$$L=-\sum_i q_i \ln q_i-(\lambda_0-1)\sum_i q_i-\lambda_v \sum_i E_v(i) q_i \tag{4.3}$$

and its unconstrained variation leads to

$$\begin{aligned}\delta L&=-\sum_i \delta q_i(1+\ln q_i)-(\lambda_0-1)\sum_i \delta q_i-\lambda_v \sum_i E_v(i) \delta q_i \\ &=-\sum_i \delta q_i\left(\ln q_i+\lambda_0+\lambda_v E_v(i)\right)\end{aligned} \tag{4.4}$$

The stationary value of the Lagrangian is obtained for

$$\ln q_i=-\lambda_0-\lambda_v E_v(i) \quad \text{or} \quad q_i=\exp\left(-\lambda_0-\lambda_v E_v(i)\right) \tag{4.5}$$

This is the end of the first stage: An explicit functional form for the distribution of final quantum states, in a form which contains two parameters whose numerical value is yet undetermined. Before we turn to their determination, note that (4.5) is not yet quite ready to be compared to experiment. By assumption, what has been measured is the distribution p_v over products' vibrational states, where, cf. (2.3), p_v is obtained by summing over all final quantum states i whose vibrational energy $E_v(i)$ is the same. Using (4.5) the form of the grouped distribution of maximum entropy is:

$$q_v=\sum_i q_i\, \delta\left(E_v(i)-E_v\right)=n_v \exp(-\lambda_0-\lambda_v E_v)=q_v^0 \exp(-\lambda_0-\lambda_v E_v) \tag{4.6}$$

Here q_v^0 is the prior distribution, $q_v^0=n_v / n$ where n_v, $\sum_v n_v=n$, is the number of final quantum states whose energy equals E_v . In going from the first to the seond line of (4.6) the, yet undetrmined, value of λ_0 has been changed by a constant amount

The result (4.6) shows explicilty how the procedure assigned equal probabilities to all the quantum states which belong to a given vibrational manifold. It did so because there was no constraint which required otherwise. Maximal entropy yeilds the most uniform distribution of final states which is consistent with the constraints. The reader is invited to derive (4.6) directly starting with the grouping form, equation (3.9), for the entropy.

The next stage is the determination of the numerical values for the two Lagrange multipliers. In principle, their values are determined by the value of the constraints. The value of the normalization condition (i.e., unity) is obviously known and so, for simplicity, we determine λ_0 once and for all by the condition $\Sigma_\nu q_\nu = 1$ or, using (4.6)

$$\exp(\lambda_0) = \sum_\nu q_\nu^0 \exp(-\lambda_\nu E_\nu) \tag{4.7}$$

This makes λ_0 a function of λ_ν and leaves us with only one undetermined multiplier. In the more general case, there will be, at this point, as many undetermined multipliers as the number of constraints which are special to the problem at hand. We discuss two approaches to the determination of the Lagrange multipliers:

• surprisal analysis

and

• surprial synthesis.

What we do not discuss, but is often important, is the error estimate [9,10] in the resulting values due to noise in the experimental data or to uncertainties in computational data.

4.2 SURPRISAL ANALYSIS

Surprisal analysis means the detrmination of the Lagrange multipliers by fitting the functional form of the distribution to the given one. If there is only one multiplier that needs to be determined, as in (4.6) then this can be done by graphical means, figure 3. Otherwise it requires a computer [11]. A trivial point, but one that is easy to overlook, is that the fit needs to be done to the logarithm, $-\ln(p_\nu / p_\nu^0)$, called the *surprisal,* rather than to the distribution directly. If one wants to use a linear least square fitting procedure on p_ν directly, one should use a weighted least squares, where the weight of each data point is proportional to its probability.Figure 4 shows a more elaborate application.

The point that the optimal distribution is the one whose surprisal is approximated as closely as possible by an expression linear in the constraints is inherent to the maximum entropy formalism. It is it which insures that the entropy is maximal. This method does give higher weight to the more probable outcomes but this is a direct implication of the search for a most uniform distribution, subject to constraints. The logarithmic plot, known as a surprisal plot, is logarthmic because this is what is needed to minimize the Lagrangian, cf. equation (4.4).

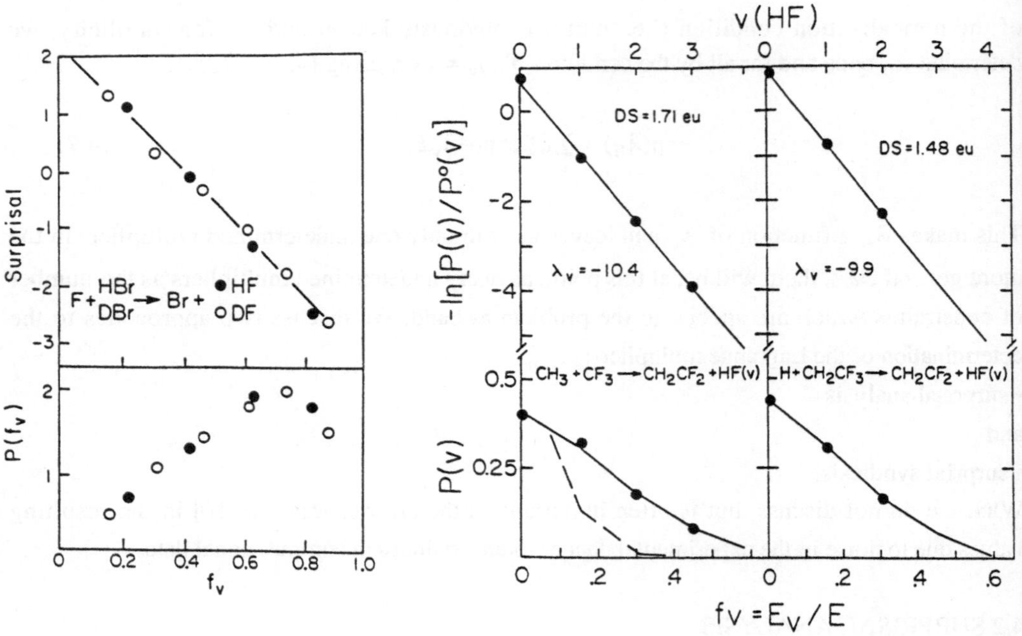

Figure 3. Applications of surprisal analysis to vibrational energy disposal in two different reactions plotted vs $f_v = E_v / E$, the fraction of the available energy which is in vibration. In each case, the bottom panel shows the meaured vibrational state distribution. A. For the exoergic reaction of F atoms with HBr or with DBr. Note how the surprisal when plotted vs. f_v is essentially invariant to isotopic substitution [1]. B. For the elimination of HF from CH_3CF_3 which is chemically activated, two about the same energy but by two different routes.Note how the surprisal is dependent only on the total energy. Both plots adapted from R. D. Levine in *Theories of Chemical Reaction Dynamics*, M. Baer, Ed, CRC press, 1984, where the sources of the experimental data are given.

We reiterate that maximum entropy is equivalent to making the best liner expansion for the surprisal with the following technical argument, which also affords a very convenient numerical procedure, particularly if more than one constraint is imposed. For this reason we consider several constraints of the generic form:

$$\langle A_r \rangle = \sum_v A_{rv} q_v \tag{4.8}$$

Here A_{rv} is the numerical value of the property A_r in the state v. Introducing a Lagrangian with a separate multiplier λ_r^T for each constraint, plus the normalization constraint, the resulting distribution of maximal entropy is

$$q_v^T = q_v^0 \exp\left(-\lambda_0^T - \sum_r \lambda_r^T A_{rv}\right) \tag{4.9}$$

The superscipt T, for 'trial', is only meant as a convenient reminder that, so far, it is a trial distribution with trial values of the Lagarnge multipliers, whose final value is yet to be determined. As before, we do impose that the trial distribution is normalized, which implies that

$$\exp(\lambda_0^T) = \sum_v q_v^0 \exp\left(-\sum_r \lambda_r^T A_{rv}\right) \tag{4.10}$$

From the last two equations it follows that

$$-\frac{\partial \lambda_0^T}{\partial \lambda_r^T} \equiv -\exp(-\lambda_0^T)\frac{\partial \exp(\lambda_0^T)}{\partial \lambda_r^T} = \sum_v A_{rv} q_v^T \tag{4.11}$$

which is an identity we shall need below.

To determine the values of the trial Lagrange multipliers from the given experimental distribution p_v we write the Lagrangian as

$$L = \sum_v p_v \ln(p_v / q_v^T) \tag{4.12}$$

It follows from (3.6) that the Lagrangian is non negative and vanishes if and only if the trial distribution provides a perfect fit to the data. In practice, the Lagrangian cannot be made to exactly vanish and its minimal value is a measure for the *quality of fit* ,[11]. By using the explicit form of q_v^T , equation (4.9), one can also verify that (4.12) is the same Lagrangian we are using throughout. Now minimize the lagrangian by varying the trial values of the Largange multipliers in q_v^T

$$\delta L = \sum_v p_v\, \delta \ln q_v^T = \sum_v p_v\left(-\delta \lambda_0^T - \sum_r \delta \lambda_r^T A_{rv}\right) = -\delta \lambda_0^T - \sum_r \delta \lambda_r^T \sum_v A_{rv} p_v$$

$$= \sum_r \delta \lambda_r^T\left[\left(\sum_v A_{rv} q_v^T\right) - \left(\sum_v A_{rv} p_v\right)\right] \tag{4.13}$$

The minimal value of the Lagrangian is achieved when the expectation value of each one of the constraints, computed for the trial distribution, equals the experimental value. In otherwords, at the minimal value of the Lagrangian, the trial distribution is of maximal entropy and satifies the constraints.

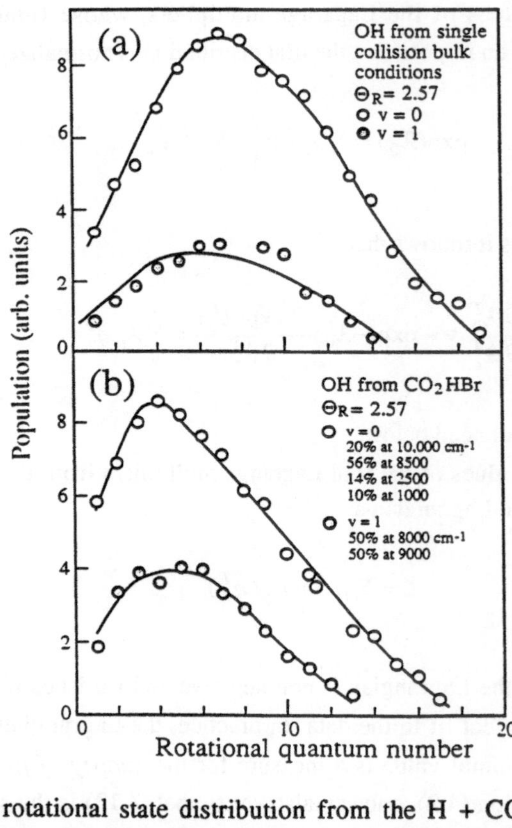

Figure 4. The OH rotational state distribution from the H + CO_2 reaction, dots, and its representation as a form of linear rotational surprisal. (a) Data from experiments on isolated collisions. (b) Data from the photodissociation of the $HBrCO_2$ van der Waals adduct. The Chattering of the H atom between its two heavier neighbors dissipates some of its energy. Source: C. Wiitig, Y. M. Engel and R. D. Levine, *Chem. Phys. Lett.* **153**, 411 (1988).

The procedure just outlined is very suitable for a numerical computation of the Lagrange multipliers because the Lagrangian is not only non negative, but it has a unique minimum. One simply iterates on the trial values of the multipliers in the direction of decreasing *L*. An example is shown in figure 4 above and in figure 5 below. In the next section we shall see how a very slight modification enables us to determine the Lagrange multipliers not from the experimental distribution but from the numerical values of the constraints.

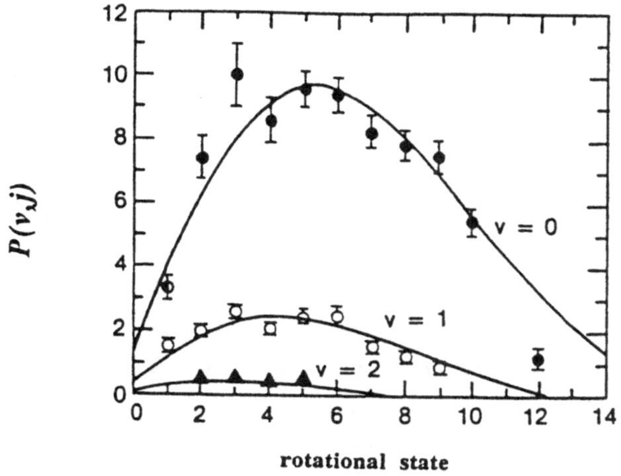

Figure 5. Experimental HD product vibrotational state distribution for the H + D_2 reaction, dots, and its fit, continous line, by a distribution of maximum entropy subject to two constraints, one on the vibrational energy, as in (4.5) and one on the rotational energy. Adapted from reference 1 where the source of the data is given.

That the minimum of the Lagrangian is unique follows from the Boltzmann-Gibbs inequality, equation (3.6). One can also take the second variation of *L* to show this. As already mentioned, much of our current interest is in the physical implications of this minimum property. The reason for this interest is that surprisal analysis itself only requires that the first variation in the Lagrangian vanishes, cf. equations (4.4) or (4.13) but does not use the sign of the second variation.

4.3 SURPRISAL SYNTHESIS

One can determine the entire distribution given only the values $\langle A_r \rangle$, $r = 1,2,..$ of the costraints. To do so, the Lagrangian (4.12) is rewritten as $L = \sum p_v \ln p_v - \sum p_v \ln q_v^T$ so that, with the intermediate steps as in (4.13)

$$\delta L = -\sum_v p_v \delta \ln q_v^T = ...$$

$$= \sum_r \delta \lambda_r^T \left[\left(\sum_v A_{rv} q_v^T \right) - \langle A_r \rangle \right] \tag{4.14}$$

The Lagrangian can therefore be minimized without any reference to the experimental distribution p_v . Only the values of the constraints are required.

Surprisal synthesis is a practical tool because the mean values $\langle A_r \rangle$ are often better determined than the raw distribution. An example drawn from experiment is the angular distribution of the products. To extract it from the actual signal one often expands it as a sum of Legendre polynomials and determines the coefficients in such an expansion by a fit. The familiar shortcoming of such a procedure is that the resulting distribution need not be positive. By definition, the coefficients that are being fitted are the mean values of the Legendre polynoials, $\langle P_l(\cos\theta) \rangle$, $l = 1,2,..$. Using (4.14) one can determine a strictly non negative angular distribution from exactly the same data set. An example drawn from computational studies is, as before, the determination of the products vibrational distribution from a classical trajectory simulation. The problem is how to assign the final states to the 'bins' of final vibrational states. A simple solution is to compute the mean products vibrational energy and, if needed, other classical averages over the final states. From these, using (4.14) one constructs a discrete vibrational distribution. A subsidiary advantage is the a classical computation of the average values of a distribution is inherently more accurate than a classical determination of a histogram of final states [12].

Surprisal synthesis is a theoretical tool because the mean values of observables can be computed in closed form [13]. An explicit, analytical, solution is, of course, possible only in very simple cases. One example when this can always be done in an approximate fashion is the, so called, *local harmonic approximation* , [13]. This can also be applied for dissipative time evolution [14].

4.4 ENTROPY DEFICIENCY

At its consrained maximum, the value of the entropy can be computed from (3.9)

$$S = \ln n - \sum_{\alpha} q_a \ln\left(q_\alpha / q_\alpha^0\right) = \ln n - \lambda_0 - \sum_{r} \lambda_r \langle A_r \rangle \tag{4.15}$$

The non negative second term, $\Sigma_\alpha q_\alpha \ln(q_\alpha / q_\alpha^0)$, is a measure of how much the entropy is below its global maximum and, as such, indicates how far one is from the strict statistical limit. For this reason it is known as the *entropy deficiency.*

Using equation (4.11) one obtains from (4.15) a quantitative measure for the importance of any given constraint

$$\lambda_r = -\partial S / \partial \langle A_r \rangle \tag{4.16}$$

A *non informative constraint* is one which does not appear in the distribution of maximal entropy. In another way of putting it, it appears but with a Lagrange multiplier which equals zero. A non informative constraint is thus a constraint whose value does not serve to lower the value of the entropy below its maximum.

5. Concluding Remarks

This tutorial review necessarily can not cover all possible applications and more advanced topics. Omitted were such questions of obvious importance as the explicit computation of the prior distribution [1,2,7,11], implications of detailed balance [15], the application to branching fractions, figure6, special quantal features, intramolecular dynamics [16] and the application to spectral fluctuations [17] and the fluctuations of unimolecular rate constants [18].

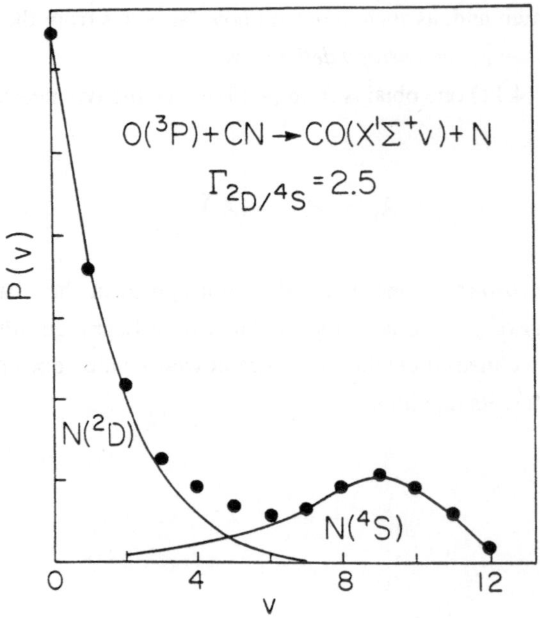

Figure 6. The CO vibrational state distribution, dots, from the O + CN reaction and its synthesis as a sum of contributions from two processes, leading to N(2D) and N(4S), lines.Adapted from R. D. Levine in *Theories of Chemical Reaction Dynamics*, M. Baer, Ed, CRC press, 1984, where the sources of the experimental data are given.

Acknowledgment

I thank my coworkers for the good scientific time we had and the US Air Force Office of Scientific Research and the Volkswagen Stiftung for their support which made it possible.

References

1. Levine, R. D. and Bernstein, R. B. (1987) *Molecular Reaction Dynamics and Chemical Reactivity*, Oxford University Press.
2. Steinfeld, J.I., Francisco, J.S. and Hase, W.L. (1989) *Chemical Kinetics and Dynamics*, Prentice Hall, New Jersey.
3. Smith, I.W.M. (1980) *Kinetics and Dynamics of Elementary Gas Reactions, Butterworths,* London.
4. Pechukas, P. (1976) Ch. 6, Statistical approximations in collision theory, in W.H. Miller (ed), *Dynamics of Molecular Collision (Part B)*, Plenum Press, New York.
5. Jaynes, E.T. (1957) Information theory and statistical mechanics, *Phys. Rev.* **106**, 620-630.
6. N.F.G. Martin and J.W. England (1981) *Mathematical Theory of Entropy*, Addison-Wesley, London.
7. Ben-Shaul, A., Levine, R.D. and Bernstein, R.B. (1972) Entropy and chemical change II. Temperature and entropy deficiency of product state distributions, *J. Chem. Phys.* **57**, 5427-5440.
8. Levine, R.D. and Bernstein, R.B. (1975) Thermodynamic approach to collision processes, in W.H. Miller (ed), *Modern Theoretical Chemistry*, Vol. III - *Dynamics of Molecular Collisions*, Plenum, New York, pp. 323-364.
9. Alhassid, Y. and Levine, R. D. (1979) Experimental and inherent uncertainties in the information theoretic approach, *Chem. Phys. Letts.* **73**, 16-24.
10. Agmon, N. Alhassid, Y. and Levine, R.D. (1979) An algorithm for finding the distribution of maximal entropy, *J. Comput. Phys.* **30**, 250-259.
11. Levine, R.D. and Kinsey, J.L. (1979) The application of information theory to molecular collisions, Ch. 22 in *Atom Molecule Collision Theory: A Guide for the Experimentalist*, Plenum, New York, pp. 693-750.
12. Levine, R.D. (1972) Classical evaluation of averaged collision rates, *J. Chem. Phys.* **56**, 1633-1700.
13. Alhassid, Y. and Levine, R.D. (1977) Entropy and chemical change III: The maximal entropy (subject to constraints) procedure as a dynamical theory, *J. Chem. Phys.* **67**, 4321-4400.
14. Ben-Nun, M. and Levine, R.D. (1992) An approximate solution of the Fokker-Planck equation for reactions on condensed phases, *Chem. Phys. Lett.* **192**, 472-480.
15. Kaplan, H., Levine, R.D. and Manz, J. (1976) The dependence of the reaction rate

constant on reagent excitation: The implications of detailed balance, *Chem. Phys.* **12**, 447-460.

16. Remacle, F. and Levine, R.D. (1993) Maximal entropy spectral fluctuations and the sampling of phase space, *J. Chem. Phys.* **99**, 2383-2400; Remacle, F. and Levine, R.D. (1993) The domain information from resonant Raman excitation profiles: A direct inversion by maximum entropy, *J. Chem. Phys.* **99**, 4908-5104.
17. Levine, R.D. (1987) Fluctuations in spectral intensities and transition rates, *Adv. Chem. Phys.* **70**, 53-70.
18. Levine, R.D. (1988) Quantal fluctuations in unimolecular rate constants, *Ber. Bunsenges. Phys. Chem.* **92**, 222-230.

TRENDS IN MOLECULAR DYNAMICS SIMULATION TECHNIQUE

J. BRICKMANN, S. M. KAST[†], H. VOLLHARDT, S. REILING[‡]
Technische Hochschule Darmstadt
Institut für Physikalische Chemie and
Darmstädter Zentrum für wissenschaftliches Rechnen
Petersenstraße 20
D-64287 Darmstadt, Germany

1. Introduction

The molecular dynamics (MD) simulation method is one of the typical statistical mechanical computer simulation techniques employed in the theoretical study of many-particle systems. The behavior of a macroscopic system, consisting of a large number of interacting particles, is usually too complicated for an analytical statistical-mechanical treatment. Computer simulations have become a valuable tool for studying structural and dynamical equilibrium and nonequilibrium properties of chemical systems. The dynamics simulation technique aims at reflecting the interaction within real systems in a mathematical *model*. Based on this model the time evolution of the particles is then numerically calculated using classical or quantum-mechanical methods. Results are obtained by *observing* and evaluating this evolution; hence the simulation technique is in principle an experimental discipline. The outcome of these computer experiments is statistically analyzed, thus giving the relevant statistical-mechanical quantities for characterizing the system. The calculated observables can be directly related to real experiments (in the case where the model scenario closely corresponds to reality) and to the results from analytic theory or simulations with reduced complexity (in the case where the model scenario was generated in order to

---

[†] Present address: Dept. of Chemistry, University of Chicago, 5735 South Ellis Avenue, Chicago, Illinois 60637, USA.

[‡] Present address: Laboratoire de Spectroscopie Moléculaire et Cristalline, Université de Bordeaux I, 351 Cours de la Libération, F-33405 Talence CEDEX, France.

E. Yurtsever (ed.), Frontiers of Chemical Dynamics, 217–253.

study the systematic response of the system to changes of individual control parameters). This interplay is schematically shown in Fig. 1.

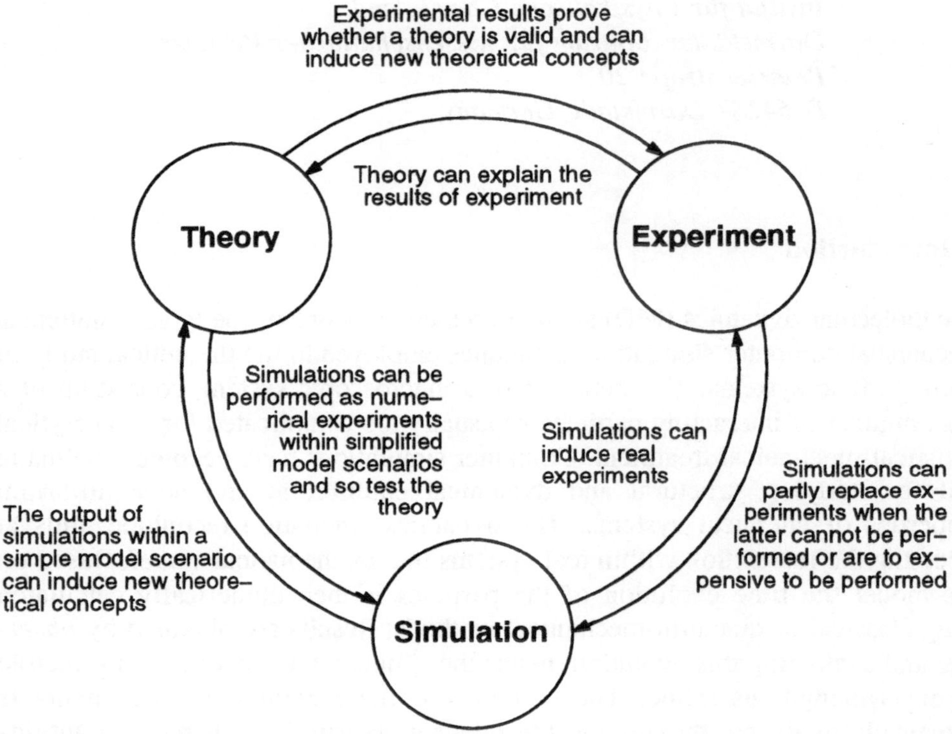

Figure 1. Theory, experiment, and simulation - the three columns of science.

Computer simulations of chemical systems supply both experiment and theory: Experimental results can be illustrated and their basis can be understood on a microscopic level; idealized theories can be tested with simplified models, for example in order to gain insight into the importance of different potential energy terms for an observed effect. Additionally, by simulating molecular scenarios, one is able to check hypotheses (or at least to falsify them) *without* the effort of an experimental study.

With current supercomputer technology, the size of chemical systems being tractable by computer simulations can be up to some 10^5 atoms, reaching time scales of several nanoseconds for complete simulations or even larger when simplified models are used [1]. The basic ideas and technologies of MD simulations are reflected in many excellent reviews and books on this subject [2-6].

The objective of this article is twofold: Firstly, we give an introduction to the statistical-mechanical background of how microscopic dynamics is evaluated with respect to macroscopic observables and we discuss the most important numerical procedures to generate and evaluate trajectories within a computer simulation. Secondly, the limitations of usual MD simulations regarding the possible simulation time and particle numbers and with respect to the simulated statistical-mechanical ensembles are reviewed together with a discussion of selected modern strategies to overcome or at least extend these frontiers. We restrict ourselves here to the case of classical MD simulations and will give some examples for the application to real systems. There have been some attempts towards the inclusion of quantum degrees of freedom in MD simulations but this topic is beyond the focus of the present article.

2. Statistical Mechanics and Computer Simulations

2.1. FROM MICROSCOPIC DYNAMICS TO MACROSCOPIC OBSERVABLES

As schematically shown in Fig. 1, there are different aspects relating theory, experiment, and simulations. MD simulations are performed in order to relate a microscopic atomistic scenario to the world of macroscopic observables under given constraints in thermodynamic equilibrium or on its way to equilibrium. Here we will focus on the simulation of equilibrated systems.

Consider, for simplicity, a one-component, equilibrated macroscopic system. The thermodynamic state of this system is defined by a small set of extensive (e.g. the particle number N or the volume V) and intensive thermodynamics variables (e.g. the temperature T or the pressure p). These observables result from the concerted dynamics of a huge number of particles. With thermodynamics measurements the effect of the microscopic dynamics is averaged over a large period of time (time average) compared to the molecular relaxation time in order to give stable equilibrium data. The conversion from microscopic dynamics to macroscopic properties is the domain of statistical mechanics [7,8]. Since the dynamics of all the particles of a macroscopic sample is far too complex to be treated explicitly, Gibbs suggested to replace the time average by an *ensemble average* in order to calculate macroscopic properties from microscopic data. In MD simulations the more natural concept of *time averaging* is the basis of the evaluation.

The physical state of an N-particle system at a given time t is fully determined by a set of $3N$ position and $3N$ momentum coordinates. This set can be interpreted as a point (or vector) Γ in a $6N$-dimensional space, the so-called *phase space*. The time evolution $\Gamma(t)$ of an individual phase space point is termed a phase space *trajectory*. Any macroscopic property A of the total system at a time t is a function of the system's current phase space point, i.e. $A(\Gamma(t))$. Measuring the property A_{obs} corresponds to an average $\langle A \rangle$ over the measuring time t_m

$$A_{obs} = \langle A \rangle = \frac{1}{t_m} \int_0^{t_m} dt' \, A(\Gamma(t')) \ . \tag{1}$$

A collection of independently moving phase space points each representing the same system is regarded as an *ensemble*. The probability density $\rho_{ens}(\Gamma)$ of these subsystems as a function of the phase space points Γ is referred to as the phase space density. The accessible phase space regions are determined by external control quantities, e.g. by fixing N, V, and the total energy E, a *microcanonical* ensemble is generated. Since each of the subsystems evolves individually in time, ρ_{ens} itself is a function of the time, i.e. $\rho_{ens}(\Gamma,t)$.

Conservation systems are those for which the Hamiltonians (sum of kinetic and potential energy) are not explicitly time dependent. For such systems the time evolution of the phase space density $\rho_{ens}(\Gamma,\mathrm{t})$ is given by the well-known Liouville equation

$$\frac{\partial}{\partial t}\rho_{ens}(\Gamma,t) = -iL\rho_{ens}(\Gamma,t) \ , \tag{2}$$

where the Liouville operator L is defined by

$$iL = \sum_i (\dot{\boldsymbol{r}}_i \nabla_{r_i} + \dot{\boldsymbol{p}}_i \nabla_{p_i}) \ , \tag{3}$$

$\boldsymbol{r}_i$ and $\boldsymbol{p}_i$ denote the coordinates and momenta of the i-th particle of a subsystem. The formal solution to Eq. (3) is given by

$$\rho_{ens}(\Gamma,t) = e^{iLt}\rho_{ens}(\Gamma,0) \ . \tag{4}$$

Analogously, the time evolution of any observable A results as

$$A(\Gamma(t)) = e^{iLt}A(\Gamma(0)) \ . \tag{5}$$

If the system is in equilibrium, the phase space density ρ_{ens} becomes stationary. This means that at the same moment when one trajectory leaves a particular phase space element, on average another one enters this element so that the probability density for finding the systems in the element remains constant. The probability fluxes into and out of an infinitesimally small phase space element must be balanced and have the same absolute value. This is the *principle of detailed balance* which is the basis of, for example, probability-theoretical descriptions of chemical systems.

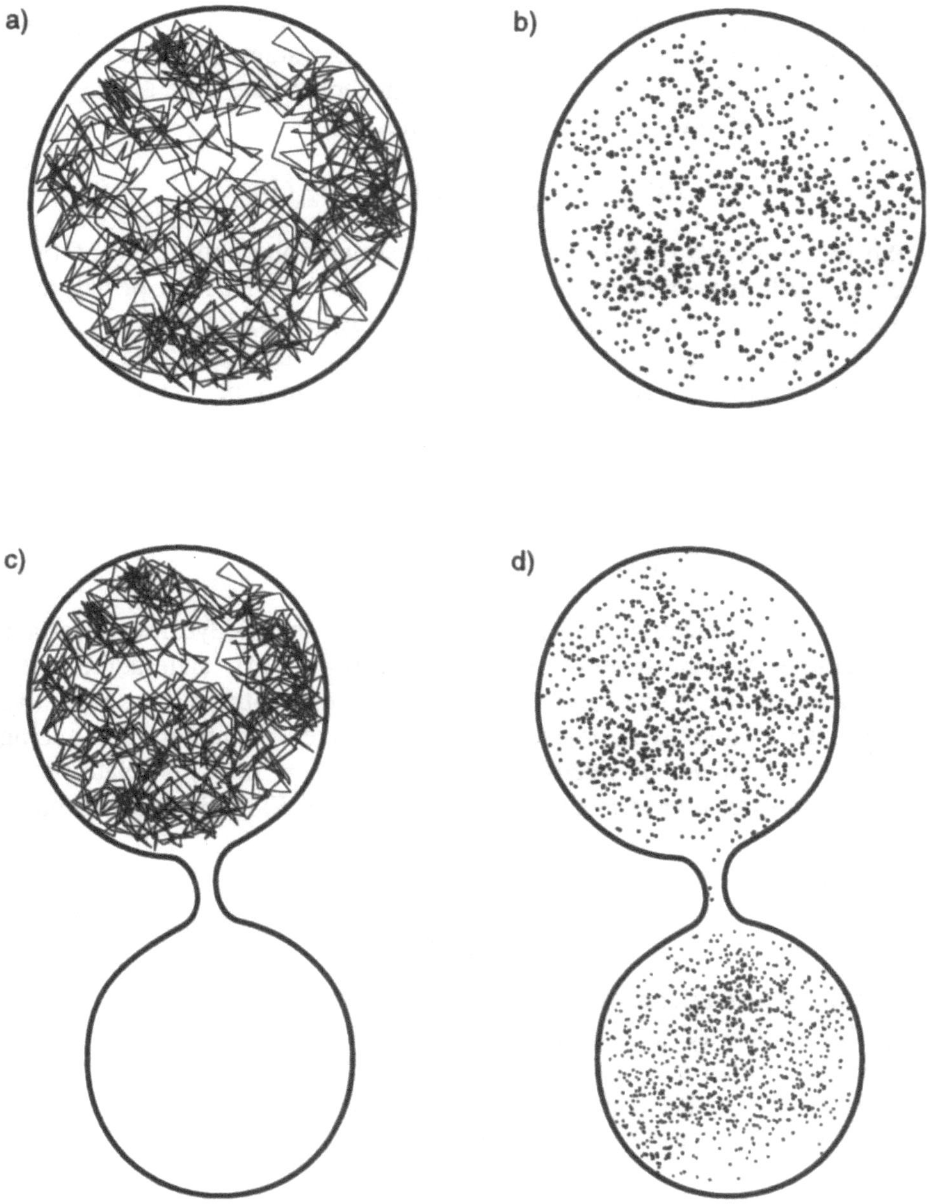

Figure 2. Phase space trajectory for an ergodic system (a) and a nonergodic system (c) (schematically). In the ergodic system the trajectory comes close to all accessible phase space points (b); in the nonergodic system this is not the case (d).

If all trajectories $\Gamma_i(t)$ which are represented by a stationary phase space density $\rho_{ens}(t_0)$ at a given time t_0 evolve in such a way that they come arbitrarily close to any point in the accessible phase space region (i.e. that part of the phase space which can be reached by the trajectory following the conservation laws of classical mechanics) within finite time the system is termed *ergodic* (as schematically depicted in Fig. 2) and one can replace the time average in Eq. (1) by an ensemble average (see Figs. 2a and 2b)

$$\langle A\rangle_{\text{time}} = \langle A\rangle_{\text{ens}} = \int_\Gamma d\Gamma\, A(\Gamma)\rho_{\text{ens}}(\Gamma) \ . \tag{6}$$

For the interpretation of the results of MD simulations knowledge of the approach towards ergodicity is absolutely necessary, at least in principle. For example, if the simulated system possesses high potential barriers between different phase space regions compared to k_BT, where k_B is the Boltzmann constant, it is rather likely that a selected trajectory may be trapped in just one region and therefore the calculated time average is not equal to the ensemble average which covers the total accessible phase space (compare Figs. 2c and 2d).

2.2. RELATIONS BETWEEN ENSEMBLES

Ensembles are characterized by a complete set of thermodynamical variables. The ensembles which are most frequently treated in MD simulations are the microcanonical (NVE), the canonical (NVT), and the isothermal-isobaric (NpT) ensemble. The variables in brackets signify the fixed quantities. The averaged quantity $\langle A\rangle_f$ (f denotes an intensive variable, e.g. $\beta = 1/k_BT$) is related to $\langle A\rangle_F$ (F denotes an extensive variable, e.g. the total energy E) by the series expansion [9]

$$\langle A\rangle_f = \langle A\rangle_{F=\langle F\rangle_f} + \frac{1}{2}\left(\frac{\partial^2}{\partial F^2}\langle A\rangle_{F=\langle F\rangle_f}\right)\langle \delta F^2\rangle_f + \dots \ . \tag{7}$$

If the number of particles of a system is very large (i.e. in the thermodynamic limit), the *fluctuation* δF^2 of an extensive quantity around the mean value becomes very small compared to the mean itself. In this limit, the correction terms containing derivatives with respect to the extensive variable in Eq. (7) become negligible and one simply has

$$\langle A\rangle_f = \langle A\rangle_{F=\langle F\rangle_f} \ . \tag{8}$$

The expectation values of any observable are identical in every ensemble. MD ensembles are usually far away from the thermodynamical limit, particularly because of the very limited number of particles and the limited simulation time. In general, it is not easy to check whether an MD ensemble is large enough in order to represent the thermodynamic limit within given error bars.

For fluctuations the ensemble equivalence (Eq. (8)) is no longer valid. One obtains the expression for the covariance of two observables A and B [9]

$$\langle \delta A\, \delta B \rangle_F = \langle \delta A\, \delta B \rangle_f + \left(\frac{\partial f}{\partial F} \right) \left(\frac{\partial}{\partial f} \langle A \rangle_f \right) \left(\frac{\partial}{\partial f} \langle B \rangle_f \right) + \dots \quad (9)$$

Here, the correction terms are of the same order of magnitude as the fluctuation itself and hence becomes non-negligible in general. One can easily show, for example, that the fluctuation of the kinetic energy, which determines the heat capacity, is *larger* in the canonical than in the microcanonical ensemble. This is reasonable, though confusing at first sight: If one interprets the *NVT* ensemble as a collection of *NVE* subensembles which are distributed according to the actual temperature, then the distribution of constant energy hypersurfaces is *folded* with the distribution of the kinetic energy on one selected hypersurface. This results in an effectively broader distribution of the kinetic energy in the canonical ensemble. Only in special cases the expression for a fluctuation is identical in any ensemble.

2.3. NUMERICAL INTEGRATIONS IN MOLECULAR DYNAMICS

2.3.1. *Principles*

Classical MD simulations are based on the numerical solution of the coupled set of Newtonian equations of motion of a (conservative) many-body system

$$m_i \ddot{\boldsymbol{r}}_i(t) = \boldsymbol{F}_i(t) = -\nabla_{r_i} V(\boldsymbol{r}_i(t)) \ , \quad (10)$$

$$\boldsymbol{p}_i(t) = m_i \dot{\boldsymbol{r}}_i(t) \ . \quad (11)$$

Here, $\boldsymbol{r}$ denotes the cartesian position vector of particle i, $\boldsymbol{F}$ the force acting on it, m its mass, V the potential, and $\boldsymbol{p}$ the momentum. The solution yields the trajectory of a single phase space point representing one subsystem of an ensemble. If the forces can be derived from a potential field (as in Eq. (10)) which is not explicitly time dependent, the total energy is a constant of motion. Analytical solutions to Eqs. (10) and (11) can be found for systems of free particles and for coupled harmonic oscillators, i.e. for normal modes of poly-atomic molecules or for the treatment of phonons in solids.

For any realistic model of chemical systems it is necessary to describe the interatomic interaction by means of more or less complex nonquadratic potentials, which is discussed in more detail in Sec. 4.4.1. The analytic solution to the Newtonian equations of motion is (in general) impossible and numerical procedures are demanded.

The integration strategies are based on a *discretization* of the continuous time t into time steps of length δt. For large systems (up to 10^5 particles) the numerical strategies have to be highly optimized in order to keep the numerical effort as small as possible. A very special type of differential equation has to be integrated. Consequently, general methods like Runge-Kutta integrators cannot efficiently be used in the field of MD simulations. The integration schemes have to take into account several necessities that are related to the available computer resources and to physical reasons:

a) *Speed and memory requirements*: The higher the integration speed, the longer might be the simulation time; the smaller the memory requirements, the larger the simulated system can be. However, the integration step is usually much less time consuming than the calculation of the intermolecular forces. Great effort is therefore spent in finding and optimizing methods for minimizing the time for the force computation.

b) *Size of the time step* δt: The larger the time step, the longer can be the simulation time. Integrators therefore should allow large time steps without a breakdown of the numerical stability. In usual (microcanonical) MD simulation, the numerical stability is tested considering the conservation of the total energy with different time steps.

c) *Time reversibility*: Since the (continuous) Newtonian equations of motion are time reversible, non-time-reversible integrators may possibly generate nonergodic trajectories and irreversible ensembles: Such systems can be irreversibly trapped in certain phase space regions far from the exact equilibrium distribution [10].

d) *Close approach to the exact trajectory*: The more closely the numerical trajectory approaches the exact (continuous) one, the more likely it is that the desired ensemble is correctly sampled (at least in the limit of long simulation times). In fact, this is not a necessary condition for a practically successful simulation: Every known integrator generates only an approximation of the exact trajectory. One has to make sure that the numerical error is under control for time intervals of the order of the molecular relaxation times of interest.

e) *Symplectic property*: If an integrator for an N-coordinate Hamiltonian system preserves the N Poincaré integral invariants, i.e. it conserves the measure in phase space (the N-th invariant is the conservation of the total phase space volume), the integrator is termed *symplectic*. It has recently been demonstrated that symplectic integrators usually lead to better energy conservation in the simulation of *NVE* ensembles than non-symplectic algorithms and to more reliable simulation results [11]. For other ensembles this property of the integrator is not important.

2.3.2. *Integration Schemes*

There are essentially two different integration schemes commonly in use, the (Gear) predictor-corrector method [12] and, most importantly, the Verlet algorithm [13] and variants that are algebraically related to the latter [14].

Predictor-corrector Methods. The predictor-corrector methods are based on the following scheme:

1. Expand the position r, velocity v, acceleration a, etc. in a Taylor series,

$$\begin{aligned} r^{p}(t+\delta t) &= r(t) + \delta t\, v(t) + (1/2)\delta t^{2} a(t) + (1/6)\delta t^{3} b(t) + \dots \\ v^{p}(t+\delta t) &= v(t) + \delta t\, a(t) + (1/2)\delta t^{2} b(t) + \dots \\ &\dots \end{aligned} \quad , \qquad (12)$$

   to obtain *predicted* ("first guess") values for the new position $r^{p}(t+\delta t)$ from the old position $r(t)$, etc.
2. With the predicted positions, calculate new, *corrected* accelerations $a^{c}(t+\delta t)$.
3. Estimate the error of the prediction step according to

$$\Delta a(t+\delta t) = a^{c}(t+\delta t) - a^{p}(t+\delta t) \quad . \qquad (13)$$

4. Use the latter expression to calculate *corrected* positions, velocities, etc.
5. Use the corrected values as predictions and repeat steps 2 to 5 up to a desired accuracy.

In principle, a very high numerical accuracy and energy stability can be achieved with predictor-corrector methods, but the technique involves at least two force calculations per time step. In practise, one has to find a compromise between accuracy of the integration and computational speed. If the number of correction iterations should be kept small, a rather small time step is required. Furthermore, the method needs the simultaneous storage of a large amount of data.

Verlet Integration. The Verlet integrator allows the use of a rather long time step in contrast to predictor-corrector strategies and involves only one force calculation per step. It is based on the following Taylor series expansion forward and backward in time

$$\begin{aligned} r(t+\delta t) &= r(t) + \delta t v(t) + (1/2)\delta t^2 a(t) + \ldots \\ r(t-\delta t) &= r(t) - \delta t v(t) + (1/2)\delta t^2 a(t) - \ldots \end{aligned} \quad . \tag{14}$$

Adding the latter two equations gives the position propagator

$$r(t+\delta t) = 2r(t) - r(t-\delta t) + \delta t^2 a(t) \quad , \tag{15}$$

whereas the subtraction yields for the velocity

$$v(t) = [r(t+\delta t) - r(t-\delta t)]/(2\delta t) \quad . \tag{16}$$

As can be seen, the velocity at a time t can only be calculated when the position of a further time step is known. This is an unavoidable feature of finite difference methods where the velocities are, implicitly or explicitly, computed from differences, not from differentiation. This property leads to several difficulties when velocity-dependent forces have to be considered, e.g. in dissipative constant-temperature integration methods [15].

In the Verlet scheme, the positions are correct up to an error of the order δt^4, but the velocities are only correct up to an order of δt^2. Nevertheless, Verlet integration usually gives better energy conservation when using a large time step than predictor-corrector methods. Due to the symmetric formulation the Verlet scheme is time reversible and conserves the measure in phase space, i.e. it is symplectic. The advantages of algorithms of this type are the small memory requirement for storing the positions and forces, their speed, and the high accuracy for large time steps.

2.3.3. *Limitations of the Simulation*

The application of molecular dynamics simulation techniques is limited mainly due to the numerical integration procedure. The largest part of current methodical research tries to overcome these limits with several algorithmic simulation "tricks". Other "strict sense" theoretical works aim at the question of how the limited knowledge that can be obtained by a simulation is transferable to much larger systems which will not ever be tractable by direct molecular simulation.

The limitations can be roughly divided into two categories: Simulation time and system size on the one hand and restrictions to ensembles that can be generated on the other hand.

a) *Time restrictions*: The restricted simulation time has essentially two pitfalls: Firstly, one has to assure that the simulated trajectory samples the accessible phase space adequately, i.e. the trajectory must be checked for ergodicity in order to calculate reliable statistical averages (see Fig. 2). This check is a major problem for systems with large (in comparison to $k_B T$) potential energy barriers between different phase space regions. The time scale of the MD simulation determines the maximum molecular relaxation time that can be studied: Only if the time correlation function of a selected quantity has decayed sufficiently close to zero within the simulation time, one is able to discuss the relaxation of the molecular process which influences this quantity.

b) *Size restrictions*: The numbers of atoms that can be treated in MD simulations are several orders of magnitude smaller than that in macroscopic systems, but one is interested in results that are valid for macroscopic systems. One possibility to overcome this problem is the introduction of *periodic boundary conditions* (see Fig. 3) in order to avoid surface effects: A simulation box or cell is periodically repeated in space. Atoms leaving the simulation box enter the box at the opposite side; each interaction site "sees" both the "real" atoms within the box and the "image" particles in the virtually repeated boxes. Special care has to be taken that the real particle and its image do not "see" each other via direct interaction. However, even if the direct interaction is excluded, it is still possible that a real molecule indirectly sees its image due to long range forces which polarize, for example, the solvent, or due to hydrodynamic interaction. These artificial interactions lead to the so-called *finite-size effects* which have not been systematically studied up to now. In any case, long-range spatial correlations can only be analyzed over a range of the order of the simulation box itself.

c) *Ensemble restrictions*: The usual MD ensemble is a subset of the microcanonical (*NVE*) ensemble with the additional constraint that the total linear momentum is conserved. As was demonstrated above, first order thermodynamic quantities are identical in any ensemble. This is not true when higher order thermodynamic derivatives are considered. Now it becomes important, what type of ensemble is simulated. Many methods have been proposed to keep, in particular, the temperature T and the pressure p constant. In the equilibration phase of a simulation the control of these variables is particularly important. If such a control cannot be imposed these quantities are only known *after* the simulation has been performed. The most important techniques for the control of external thermodynamic variables will be discussed in Sec. 3.

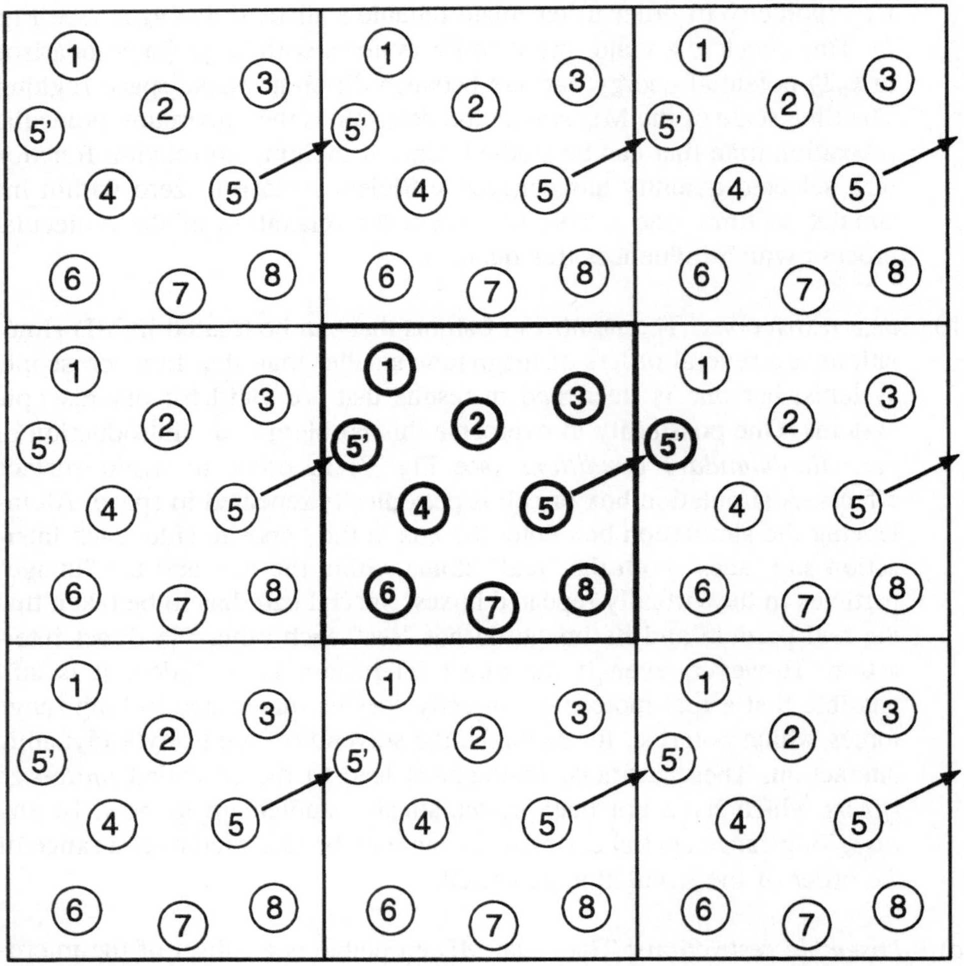

Figure 3. Schematical (2D) representation of periodic boundary conditions: When atom 5 leaves the central simulation cell to the virtual position 5', its image from the left side re-enters the central cell.

2.3.4. *Finite Systems*

The simulation restrictions mentioned above can be overcome in part by several techniques, the most actual of which will be discussed in Sec. 3. Here, we give two examples of different approaches related to the question "How can information about finite systems be transferred to macroscopic reality?". There are in principle two possible strategies:

1. One can try to find out a formalism for an extrapolation of the results obtained from finite systems to systems with many particles (in the order of Avogadro's number, i.e. in the thermodynamic limit).

2. One can study such systems for which experimental results are available also for systems with a very limited number of particles (i.e. isolated clusters).

A new statistical-mechanical approach towards the answer of the first question has recently been presented by Lustig [16]. In his work, exact expressions were derived for thermodynamic quantities up to fourth order for systems containing a finite number of particles under the conditions of the usual MD ensemble. This is the first time that the generally accepted *assumption* is systematically checked, as to whether the formulas valid in the thermodynamic limit can be used to evaluate properties of finite systems. The proposed method allows in principle to establish relations between the quantities calculated for finite systems (within a simulation) and the results in the thermodynamic limit.

Discrepancies between finite systems and the thermodynamic limit do not occur, when the experimental reality deals with finite systems as well: One may use *clusters* of atoms or molecules as finite model systems and study phase transitions or investigate the influence of the potential parameters on the thermodynamic behavior. In this case, the model systems are simple enough to perform MD simulations over a rather long period of time, but are nevertheless complicated enough to enlighten several aspects of cooperative motion or chaotic dynamics and their connection to the topology of the underlying potential surface. The systematic variation of the cluster size allows the study of several phenomena as function of the particle number and therefore, at least in principle, the extrapolation to macroscopic systems. An example for this type of MD applications has been recently summarized by Berry [17].

3. Extension of the Limits of Simple MD Simulations

3.1. SAMPLING FROM DIFFERENT ENSEMBLES

3.1.1. *Constant-temperature Simulations*

Simulations in the canonical (*NVT*) ensemble, i.e. at a constant temperature, require coupling the equations of motion of a simulated system to an external heat reservoir which allows the system to move in between phase space hypersurfaces of different energy. The canonical phase space density reads

$$\rho_{NVT}(\boldsymbol{P},\boldsymbol{Q}) \propto \exp[-\beta H(\boldsymbol{P},\boldsymbol{Q})] \tag{17}$$

where the $\boldsymbol{P}$, $\boldsymbol{Q}$ are momenta and coordinates of a system and $\beta = 1/k_BT$. During the course of an MD simulation, typically the kinetic (velocity space) temperature T_k according to the equipartition theorem

$$\left\langle \sum_{i=1}^{N} \frac{\boldsymbol{p}_i^2}{2m_i} \right\rangle = \frac{3}{2} N k_B T_k \tag{18}$$

can be easily monitored. In contrast, the position space temperature, which should be equal to the kinetic temperature if the canonical density is achieved, must be derived from inverting simulated distributions. In fact, the truncation error of the numerical integration methods leads to deviations between kinetic and position temperature if not special (non-MD) techniques are applied to generate the canonical ensemble. This is discussed in more detail below.

In the past two decades many different methods for conducting MD simulations at constant temperature have been proposed. They vary in the way that the system and the heat bath are coupled. Only a short outline of the existing techniques can be given here. The strategies can be roughly divided into deterministic and stochastic methods.

Deterministic Methods. Different authors constrained the kinetic energy to a fixed value corresponding to the desired kinetic temperature. In this way the natural fluctuations of the kinetic energy which determine the system's heat capacity are ignored, although the canonical distribution in configuration space is obtained [18-21]. Berendsen *et al.* [22] developed a velocity scaling scheme to adjust the kinetic temperature rapidly to the desired value. However, the method does not generate states in the canonical ensemble. Furthermore, local temperature differences may be preserved over rather long times resulting in an inefficient equilibration procedure.

The most widely used deterministic method is that of Nosé [23] and Hoover [24], in which the simulated system is extended by an additional degree of freedom, s. The potential associated with s (f denotes the system's number of degrees of freedom) is given as

$$V_s = (f+1) k_B T \ln s \tag{19}$$

while the kinetic energy reads

$$K_s = p_s^2/(2Q) \tag{20}$$

The "mass" Q is an adjustable parameter. The extra degree of freedom acts on the particle velocities via

$$v_i = p_i/(m_i s) \quad . \tag{21}$$

It can be shown that the total system including the extra degree of freedom, characterized by the Hamiltonian $H = V + V_s + K + K_s$, is microcanonical, whereas the simulated system alone obeys the canonical phase space density. While in older applications of the Nosé-Hoover scheme problems occurred with regard to ergodicity and strong dependencies on the coupling parameter [25], these problems have in part been resolved in recent work [26]: The extra degree of freedom is itself thermalized by additional Nosé terms, thus damping down eventual energy oscillations between system and reservoir. A recent review of deterministic methods has been given by Nosé [27].

Stochastic Methods. These techniques provide an alternative approach to generating canonical MD ensembles. Andersen [28] proposed a procedure where the velocities of randomly selected system particles are chosen at distinct times from a Maxwellian distribution at the desired temperature. This technique and related approaches [29-31] all generate trajectories which sample the canonical ensemble, but the coupling parameters are not easy to estimate for a given application. Other strategies use Brownian dynamics algorithms to integrate the strict Langevin equations of a many-body system [32,33]. The canonical ensemble is sampled, but the computational effort is very high.

Recently, a new stochastic approach to constant temperature dynamics has been developed in our group by Kast *et al.* [15,34]. The strategy is based on central impulsive collisions between system particles and imaginary heat bath particles of finite mass, therefore offering a rather intuitive approach to the system-reservoir coupling problem. The method is an adaptation of the Rayleigh model [35] of an ensemble of frictionless pistons subject to one-dimensional collisions with Maxwellian distributed heat bath particles. The technique is computationally efficient, easy to implement into existing MD code, and does not suffer the difficulties observed with the deterministic thermalization of small systems. The new algorithm can be characterized as follows.

The force vector acting on atom i at time step n

$$\boldsymbol{F}_{i,n} = \boldsymbol{F}_{i,n}^{(s)} + \boldsymbol{F}_{i,n}^{(c)} \tag{22}$$

is the sum of the systematic part

$$\boldsymbol{F}_{i,n}^{(s)} = -\nabla_{r_{i,n}} V(\boldsymbol{r}_1, \ldots, \boldsymbol{r}_N) \quad , \tag{23}$$

where N is the total number of atoms in the system, and the collision part

$$F_{i,n}^{(c)} = m_i \frac{2\alpha}{\delta t}(u_{i,n} - v_{i,n-1}) = \frac{2m_i}{\delta t}\frac{m_r}{1+m_r}(u_{i,n} - v_{i,n-1}) \;, \tag{24}$$

where $\alpha = m_{b,i}/(m_i+m_{b,i}) = m_r/(1+m_r)$ with $m_r = m_{b,i}/m_i$, $m_{b,i}$ being the mass of the heat bath particles. Note that only the velocity of step n-1 is available when the forces at step n are calculated. The velocity vector of the bath particles is a random number vector denoted by $\boldsymbol{u}_i$. The velocity distribution $f_{b,i}(u_i)$ is independently the same for each spatial direction and is taken to be Maxwellian at the bath temperature T_b:

$$f_{b,i}(u_i) = \left(\frac{m_{b,i}}{2\pi k_B T_b}\right)^{1/2} \exp\left(\frac{-m_{b,i}u^2}{2k_B T_b}\right) \;, \tag{25}$$

The bath particles do not interact with each other. The random variables u_i are taken to be *independent*, i.e. the bath is assumed to be infinitely large.

For Verlet-type algorithms, the method has been analytically studied for the cases of the free particle and the harmonic oscillator with respect to the resulting phase space density and to time correlation functions by means of solving the set of difference equations. Additionally, numerical studies were carried out on gaseous hydrogenfluoride and liquid argon. The canonical density is achieved while the influence on time correlation functions may be kept small.

From the velocity time correlation functions of a free particle subject to the heat bath two limiting cases to characterize the dynamics can be derived: Firstly, if the mass ratio between system particles and heat bath particles and the time step both tend to zero, the discrete correlation function approaches the continuous correlation function from Langevin dynamics of a free particle. Secondly, for a critical mass ratio $m_r = (3-8^{1/2})/(8^{1/2}-2) \sim 0.2071$ the dynamics switches from monotonically decreasing correlation functions to oscillatory behavior, since, for larger mass ratios, the particle's momentum changes its sign on average and the energy transfer becomes less effective. In the region of mass ratios around the critical value the dynamics is similar to that induced by the Andersen thermostat. This means, one is able to switch between Langevin-type and Andersen-type thermalization by means of only one control parameter, the mass ratio, without using different integration schemes.

In the case of the harmonic oscillator, the effect of the truncation error of Verlet integration on the resulting kinetic and position temperature could be analytically calculated: With $\varepsilon = 4\pi^2(\delta t/T)^2$ (T is the vibration period length of the oscillator) as a measure for the *graining* of the vibration period. Kast and Brickmann [34] could demonstrate that the simulated velocity v_n at a time step n is related to the exact (analytic) velocity v_n^* by

$$v_n = v_n^* \frac{\sin\sqrt{\varepsilon}}{\sqrt{\varepsilon}} , \tag{26}$$

i.e., the velocities are systematically *underestimated* compared to the analytic solution, whereas the turning points of a period are systematically *overestimated.* This yields an effectively too high kinetic temperature whereas the position temperature may deviate in both directions from the bath temperature.

These effects can be efficiently minimized to generate the canonical ensemble with a high accuracy at the desired temperature by applying a dynamic control circuit on the heat bath temperature: During intervals of some 10^2 time steps the average kinetic temperature of the simulated system is computed from the beginning of each interval up to the current step. The difference between this average system temperature and the desired temperature is multiplied by an adjustable parameter and added to or, respectively, subtracted from the heat bath temperature depending on whether the system is too cold or too hot. Therefore the bath temperature fluctuates strongly at the beginning of each cycle, while the fluctuations damp down toward the end. In this way, the system's modes are effectively uncoupled from the oscillatory "bath mode" and the system temperature can be accurately adjusted.

The method has been and is used in many MD simulations performed in the group of the authors, e.g. for the investigation of carbohydrate dynamics in solution [36], the study of a phospholipid bilayer-water interface (see Sec. 4), or simulations of xenon diffusion in zeolites [37].

Hybrid Monte-Carlo (MC) Methods [38,39]. The problems with finite time steps cannot be overcome with straightforward dynamics. The hybrid MC methods solve this problem in the following manner: Starting with a given system configuration and Maxwellian velocity distribution, several MD steps with a time-reversible and symplectic algorithm are performed in order to induce a global configuration change. After the MD period the resulting configuration is accepted or rejected with the usual Metropolis MC criterion [3]. After decision the velocities are updated according to a Maxwellian distribution, and the procedure is repeated. The MC step guarantees that the resulting (canonical) ensemble is free of integration errors. Therefore the MD time steps can be made larger than in usual MD simulations.

The hybrid MC strategy is particularly suited for bulk polymer simulations with full-detailed models where usual MC suffers from too high a rejection rate when global changes are induced randomly [39]. However, one has critically to balance out the length of the MD period and the time step length with regard to an acceptable rejection rate. It was found [39] that, compared to usual MD simulations, a speedup factor of approximately 1.5 can be achieved while the MD periods can be made arbitrarily large. It is not clear up to now whether the method can be used to study not only static ensemble quantities but dynamic properties.

3.1.2. *Constant-pressure Simulations*

MD simulations at a constant pressure are not only neccessary for the equilibration phase of a simulation of large systems where the initial configuration usually is far from equilibrium, but are useful e.g. for the simulation of phase transitions in solids or the calculation of some special thermodynamic quantities, for example the adiabatic compressibility. We only review the most popular methods here and describe the major difficulties.

The instantaneous pressure at time step n of a system is given by

$$P_n = \frac{N}{V} k_B T_n + \frac{1}{3V} \sum_{i=1}^{N} \boldsymbol{r}_{i,n} \cdot \boldsymbol{F}_{i,n} \ , \tag{27}$$

where V is container volume, T_n is the instantaneous temperature. Ensemble averaging of Eq. (27) gives the thermodynamic pressure. We here discuss the constant-pressure techniques on the example of a monoatomic fluid in a cubic box with volume V and periodic boundary conditions ($0 \le \boldsymbol{r}_i < V^{1/3}$). The corresponding Hamiltonian reads, with u being the interparticle potential:

$$H_1(\boldsymbol{r}^N, \boldsymbol{p}^N) = (2m)^{-1} \sum_i \boldsymbol{p}_i \boldsymbol{p}_i + \sum_{i<j} u(r_{ij}) \ . \tag{28}$$

Andersen's Method [28]. The cartesian coordinates of the particles are replaced by fractional coordinates

$$\boldsymbol{r}_i \rightarrow \rho_i = \frac{\boldsymbol{r}_i}{V^{1/3}} \tag{29}$$

When simulating at constant pressure, the box volume is variable in the equations of motion. Associated with the box volume is a mass (i.e. a piston acting on the system). The new Hamiltonian is

$$\begin{aligned} H_2(\rho^N, \pi^N, V, \Pi) = {} & (2mV^{2/3})^{-1} \sum_i \pi_i \pi_i + \sum_{i<j} u(V^{1/3} \rho_{ij}) \\ & + (2M)^{-1} \Pi^2 + P_{ex} V \end{aligned} \tag{30}$$

with the particle momenta

$$\pi_i = m V^{2/3} \dot{\rho}_i \tag{31}$$

and the momentum of the piston

$$\Pi = M\dot{V} \tag{32}$$

The motion of the piston is given as

$$\frac{Md^2V}{dt} = -P_{ex} + \frac{\frac{2}{3}\sum_i \frac{p_i p_i}{2m} - \frac{1}{3}\sum_{i<j} r_{ij} u'(r_{ij})}{V} . \tag{33}$$

In the limit $M \rightarrow \infty$ and $dV/dt \rightarrow 0$ the equations of motion of the original system are restored. In fact, without using one of the constant-temperature simulation methods described above, a somewhat unusual isobaric-isenthalpic ensemble is generated. When a thermostat is applied, averages of observables are indeed those of an isobaric-isothermal ensemble. For fluctuations (or autocorrelation functions), the equivalence has not been proved.

Berendsen's Method [22]. In this approach, an extra term is added to the equations of motion to couple the system to an external "pressure bath":

$$\mu = \underline{1} - \frac{\beta \delta t}{3\tau_p}(P_0 \underline{1} - P) \tag{34}$$

where τ_p is the compressibility, β is a coupling constant, μ is a scaling tensor, and P_0 is the target pressure. At any time step the box volume is scaled in order to induce a pressure change. After calculation of the pressure tensor P and the scaling tensor μ, the new coordinates of the particles and the new volume of the box are given as

$$\boldsymbol{r}_i' = \mu \boldsymbol{r}_i , \tag{35}$$

$$V' = |\mu| V . \tag{36}$$

Extension to Nonrectangular Systems. Parinello and Rahman [40] extended Andersen's method to systems with nonrectangular simulation boxes. The fractional coordinates are now given by

$$\boldsymbol{r}_i = \boldsymbol{u}_1 s_{i1} + \boldsymbol{u}_2 s_{i2} + \boldsymbol{u}_3 s_{i3} = h \boldsymbol{s}_i \tag{37}$$

where $\boldsymbol{u}$ are the unit vectors which define the box. With the tensor

$$h = (\boldsymbol{u}_1, \boldsymbol{u}_2, \boldsymbol{u}_3) \tag{38}$$

we now obtain the equations of motion of the simulation box

$$\frac{M d^2 h_{ab}}{dt^2} = [(S - P_{ex}\underline{1})\, V [h^{-1}]^T]_{ab} \tag{39}$$

with S being the stress tensor which describes the deformation forces acting on the simulation box

$$S_{ac} = \frac{1}{V}\left\{ m \sum_i [h\dot{s}_i]_a^t [h\dot{s}_i]_c + \sum_{i>j} [u'(r_{ij}) e_{ij}]_a \; [h(s_i - s_j)]_c \right\} . \tag{40}$$

Rotation of the Simulation Box. In the *NpT* ensemble the momentum and the angular momentum are no longer conserved. As a consequence, the simulation box might rotate [41]. The tensor h has 9 components amongst which 6 components describe the box (3 axes lengths, 3 angles), 3 components describe the box orientation in space (Euler angles). The rotation of the MD box ("superfluous rotation") has no effect on the motion of the particles, but makes the analysis of the resulting trajectory more complicated. According to Nosé [41], the tensor h can be subdivided into a symmetric and an asymmetric tensor, formally $h = sym + asym$. The asymmetric tensor describes the rotation of the simulation box. By introducing the constraint $asym = 0$, the rotation of the box can be supressed. The new equation of motion then reads

$$\frac{M d^2 sym_{ab}}{dt^2} = \frac{[(S - P_{ex} 1)\, V [h^{-1}]^T]_{ab} + [(S - P_{ex} 1)\, V [h^{-1}]^T]_{ba}}{2} . \tag{41}$$

Comparison. Andersen's method is characterized by the facts that the properties of the generated ensembles are to some extent known, but the implementation is complicated. Here, the pressure shows damped oscillatory behavior when approaching its equilibrium value. In contrast, Berendsen's method is easy to implement and numerically more stable, but the ensemble properties are not clear.

3.2. EXPANDING SIZE AND TIME LIMITS

Several techniques have been developed to expand the possible number of particles that can be simulated in order to study long range spatial correlations, and to lengthen the simulation time for investigation into slow molecular relaxation processes. We will not discuss here the standard strategies like, e.g., the united-atom approach, constraint dynamics and recent improvements [42], or methods to coarse-grain a model, e.g. Brownian dynamics simulations. Instead, we will focus on some new developments.

3.2.1. *Long-range Interactions*
The biggest numerical effort of an MD simulation is related to the calculation of long-range interactions (mainly Coulombic and to some extent Lennard-Jones interactions). If all interactions are calculated, the computational expense increases with the square of the particle number as far as only pair potentials are used. Since the common intermolecular potentials approach to zero for large distances one may calculate the interaction energy and the forces within a limited region around the sites only. We only treat some aspects of this strategy here and avoid a discussion of Ewald summation techniques [43] to compute exact Madelung energies for periodic structures.

Shifted-force Potentials. If one truncates the interatomic potential at a certain (spherical) cut-off distance r_{cut} singularities occur for the forces at this distance resulting in numerical instability of the integration procedure. Several possibilities exist to modify the truncated potential in such a manner that both the potential and the force decay to zero when approaching r_{cut}, leading to so-called shifted-force (SF) potentials. Dufner *et al.* [44] investigated the properties of two different SF potentials for static ionic crystals with regard to the convergence to the Madelung energy when different cut-off radii are used, and to the convergence behavior of differently complex structures. Two types of SF potentials are presented in some detail. These potentials are the CHARMM cut-off potential V_{CHARMM} for Coulomb interactions

$$V_{\mathrm{CHARMM}}(r_{ij}) = \begin{cases} \dfrac{q_i \cdot q_j}{r_{ij}} \left[1 - \left(\dfrac{r_{ij}}{r_{cut}} \right)^2 \right]^2, & r_{ij} < r_{cut} \\ 0 \,, & r_{ij} \geq r_{cut} \,. \end{cases} \tag{42}$$

used within the CHARMM program [45] and the Schrimpf cut-off potential V_{SCHRIMPF}

$$V_{\mathrm{SCHRIMPF}}(r_{ij}) = \begin{cases} \dfrac{q_i \cdot q_j}{r_{ij}} \left[1 - \dfrac{4}{3}\dfrac{r_{ij}}{r_{cut}} + \dfrac{1}{3}\left(\dfrac{r_{ij}}{r_{cut}}\right)^4 \right], & r_{ij} < r_{cut} \\ 0\,, & r_{ij} \geq r_{cut} \end{cases} \tag{43}$$

developed in the group of the authors [46]. It was found that one has to select different shifted-force potentials according to the question of whether either the energy or the force should be accurately calculated with respect to the exact Madelung value. The CHARMM potential is better suited for energy convergence whereas the SCHRIMPF potential has advantages when a high force accuracy is required, i.e. the latter is better for MD simulations. However, the actual convergence behavior strongly depends on the complexity of the simulated system. The more complex the structure the smaller may be the cut-off for obtaining reliable results. It was demonstrated that the usual cut-offs of the order 10 Å are sufficient only if the structure is not too simple. However, this is the case for most current MD simulations.

The Cell-multipole Method. Whilst cut-off techniques represent one set of possibilities for a reduction of computational effort, the cell multipole method (CMM) developed by Ding *et al.* [47] is another. This technique reduces the computational effort for the calculation of long-range interactions without using a cut-off, in a similar way to the fast multipole algorithm by Board *et al.* [48]. In CMM, atoms are grouped in cells and the electrostatics of these atoms in the cell are described by a multipole expansion of the cell. Only the interactions with the nearest cells are calculated on an atom-atom basis ("near potential"), the "far interactions" are calculated by the cell- multipole expansion [49,50].

$$\begin{aligned} V(\boldsymbol{r}) &= V_{near}(\boldsymbol{r}) + V_{far}(\boldsymbol{r}) \\ &= \sum_{atoms} \frac{q_j}{|\boldsymbol{r} - \boldsymbol{r}_{atom}|} + \sum_{cells} V_{cell}^{pole}(\boldsymbol{r} - \boldsymbol{r}_{cell}) \end{aligned} \tag{44}$$

$$V_{cell}^{pole}(\boldsymbol{r} - \boldsymbol{r}_{cell}) = \frac{Z}{\boldsymbol{R}} + \frac{\mu_\alpha \boldsymbol{R}_\alpha}{\boldsymbol{R}^2} + \frac{Q_{\alpha\beta}\boldsymbol{R}_\alpha \boldsymbol{R}_\beta}{\boldsymbol{R}^4} + \frac{O_{\alpha\beta\gamma}\boldsymbol{R}_\alpha \boldsymbol{R}_\beta \boldsymbol{R}_\gamma}{\boldsymbol{R}^6} + \ldots\ldots \tag{45}$$

where Z denotes the charges, μ the dipole moments, Q the quadrupole moments, and O the octopole moments with $R = r - r_{\mathrm{cell}}$. As the distance from an atom increases, the size of these cells is increased by grouping cells into bigger cells (hierarchical cell clustering).

A further reduction of computational effort is achieved when the interactions are not calculated between a certain atom and a cell, but only between cells. The resulting interaction for the atoms in the cell is described by a Taylor series expansion of the cell-cell interaction.

$$\sum_{\text{cells}} V_{\text{cell}}^{\text{pole}}(r - r_{\text{cell}}) = V^{(0)} + V_{\alpha}^{(1)} r_{\alpha} + V_{\alpha\beta}^{(2)} r_{\alpha} r_{\beta} + \ldots \tag{46}$$

with r and r_{cell} being distances of atoms or cells to the center of a certain cell. CMM is especially suited for parallel computer architectures and obtains large speedups when the systems to be simulated are sufficiently large [48].

3.2.2. *Multiple-time-step Methods*

The length of a time step used in an MD simulation is limited by the fastest modes of the molecular systems (usually bond stretching vibrations). Typically, one is interested in far slower modes, i.e. much computing time is spent in the calculation of small-amplitude motion while the interesting modes remain almost unchanged. Therefore, it is desirable to subdivide a system into several differently fast modes each of which are integrated with a different time step. Early attempts to achieve this goal reached a speedup of approximately 2-3 [51]. A new formalism was recently presented by Askar [52]. It has been presented at this summer school as well and so we will not review it here.

Recently, Tuckerman *et al.* [53] derived a multiple-time-step (MTS) scheme which is based on the Trotter factorization [54] of the classical Liouville propagator (4)

$$\begin{aligned} e^{i(L_1+L_2)t} &= \left[e^{i(L_1+L_2)t/P}\right]^P \\ &= \left[e^{iL_1(\delta t/2)} e^{iL_2\delta t} e^{iL_1(\delta t/2)}\right]^P + O(t^3/P) \end{aligned} \tag{47}$$

which is here formulated for a Liouville operator that is decomposed into two parts L_1 and L_2. This factorization, applied to a phase space point, yields exactly time-reversible algorithms. For example, with the one-particle Liouville operator

$$iL_1 = F(x)\frac{\partial}{\partial p} \;;\quad iL_2 = \dot{x}\frac{\partial}{\partial x} \tag{48}$$

and the Trotter factorization

$$e^{\delta t(\dot{x}\partial/\partial x + F(x)\partial/\partial p)} \approx e^{(\delta t/2)F(x)\partial/\partial p} e^{\delta t\,\dot{x}\partial/\partial x} e^{(\delta t/2)F(x)\partial/\partial p} \tag{49}$$

and by using the theorem

$$e^{y\partial/\partial x} f(x) = f(x+y) \tag{50}$$

where y is taken to be independent of x, one can derive the velocity Verlet integrator by operating with the right hand side of Eq. (49) on an initial state $\{x(0),p(0)\}$:

$$\begin{aligned} & e^{(\delta t/2)F(x)\partial/\partial p} e^{\delta t\, \dot{x}\partial/\partial x} e^{(\delta t/2)F(x)\partial/\partial p} \{x(0),p(0)\} \\ & = e^{(\delta t/2)F(x)\partial/\partial p} e^{\delta t\, \dot{x}\partial/\partial x} \{x(0),p(\delta t/2)\} \\ & = e^{(\delta t/2)F(x)\partial/\partial p} \{x(\delta t),p(\delta t/2)\} \\ & = \{x(\delta t),p(\delta t)\} \end{aligned} \tag{51}$$

where

$$p(\delta t/2) = p(0) + (\delta t/2)\, F(x(0)) \ , \tag{52}$$

$$\begin{aligned} p(\delta t) & = p(\delta t/2) + (\delta t/2)\, F(x(\delta t)) \\ & = p(0) + (\delta t/2)[F(x(0)) + F(x(\delta t))] \ , \end{aligned} \tag{53}$$

and

$$x(\delta t) = x(0) + (\delta t/m)[p(0) + (\delta t/2)\, F(x(0))] \tag{54}$$

with the particle's mass m.

The Liouville operator may also be decomposed into separate fractions each belonging to differently hard forces. In this way, one generates MTS integrators for a broad variety of systems. However, the numerical stability must be critically checked for any decomposition. Algorithms of this type were exemplarily applied to systems of disparate masses, short-range and long-range forces, and to systems containing stiff oscillators in a soft bath. A generalized procedure for the simulation of macromolecules has been recently reported [55] which obtains a speedup of 4-5. In a simulation of solid C_{60} the speed-up even reaches 20-40 compared to usual simulations [56]. A particularly important application of the MTS scheme is the implementation into Car-Parrinello programs [57,58] where the time step is one order of magnitude smaller than in classical MD simulations.

4. Application of Large Scale MD Simulations: Structure and Dynamics of Biomembranes

4.1. INTRODUCTION

Biological membranes play an important role for the life of almost every organism. Membranes enclose every living cell and partition the interior into different reaction sites. They regulate the ion concentration inside the cells and influence the transport of small molecules between cells. Their complex structure can be treated as a two-dimensional (2D) liquid containing lipids, proteins and steroids (Fig. 4). Unlike proteins, bilayers have no boundaries in two dimensions. Hence, a realistic model must be sufficiently sized in order to avoid arteficial effects due to long-ranging correlations which will be enhanced by periodic boundary conditions (see Sec. 2.3.3.).

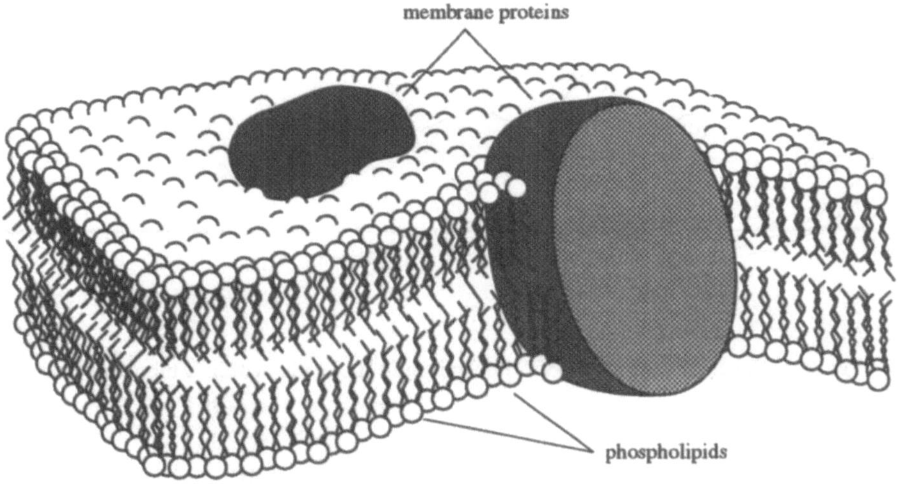

Figure 4. Schematic structure of a biomembrane.

4.2. STRUCTURE OF PHOSPHOLIPIDS

Phospholipids have the general structure given in Fig. 5. The head group consists of a polar fragment (e.g., ethanolamine, choline, serine). It is attached to a phosphate group, which is itself attached to a backbone fragment via an ester linkage. The backbone consists of glycerol or propanediol. The remaining

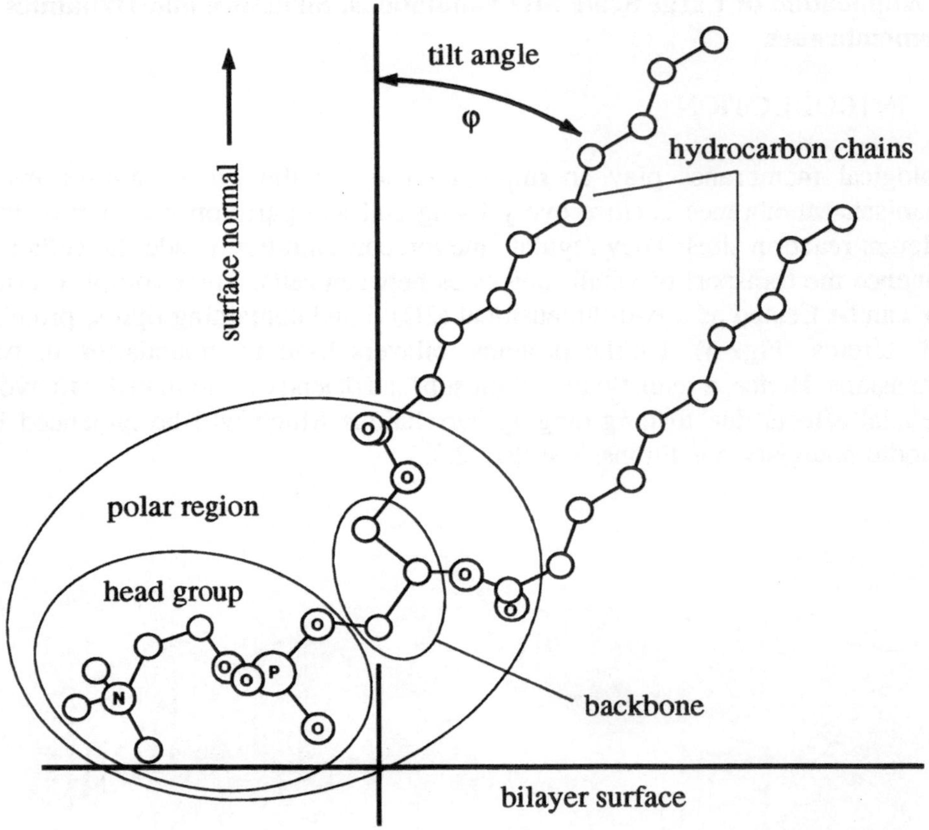

Figure 5. Schematic structure of a phospholipid molecule within a biomembrane.

one or two hydroxyl groups are then attached to fatty acids through ester linkages. They form the nonpolar portion of the molecule, whereas the head group and the backbone (including the ester linkage) constitute the polar region. Thus, these molecules are amphiphilic. This property allows them to form a variety of molecular assemblies. Besides micelles, tubes, and monolayers, the bilayer phase is the most important. It consists of two monolayers connected with the lipid tails pointing toward each other and the head groups forming the surface. When numerous bilayers are stacked they form a multilamellar phase.

4.3. EARLY MODELS

4.3.1. *Bilayer*

One of the first MD simulations of a bilayer system was performed by Ploeg *et al.* [59] The bilayer contained 2 x 16 decanoate molecules and was 2D periodic. The authors observed from their 80 ps simulation a collective tilt of the hydrocarbon chains extending over the whole simulation cell. This long-range correlation was enhanced by the periodic boundary conditions. Therefore, they performed a second MD run with a larger system, consisting of 2 x 64 decanoate molecules. This time, the collective tilt did not extend over the whole unit cell.

Egberts *et al.* [60] extended this simple model significantly. They not only added water and counterions to their system, but they also used a mixture of decanole and sodium decanoate. This time, periodic boundary conditions were applied in three dimensions, i.e. the model system was a multilamellar system. The simulation was performed under constant temperature and constant pressure (*NpT*) conditions at 300 K and 1 atm. The initial simulation cell contained only 20 decanole molecules, 12 decanoate ions, 12 sodium ions and 128 SPC [61] water molecules. Due to the long-ranging correlations which have been observed previously [59], this small system tends to crystallize already in the equilibration run. The authors therefore decided to increase the system size to 76 decanole molecules, 52 decanoate ions, 52 sodium ions and 526 water molecules. Now, the system stabilizes at a liquid-crystalline state.

4.3.2. *Monolayer*

One approach to reduce the required computer resources, and thus increase the system size to overcome finite size effects (see Sec. 2.3.3.), is to use monolayers as model membranes. Böcker *et al.* [62,63] simulated a monolayer consisting of 64 *n*-hexadecyltrimethylammonium chloride ($C_{16}TAC$) molecules and 1632 TIP4P [64] water molecules to investigate the structural and dynamic properties of a membrane-water interface. The water phase was connected to a "Lennard-Jones" wall by applying a van-der-Waals-type interaction to the water oxygens depending on their distance from the wall. Like Egberts *et al.* [60] they had to take care preparing the initial configuration. All simulation runs were started with an area per head group of 45 Å^2 according to experimental results [65,66]. In the first simulation, the amphiphilic molecules had an initial tilt angle of 0°, which resulted in a lower density for the hydrocarbon groups compared to that of the head groups. For that reason, the tails collapsed and turned their ends in the direction of the head groups in order to achieve a higer density. In the final simulation run, the authors used a tilt angle of 44° in the starting configuration. It turned out that the average tilt angle fluctuated slightly around the initial value during the whole simulation. Böcker *et al.* predicted the results of a neutron reflectivity study of monolayers of *n*-tetradecyltrimethylammonium bromide (C_{14}TAB) and *n*-octadecyltrimethylammonium bromide (C_{18}TAB) [67]. The calculated density profiles, perpendicular to the monolayer, of water, head groups, and chains are in excellent agreement with the ones from the neutron reflectivity study.

4.3.3. *Micelle*

Additionally, Böcker *et al.* [68] performed an MD simulation of a micelle in water. The system contained 30 *n*-decyltrimethylammonium chloride (C_{10}TAC) molecules and 2166 TIP4P water molecules using the same model potentials as in the previous study. After 275 ps simulation time, the final shape was slightly prolate ellipsoidal. It was found that the head groups are located in a molecularly sharp interfacial region, which is in very good agreement with experimental studies [65] of a similar system.

4.4. PHOSPHOLIPID MODELS

4.4.1. *Parameterization*

The first available force fields for biomolecules were developed for the simulation of proteins [45,69-79]. Schlenkrich *et al.* [80] extended the version 22 of the CHARMM force field [45] with parameters for phospholipids. This new version of the CHARMM force field includes effective potential functions, which are important for the consideration of polarization effects due to the water molecules.

The CHARMM force field is subdivided into three parts: one for intermolecular interactions, one for the vibrational modes, and one for the torsional flexibility (see Fig. 6). The functional form reads

$$V = \sum_{\text{bonds}} k_b (r - r_0)^2 + \sum_{\text{angles}} k_\alpha (\alpha - \alpha_0)^2 + \sum_{\text{impropers}} k_\phi (\phi - \phi_0)^2 + \sum_{j>i} \left(\frac{A_{ij}}{r_{ij}^{12}} - \frac{B_{ij}}{r_{ij}^{6}} + \frac{q_i\, q_j}{4\pi\varepsilon_0 r_{ij}} \right) + \sum_{\text{torsions}} k_\tau [1 + \cos(n\tau - \delta)] \qquad (55)$$

(r: bond length, r_0: equilibrium bond length, k_b: force constant (bond), α: bending angle, α_0: equilibrium bending angle, k_α: force constant (bending), τ: torsional angle, n: multiplicity, δ: phase angle, k_τ: force constant (torsion), ϕ: improper torsional angle, k_ϕ: force constant (improper torsion), ϕ_0: equilibrium improper torsional angle, r_{ij}: distance, A_{ij}, B_{ij}: Lennard-Jones parameters, q_i, q_j: charges).

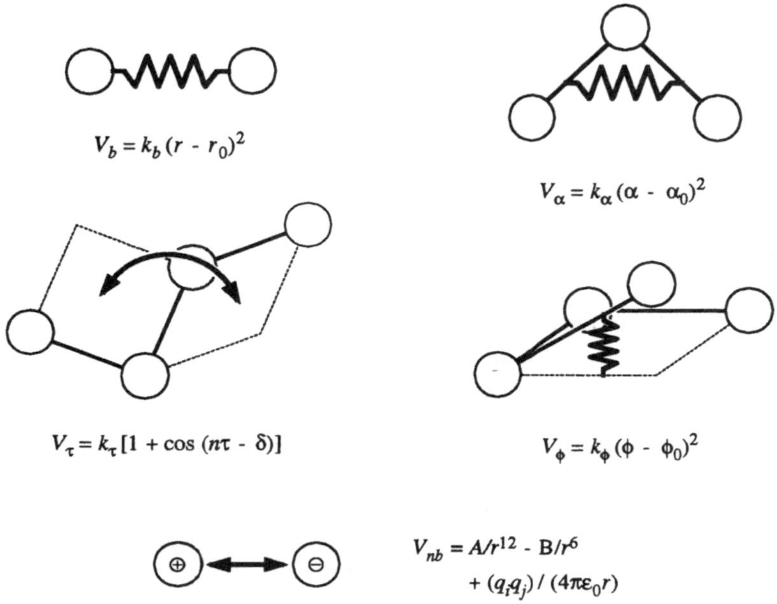

$V_b = k_b (r - r_0)^2$
$V_\alpha = k_\alpha (\alpha - \alpha_0)^2$
$V_\tau = k_\tau [1 + \cos (n\tau - \delta)]$
$V_\phi = k_\phi (\phi - \phi_0)^2$
$V_{nb} = A/r^{12} - B/r^6 + (q_i q_j) / (4\pi\varepsilon_0 r)$

Figure 6. Functional form of the CHARMM force field.

The vibrational part of the force field is represented by harmonic potentials for internal coordinates (bond lengths, valence angles, and out-of-plane displacements). These terms essentially represent the vibrational frequencies as experimentally observed. The intermolecular interactions are treated by Coulombic and Lennard-Jones (12,6) pairwise interactions. These terms are essential for the interaction of the phospholipids with their solvent and for the heat of solvation. The third and last contributions are the potentials associated with the dihedral angles, representing most of the molecular flexibility.

In the CHARMM22 force field, the partial charges were derived to treat the polarization effects arising from the solvent in an effective manner and to reproduce the experimentally measured free energies of solvation. The harmonic force constants and equilibrium distances and angles were fitted to the experimentally observed vibrational frequencies. These local parameters should not change when larger molecules are composed, as there are no known electron delocalization effects for these compounds. The new parameters derived by Schlenkrich *et al.* are based on accurate ab-initio calculations of small molecules. Quantum-chemical calculations were performed with these fragments to obtain relative energies of conformers and barrier heights. The obtained data was used to modify the force field for the best possible agreement.

4.4.2. *Verification*

Model Systems. We performed MD simulations of phospholipid crystals in order to show the capability of the extended CHARMM22 force field to describe large lipid bilayers. The force field has been tested with smaller systems like the unit cells of phospholipid crystals [81]. We have expanded these tests with larger systems under more realistic conditions. For that reason simulations were carried out with the experimentally obtained crystal structures of 2,3-dilauroyl-DL-glycerol-1-phosphorylethanolamine acetic acid (DLPE) [82], 2,3-dimyristoyl-DL-glycerol-1-phosphorylcholine dihydrate (DMPC) [83], and 3-lauroylpropanediol-1-phosphorylcholine monohydrate (LPPC) [84] as starting points (simulation cell parameters are given in Tab. I). Each system was simulated in two different ways, in the microcanonical ensemble (*NVE*) and under constant temperature and constant pressure conditions (*NpT*). The latter was applied to consider the natural circumstances, i.e. room temperature and atmospheric pressure. The constant pressure simulations were also employed to investigate the stability of the box size and shape.

TABLE I. Simulation cell parameters of the model systems used in the simulations.

	DLPE	DMPC	LPPC
molecules	96 lipid 96 acetic acid	96 lipid 192 water	96 lipid 96 water
atoms	10080	11904	7008
volume [nm^3]	88.568	102.558	61.248
a [Å]	47.73	52.32	24.83
b [Å]	46.64	35.68	57.18
c [Å]	39.81	55.40	43.76
α [°]	90.00	90.00	90.00
β [°]	92.02	97.40	99.66
γ [°]	90.00	90.00	90.00

From the previously reported simulations it is known that the molecular motions in amphiphilic bilayers are marked by long-ranging correlations [59,60,63]. Therefore, each simulation cell consists of 24 unit cells (96 lipid molecules) arranged in a 6 x 4 matrix. We used the CHARMM22 force field with the above mentioned extensions for phospholipid molecules. Two types of molecular dynamics simulations were carried out. In the first simulation the systems were heated near room temperature in a short time interval. After equilibration the systems were simulated in the microcanonical ensemble over a period of approximately 100 picoseconds. In the second simulation the systems were coupled to an external heat bath to keep the temperature constant [15,34] (see Sec. 3.1.1.). In addition we employed the Berendsen method [22] (see Sec. 3.1.2.) to keep the pressure constant near the atmospheric pressure. Because of the low frequency fluctuations of the lattice parameters the systems were simulated for more than 200 picoseconds after equilibration. The simulation conditions are summarized in Tab. II.

TABLE II. Simulation conditions for the model systems.

		DLPE	DMPC	LPPC
time [ps]	*NVE*	107	104	100
av. temperature [K]	*NVE*	285.6	294.0	292.6
time [ps]	*NpT*	260	250	220
temperature [K]	*NpT*	293.3	293.8	293.4
pressure [bar]	*NpT*	1.0	1.8	1.0

Results. We have concentrated our analysis on molecular conformations and their changes with respect to crystal structure. Fig. 7 shows the average structures of the four different lipid molecules calculated from the *NVE* simulations. The atoms are color-coded with respect to the mean fluctuations around their equilibrium position. This reveals their thermal motions. The terminal methyl groups of the hydrocarbon chains of DLPE and DMPC experience the largest fluctuations, which is in good agreement with X-ray structures [82] where the tails show nearly fluid behavior. The distributions of the dihedral angles show that all angles oscillate roughly ±20° around their mean value. Besides a few exceptions, the torsion angles calculated from the crystal structures lie inside these distributions. None of them show conformational transitions. The tilt angle (Fig. 5) and its distribution reveal the flexibility of the hydrocarbon chains. The LPPC chains show the sharpest distributions of the tilt angles due to their interdigitated packing. The distribution of the angles gets broader in the *NpT* simulations in the case of DPLE and DMPC due to the additional degrees of freedom. Simulation results and corresponding crystal data are compared in Tabs. III and IV.

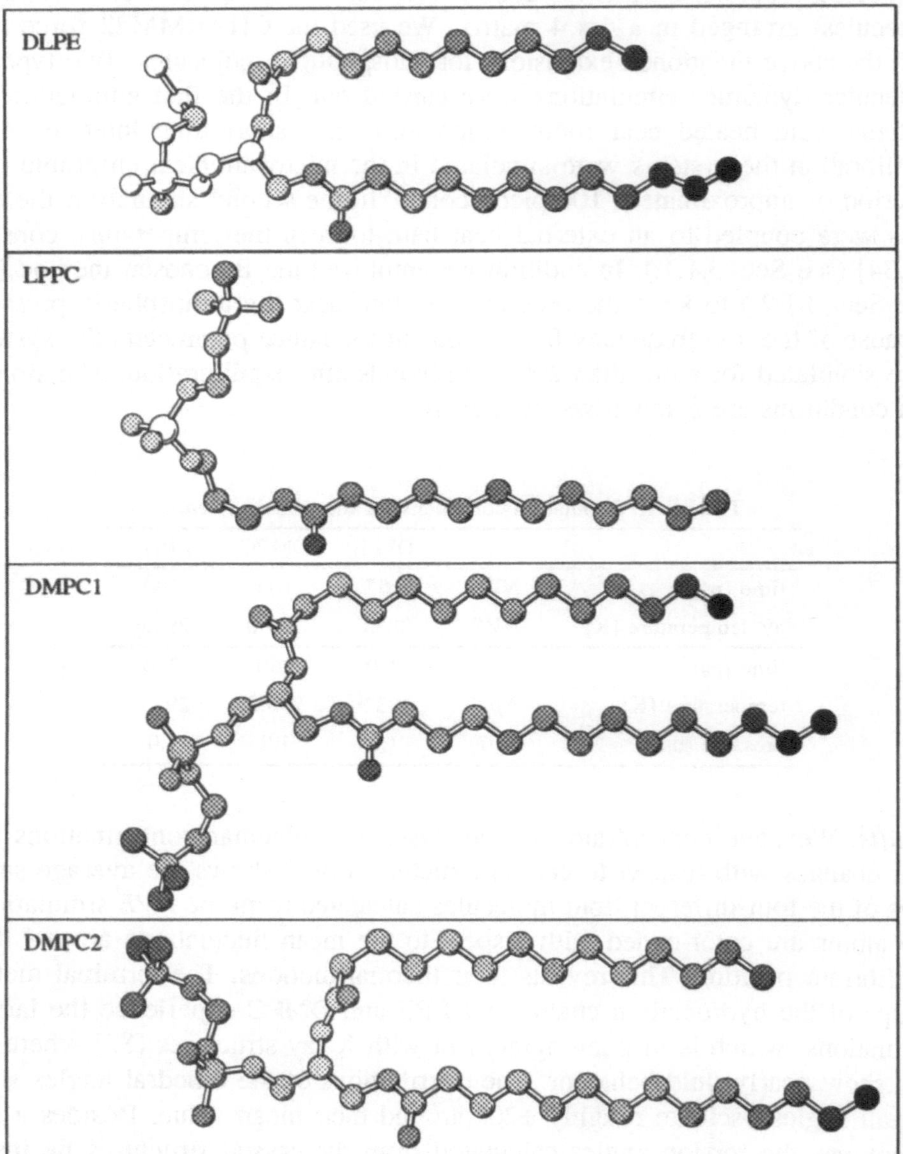

Figure 7. Average structures of different lipid species as calculated from the NVE simulations. DMPC1 and DMPC2 refer to the two distinguishable molecules within the DMPC unit cell. Atomic RMS fluctuations are color coded: white = 0.27 Å, black = 0.95 Å.

The largest changes in the box size occurred in the initial phase of the *NpT* simulation of DLPE and LPPC. In both cases the drifts happened during the first two picoseconds. After the equilibration phase the box sizes showed no further changes.

TABLE III. RMS deviations between the crystal structures and the averages structures (r: bond length, α: valance angle, θ: dihedral angle).

		DLPE	DMPC	LPPC
$\sigma(r)$ [Å]	*NVE*	0.06	0.12	0.004
$\sigma(\alpha)$ [°]	*NVE*	6.1	11.0	0.8
$\sigma(\theta)$ [°]	*NVE*	7.1	9.2	5.5
$\sigma(r)$ [Å]	*NpT*	0.06	0.12	0.005
$\sigma(\alpha)$ [°]	*NpT*	6.1	11.0	0.8
$\sigma(\theta)$ [°]	*NpT*	9.0	9.4	6.9

TABLE IV. Changes in the box shape and size during the *NpT* simulations. The values in parentheses are the differences compared to the crystal values.

	DLPE		DMPC		LPPC	
volume [nm^3]	93.33	(+5.4%)	101.97	(-0.6%)	58.37	(-4.7%)
a [Å]	47.56	(-0.4%)	53.20	(-1.7%)	24.95	(+0.5%)
b [Å]	46.97	(+0.7%)	35.33	(-1.0%)	54.46	(-4.8%)
c [Å]	41.79	(+5.0%)	54.73	(-1.2%)	43.57	(-0.4%)
α [°]	89.78	(-0.2%)	89.98	(-0.0%)	89.99	(-0.0%)
β [°]	91.19	(-0.9%)	97.54	(+0.1%)	98.66	(-1.0%)
γ [°]	90.68	(+0.8%)	90.22	(+0.2%)	90.01	(+0.0%)

The introduced results show that the extended CHARM22 force field is reliable for molecular dynamics simulations of phospholipids in large aggregates. Moreover, it can be used for simulations under constant pressure conditions. This is important for the generation of hydrated bilayers to be used as models for biomembranes. In this case the area per head group is often unknown.

4.5. SIMULATIONS OF PHOSPHOLIPID BILAYERS

4.5.1. *Preparation of a Starting Point*

The generation of an initial configuration is one of the main problems in simulation of biological membranes. The structural properties of membranes under physiological conditions are not well known and differ from related crystal structures. No general rule exists for the generation of an initial structure.

We used the crystal structure of DLPE [82] to prepare an initital configuration [85]. Firstly, we added a water box of 1485 TIPS3P [64] molecules on top of the bilayer surface. We applied constant temperature (320 K) and constant pressure (1 bar) conditions in order to give the system the freedom to change its state, e.g., to adjust the area per head group. After several hundreds of picoseconds simulation time and an increase of the temperature up to 380 K the system was left nearly unchanged. No penetration of the water molecules into the head group area was detected. We conclude that:

1. the phase transition is too slow to occur on the time scale of a typical molecular dynamics simulation,
2. the hydrogen bonding network between the ethanolamine head groups is too strong to be disrupted by the water molecules at these moderate conditions.

Therefore, we performed a second attempt. This time, a modified crystal structure was used as starting point. In order to interrupt the strong hydrogen bonding network, all L-enantiomeres, which are not found in nature, were removed from the crystal structure. Again, a water box with 2112 molecules was set on top of the bilayer. We applied constant temperature (320 K) and constant pressure (1 bar) conditions in order to remove the introduced holes in the bilayer. The system equilibrates rapidly and after 150 ps the area per head group remains constant at 47 Å^2 over a period of 100 ps. This value is in good agreement with experimental values [86].

5. Conclusion and Outlook

It has been demonstrated above that the molecular dynamics simulation technique is a powerful instrument for the study of thermodynamic and kinetic properties of matter. It has also been shown that there are still some restrictions to the applicability of this method. Using standard techniques the time scale is limited to a few nanoseconds and the number of particles in the simulation cell should not be larger than a several thousand in order to keep the numerical effort within reasonable limits. The approach of new computer architecture (parallel computing) on one side and the systematic development of algorithms (cell-multipole and multiple-time-step methods, mixed MD-Monte Carlo simulations) may help to overcome most of these restrictions in the very next future.

6. References

1. McCammon, J. A. and Harvey, S. C. (1987) *Dynamics of Proteins and Nucleic Acids*, Cambridge University Press, Cambridge.
2. Van Gunsteren, W. F. and Berendsen, H. J. C. (1990) *Angew. Chem.* **102**, 1020.
3. Allen, M. P. and Tildesley, D. J. (1990) *Computer Simulation of Liquids*, Clarendon Press, Oxford.
4. Heermann, D. W. (1986) *Computer Simulation Methods in Theoretical Physics*, Springer-Verlag, Berlin.
5. Ciccotti, G. and Hoover, W. G. (1986) *Molecular Dynamical Simulation of Statistical-Mechanical Systems*, North-Holland, Amsterdam.
6. Hoover, W.G. (1986) *Molecular Dynamics*, Springer-Verlag, Berlin.
7. McQuarrie, D. A. (1976) *Statistical Mechanics*, Harper and Row, New York.
8. Chandler, D. (1987) *Introduction to Modern Statistical Mechanics*, Oxford University Press, New York.
9. Lebowitz, J. L., Percus, J. K., and Verlet, L. (1967) *Phys. Rev.* **153**, 250.
10. Toxvaerd, S. (1991) *Molec. Phys.* **72**, 159.
11. Gray, S. K., Noid, D. W., and Sumpter, B. G. (1994) *J. Chem. Phys.* **101**, 4062.
12. Gear, C. W. (1971) *Numerical initial value problems in ordinary differential equations*, Prentice-Hall, Englewood Cliffs, NJ.
13. Verlet, L. (1967) *Phys. Rev.* **159**, 98.
14. Berendsen, H. J. C. and Van Gunsteren, W. F. (1985) *Proceedings of the Enrico Fermi Summer School, Molecular dynamics simulation of statistical mechanical systems*, Soc. Italiana di Fisica, Bologna.
15. Kast, S. M., Nicklas, K., Bär, H.-J., and Brickmann, J. (1994) *J. Chem. Phys.* **100**, 566.
16. Lustig, R. (1994) *J. Chem. Phys.* **100**, 3048; and subsequent articles.
17. Berry, R. S. (1994) *J. Phys. Chem.* **98**, 6910.
18. Woodcock, L. V. (19971) *Chem. Phys. Lett.* **10**, 257.
19. Evans, D. J., Hoover, W. G., Failor, B. H., Moran, B., and Ladd, A. J. C. (1983) *Phys. Rev. A* **28**, 1016.
20. Evans, D. J. and Morriss, G. P. (1984) *Comput. Phys. Rep.* **1**, 297.
21. Esparza, C. H. and Kronmüller, H. (1989) *Molec. Phys.* **68**, 1341.
22. Berendsen, H. J. C., Postma, J. P. M., Van Gunsteren, W. F., DiNola, A., and Haak, J. R. (1984) *J. Chem. Phys.* **81**, 3684.
23. Nosé, S. (1984) *Molec. Phys.* **52**, 255.
24. Hoover, W. G. (1985) *Phys. Rev.* **A 31**, 1695.
25. Toxvaerd, S. and Olsen, O. H. (1990) *Ber. Bunsenges. Phys. Chem.* **94**, 274.
26. Martyna, G. J., Klein, M. L., and Tuckerman, M. (1992) *J. Chem. Phys.* **97**, 2635.
27. Nosé, S. (1991) *Prog. Theor. Phys. Suppl.* **103**, 1.
28. Andersen, H. C. (1980) *J. Chem. Phys.* **72**, 2384.
29. Tanaka, H., Nakanishi, K., and Watanabe, N. (1983) *J. Chem. Phys.* **78**, 2626.

30. Ciccotti, G. and Tenenbaum, A. (1980) *J. Stat. Phys.* **23**, 767.
31. Bonomi, E. (1985) *J. Stat. Phys.* **39**, 167.
32. Schneider, T. and Stoll, E. (1978) *Phys. Rev.* **B 17**, 1302.
33. Van Gunsteren, W. F., Berendsen, H. J. C., and Rullmann, J. A. C. (1981) *Molec. Phys.* **44**, 69.
34. Kast, S. M. (1994) Dr.-Ing. thesis, Technische Hochschule Darmstadt; Kast, S. M. and Brickmann, J. (submitted for publication) *J. Chem. Phys.*
35. Strutt, J. W. (Baron Rayleigh) (1891) *Phil. Mag.* **32**, 424; (1902) *Scientific Papers*, Vol. 3, Cambridge University Press, London.
36. Reiling, S. (1994) Dr.-Ing. thesis, Technische Hochschule Darmstadt.
37. Schrimpf, G. (1993) Dr.-Ing. thesis, Technische Hochschule Darmstadt.
38. Duane, S., Kennedy, A. D., Pendleton, B. J., and Roweth, D. (1987) *Phys. Lett.* **B 195**, 216.
39. Forrest, B. M. and Suter, U. W. (1994) *J. Chem. Phys.* **101**, 2616.
40. Parinello, M. and Rahmann, A. (1980) *Phys. Rev. Lett.* **45**, 1196.
41. Nosé, S. and Klein, M. L. (1983) *Mol. Phys.* **50**, 1055.
42. Nicklas, K. (1993) Dr.-Ing. thesis, Technische Hochschule Darmstadt.
43. Ewald, P. P. (1921) *Ann. Phys.* **64**, 253.
44. Dufner, H., Schlenkrich, M., and Brickmann, J. (manuscript in preparation).
45. Brooks, B. R., Bruccoleri, R. E., Olafson, B. D., States, D. J., Swaminathan, S., and Karplus, M. (1983) *J. Comp. Chem.* **4**, 187; Nilsson, L. and Karplus, M. (1986) *ibid.* **7**, 591.
46. Schrimpf, G., Schlenkrich, M., Brickmann, J., and Bopp, P. A. (1992) *J. Phys. Chem.* **96**, 7404.
47. Ding, H.-Q., Karasawa, N., and Goddard III, W. A. (1992) *J. Chem. Phys.* **97**, 4309.
48. Board, J. A., Causey, J. W, Leathrum Jr., J. F., Windemuth, A., and Schulten, K. (1992) *Chem. Phys. Lett.* **19**, 89.
49. Appel, A. W. (1985) *SIAM J. Sci. Stat. Comput. Chem.* **6**, 85.
50. Greengard, L. and Rokhlin, V. I. (1987) *J. Comput. Phys.* **73**, 325.
51. Telemann, O. and Jönsson, B. (1986) *J. Comput. Chem.* **7**, 58.
52. Askar, A., Space, B., and Rabitz, H. (1995) Long Time Scale Molecular Dynamics and the Subspace Method for Molecular Dynamics, this volume.
53. Tuckerman, M., Berne, B. J., and Martyna, G. J. (1992) *J. Chem. Phys.* **97**, 1990.
54. Sexton, J. C. and Weingarten, D. H. (1992) *Nucl. Phys. B Proc. Suppl.* **26**, 613.
55. Humphreys, D. D., Friesner, R. A., and Berne, B. J. (1994) *J. Phys. Chem.* **98**, 6885.
56. Procacci, P. and Berne, B. J. (1994) *J. Chem. Phys.* **101**, 2421.
57. Tuckerman, M. and Parrinello, M. (1994) *J. Chem. Phys.* **101**, 1302.
58. Tuckerman, M. and Parrinello, M. (1994) *J. Chem. Phys.* **101**, 1316.
59. Van der Ploeg, P. and Berendsen, H. J. C. (1982) *J. Chem. Phys.* **76**, 3271.
60. Egberts, E. and Berendsen, H. J. C. (1988) *J. Chem. Phys.* **89**, 3718.
61. Berendsen, H. J. C., Postma, J. P. M., Van Gunsteren, W. F., and Hermans, J. (1981) in B. Pullmann (ed), *Intermolecular Forces*, Reidel, Dordrecht, p. 331.
62. Böcker, J. (1993) Dr.-Ing. thesis, Technische Hochschule Darmstadt.
63. Böcker, J., Schlenkrich, M., Bopp, P. A., and Brickmann, J. (1992) *J. Phys. Chem.* **96**, 9915.
64. Jorgensen, W. L., Chandrasekharm, J., Madura, J. D., Impey, R. W., and Klein, M. L. (1983) *J. Chem. Phys.* **79**, 926.

65. Bradly, J. E., Lee, E. M., Thomas, R. K., Willatt, A. J., Penfold, J., Ward, R. C., Gregory, D. P., and Waschkowski, W. (1988) *Langmuir* **4**, 821.
66. J. B. Rijnbout, J. Colloid Interface Sci. **62**, 81 (1977).
67. Lu, J. R., Simister, E. A., Thomas, R. K., and Penfold, J. (1993) *J. Phys. Chem.* **97**, 6024.
68. Böcker, J., Bopp, P. A., and Brickmann, J. (1994) *J. Phys. Chem.* **98**, 712.
69. AMBER 3.1, Singh, U. C, Weiner, P. K., and Kollman, P. A. (1988) Department of Pharmaceutical Chemistry, School of Pharmacy, University of California, San Fransisco.
70. Weiner, S. J., Kollman, P. A., Case, D. A., Singh, U. C., Ghio, C., Algona, G., Profeta Jr., S., and Weiner, P. K. (1984) *J. Am. Chem. Soc.* **106**, 765.
71. Weiner, S. J., Kollman, P. A., Nguyen, D. T., and Case, D. A (1986) *J. Comp. Chem.* **7**, 230.
72. Momany, F. A., McGuire, R. F., Burgess, A. W., and Scheraga, H.A. (1975) *J. Phys. Chem.* **79**, 2361.
73. Némethy, G., Pottle, M. S., and Scheraga, H. A. (1983) *J. Phys. Chem.* **87**, 1883.
74. Sipple, M. J., Némethy, G., and Scheraga, H. A. (1984) *J. Phys. Chem.* **88**, 6231.
75. Van Gunsteren, W. F. and Berendsen, H. J. C. (1987) *Groningen Molecular Simulation (GROMOS) Library Manual*, Biomos, Groningen.
76. Hermans, J., Berendsen, H. J. C., Van Gunsteren, W. F., and Postma, J. P. M., *Biopolymers* **23**, 1513.
77. Discover Molecular Modeling System, BIOSYM Technologies, Inc., 10065 Barnes Canyon Road, Suite A, San Diego, CA 92121.
78. Hagler, A. T., Lifson, S., and Dauber, P. (1979) *J. Am. Chem. Soc.* **101**, 5122.
79. Daubler-Osguthorpe, P., Roberts, V. A., Osguthorpe, D. J., Wolff, J., Genest, M., and A. Hagler, T. A. (1988) *Proteins: Structure, Function, and Genetics* **4**, 31.
80. Schlenkrich, M., Brickmann, J., MacKerrel, A., and Karplus, M. (manuscript in preparation).
81. Schlenkrich, M. (1993) Dr. rer. nat. thesis, Technische Hochschule Darmstadt.
82. Hitchcock, P. B., Mason, R., Thomas, K. M. and Shipley, G. G. (1974) *Proc. Nat. Acad. Sci. USA* **71**, 3036; Elder, M., Hitchcock, P., Mason, R. and Shipley, G. G. (1977) *Proc. R. Soc. Lond. A.* **354**, 157.
83. Pearson, R. H. and Pascher, I. (1979) *Nature* **281**, 499.
84. Hauser, H., Pascher, I. and Sundell, S. (1980) *J. Mol. Biol.* **137**, 249.
85. Vollhardt, H. and Brickmann, J. (manuscript in preparation).
86. McIntosh, T. J. and Simon, S. A. (1986) *Biochemistry* **25**, 4948.

SUBSPACE MOLECULAR DYNAMICS FOR LONG TIME PHENOMENA

Attila ASKAR
Koç University
Istinye, 80860 Istanbul
Turkey

Abstract

This paper presents an analysis of molecular dynamics towards an understanding of the contribution of slow and fast modes and proposes an integration method, the "Subspace Molecular Dynamics". The analysis of exact molecular dynamics for biological polymers reveals that only a very small proportion of the slow motion modes, of the order to a few percent to ten percent, participate in the fundamental dynamics. Furthermore, the calculations presented in this paper indicate that the system remains in subspaces for relatively long times of the order of picoseconds. Both of these observations suggest that one could implement these ideas in working directly with the subspace. The major difficulty is that although the system evolves in a low dimensional subspace, the basis vectors defining this subspace rotate in time.

1. Introduction.

The fundamental biological polymers are the DNA, which is at the initiation of life through replication and transcription and proteins which are the basic polymers for sustaining life through energy storage, transport, use and enzymatic catalysis. The DNA is a nucleic acid with a basic double helical structure. The proteins are polypeptide chains with amino acid residues with the alpha helices and the beta sheets as the basic structures [1]. Several considerations guide the formulation of molecular dynamics equations and the choice of solution methods as discussed below.

1.1 Considerations in formulating the molecular dynamics simulations.

- **Coordinates.** Three main coordinate systems as cartesian, internal and modal are distiguished [1-2]. The cartesian coordinates are most convenient for writing the kinetic energy and the configuration of the molecule is immediately available. Conversely, the potential energy as well as the physical distortions are most easily expressed in internal coordinates while the kinetic energy takes a complicated form. Modal coordinates correspond to the local eigenvalues of the system and permit a classification of the various time and correlation length scales. The low and high frequencies respectively coincide to the global and localized deformation modes.

E. Yurtsever (ed.), Frontiers of Chemical Dynamics, 255–266.

• **Time scales.** The short time phenomena permit the study of the thermodynamic statistical averages with the system oscillating about a mechanical or thermal equilibrium state. In this case, a linear analysis accompanied by a modal representation permits a reasonable description, although non linear effects can not be completely ignored. A second important class of events. Transitions from a stable configuration to another constitute. The transition process itself has a short duration, while occurring infrequently. Such processes are called rare events. Therefore calculations have to be carried for an extended period of time in order to catch these rare events. Finally, the system may undergo extreme changes in a quasi-static manner starting from an initial state and takes a very long time as compared to the oscillations it experiences. Examples of such long time processes are the folding in proteins and denaturation in DNA and . The time scales vary from about 10^{-15} to 10^{-3} [1].

• **Environment.** The molecular dynamics is influenced strongly by its environment. Significant differences exist in the behaviour of the molecular system in vacuo as compared to the wet case where the surrounding solvent is considered [3-6]. The solvent can be accounted for in varying degrees of details, starting with the macroscopic description of the as a continuum fluid to the atomistic picture.

• **Dynamical equations.** Newton's equations are the starting point for the simulation of molecular dynamics. These equations describe the dry molecular system as well as the wet case where the solvent atoms are also an integral part of the systems. For the dynamics of the molecular system within a solvent, the Langevin equation provides a convenient tool for a semi phenomenological description. The random forces account for the thermal excitation input on the system from the surrounding solvent. The Langevin equation can also be used for studying a local region in a large molecule where the exterior region is treated as a heat bath. The Langevin equation comprises a damping term which is taken as proportional to the velocity of the particles. This term describes the momentum exchange between the molecule and the surrounding solvent or the portion of the molecular system under consideration with the rest of the molecule. In heavily viscous solvents, the oscillations may be overdamped and the neglect of the acceleration term becomes justifiable. This singular perturbation limit to the Langevin equation is called the Brownian dynamics equation [5,6].

1.2. Approximate solution methods.

Major difficulties exist for numerical simulations due particularly to the large number of particles leading to several thousands degrees of freedom, the large range of the time scales as a result of the vastly different magnitudes in the interaction potentials, the extremely large excursions the particles may undergo leading to strong nonlinearities. The various approaches towards surmounting these difficulties are described below.

• **Constrained Dynamics** reduces the degrees of freedom of the system e. g. by freezing the bond lengths and the bending angles,[7-11]. **Lumping** consists of grouping parts of the molecule into a single mass unit or making continuum assumptions again with the goal of reducing the degrees of freedom in the system [12-19].

assumptions again with the goal of reducing the degrees of freedom in the system [12-19].

- **Residue clustering** splits the molecule into subsystems, i. e. smaller degrees of freedom which are reunited by putting back the interaction between them [16,17]. **The static reduction** is a mathematical scheme for treating many degree of freedom systems that can be separated into groups and where the effect of a group on the other is treated in a quasi- statical manner [18-19]. **DISCOS** (Dynamic Interaction Simulation and Control of Structures) is similar in spirit to residue clustering in considering subsystems while elements of finite body dynamics are utilized [20,21].

- **The subspace dynamics** is the subject of the presentation in this paper [22,23]. The analysis of exact calculations as well as physical intuition indicate that only a very small fraction, typically from a few per cent to ten per cent of the local eigenmodes participate significantly in the dynamics of large molecular systems [24-30]. The basic messages of this paper are to report that the dynamics takes place in a rather small subspace even when the molecular system undergoes extreme changes; the system remains in the subspace for a long time, of the order of picoseconds; the high frequencies in the spectrum are invariant from the configuration of the system; and nevertheless, the subspace rotates. The first three aspects are quite favourable for the design of a numerical algorithm while the last is a serious difficulty that needs to be overcome.

2. Potential energy and the dynamical equations

The kinematics of a polymer chain is described in terms of the bond lengths b_{jk}, bending angles θ_{jkl}, the dihedral (or torsion) angles ϕ_{jklm} , various long range unbinding interaction lengths as well as hydrogen bonds bond lengths d_{jk}. Here, jk, jkl, and jklm indicate respectively two, three and four consecutive atoms in the polymer. The kinematic variables are defined in terms of the coordinates $\{\mathbf{x}_j\}$ of the particles forming the polymer as [1,2,15]:

$$b_{jk} = |\mathbf{x}_j - \mathbf{x}_k| \qquad \cos\theta_{jkl} = \mathbf{c}_{jk} \cdot \mathbf{c}_{kl} \qquad \sin\phi_{jklm} = \det[\mathbf{c}_{jk}, \mathbf{c}_{kl}, \mathbf{c}_{lm}] \tag{2.1}$$

Above, det denotes the triple product of the vectors appearing as arguments and are calculated as a determinant and $\mathbf{c}_{jk}$ are the unit vectors along the bonds between the particles at $\mathbf{x}_j$ and $\mathbf{x}_k$: $\mathbf{c}_{jk}=(\mathbf{x}_k - \mathbf{x}_j) / b_{jk}$. The long range unbinding interaction lengths as well as hydrogen bonds bond lengths are defined just as b_{jk}, except that the particles $\mathbf{x}_j$ and $\mathbf{x}_k$ are not necessarily near neighbors. The potential energy is taken typically as:

$$V = \sum K_{ij}(b_{ij} - B^o_{ij})^2 + \sum K_{Bijk}(\theta_{ijk} - \Theta^o_{ijk})^2 + \sum K_{Tijkl}[1 - \cos 3(\phi_{ijkl} - \Phi^o_{ijkl})] + \sum U(d_{ij}) + \sum H(d_{ij}) \tag{2.2}$$

The values of coefficients K_{ij}, K_{Bijk} and K_{Tijkl} are known for various bonds along with the equilibrium values B^o_{ij}, Θ^o_{ijk} and Φ^o_{ijkl}. The form of the unbinding and hydrogen bond, U and H interactions can vary from model to model and between pairs of atoms. In the present work, in order for observing large deformations, the interactions are limited to the first three nearest neighbors[23].

The dynamical equations in the Cartesian coordinates are given by the Newton equations

$$\ddot{\mathbf{x}} = \mathbf{F}(\mathbf{x}) \qquad (2.3)$$

where for simplicity in the presentation the masses are taken as normalized and $F_j = -dV/dx_j$. The vector $\mathbf{x}$ comprises the position vectors of the n particles. A common procedure for the integration of these equations is the Verlet algorithm. This algorithm proceeds by discretizing the position coordinate at time t+dt in terms of the past values as[3]:

$$\mathbf{x}(t+\delta t) = 2\,\mathbf{x}(t) - \mathbf{x}(t-\delta t) + \delta t^2\,\mathbf{F}(t) \qquad (2.4)$$

The velocities do not appear in the equations. These may be needed particularly for calculating the kinetic energy and are evaluated as: $\mathbf{v}(t) = (\mathbf{x}(t+\delta t) - \mathbf{x}(t-\delta t))/2$.

An alternative to the algorithm above is the so-called velocity Verlet algorithm which integrates the equation $\dot{\mathbf{v}} = \mathbf{F}(\mathbf{x})$ for $\mathbf{v}$ [5]. This latter form is more stable as the former involves the ratio of two small quantities for evaluating the velocities. In spite of its simplicity, the algorithm above is limited to small times as δt is determined by the smallest of the time scales for stability and accuracy requirements. Considering that there may be time scales differing by up to 3 to 12 orders of magnitude, there would be need to carry calculations by this many time steps, thereby leading to impossibly long computational tasks. More precisely, for time typical steps of the order of 0.01 fsec, a long time calculation to a milisecond, would require to run the calculations for 10^{14} time steps. The impossibility of the task with even the most advanced computers leads us to exploring accurate approximations.

3. Subspace Molecular Dynamics

The "Subspace Molecular Dynamics" method is aimed at removing the high frequency components from the integration of the dynamical equations. This permits to reduce the size of the system and use long time steps. For motivating the method, some observations from exact molecular dynamics are presented. Fig.1a shows the mode density for a representative atom along the backbone of a protein and the radius of gyration [24] and Fig. 1b projections for the motion of particles of a protein on eigen modes[25]. The plots in these figures reveal two fundamental phenomena: the high frequency components are much smaller than the low frequency ones and the low frequency modes are much more densely populated. Moreover, further results show that only a few percent of the modes corresponding to the lowest ones carry substential amplitude[24-30]. One can therefore suggest that the fundamental dynamics is described by a small set of modes corresponding to the slow motion. This observation is suggestive for the term "Subspace Molecular Dynamics" and the algorithm to be presented.

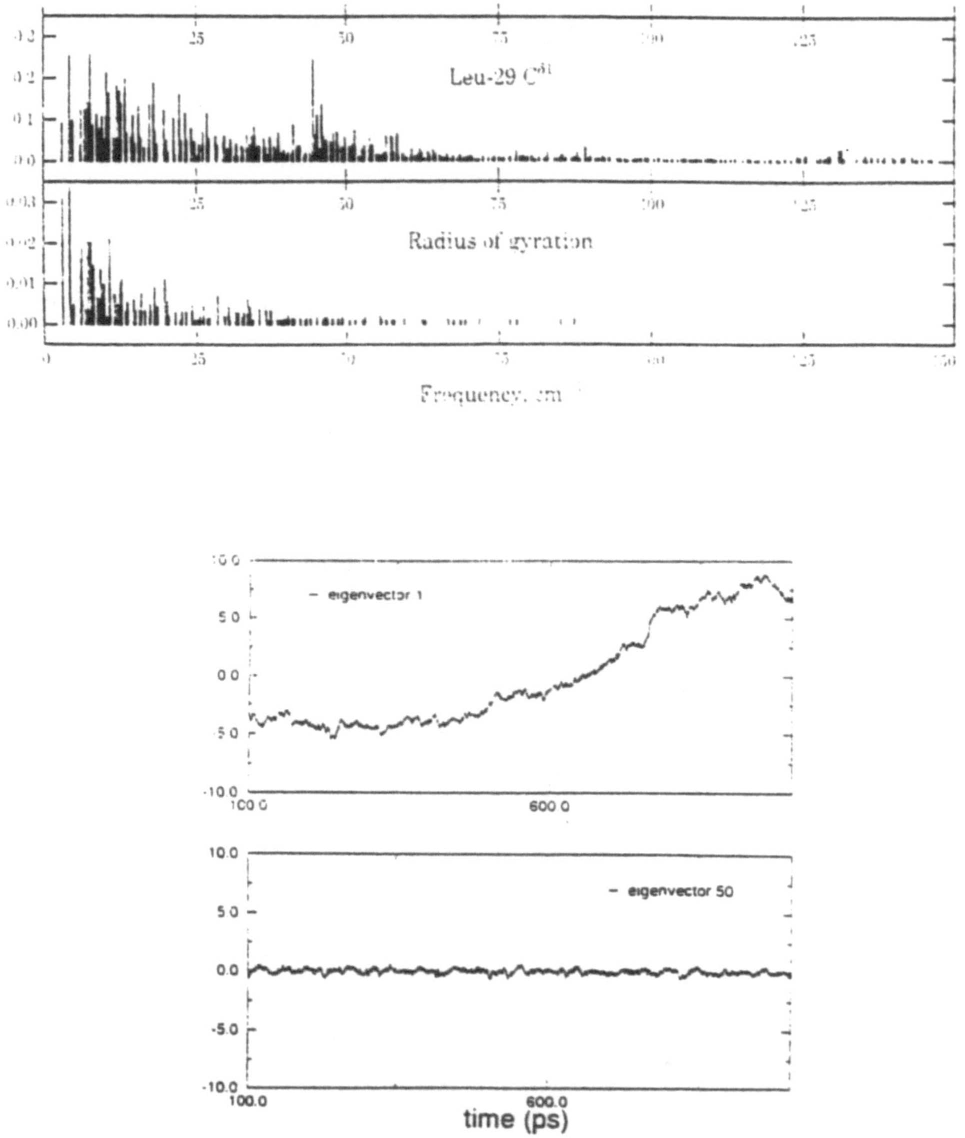

Fig. 1. Results from exact moleculacular dynamics calculations. (a) Mode density for a representative atom and radius of gyration along the backbone of a protein [24]; (b) Projections for the motion of particles of a protein[25].

Towards the selection of the subspace, the Jacobian of the potential energy is considered at a reference configuration **x**=**X**:

$$J_{ij} = d^2V/dx_i dx_j \tag{3.1}$$

coordinates vector. To motivate the general idea in the subspace method and suggest a hyerarchy of approximations based on the general concept, consider a splitting of the modal vector into slow and rapid constituants $\mathbf{z_s}$ and $\mathbf{z_r}$ respectively as:

$$\mathbf{u} = \mathbf{Q}\,.\,\mathbf{z} = \mathbf{P}\,.\,\mathbf{z_s} + \mathbf{R}\,.\,\mathbf{z_r} \tag{3.2}$$

Above, with the vectors $\mathbf{z_s}$ and $\mathbf{z_r}$ being respectively of dimensions m and 3n-m, the projection operators **P** and **R** are rectangular matrices with n rows and respectively with m and 3n-m columns. Following $\mathbf{Q}^T.\,\mathbf{Q} = \mathbf{I}$, they have the properties

$$\mathbf{P}^T.\,\mathbf{P} = \mathbf{I} \qquad \mathbf{P}^T.\,\mathbf{R} = 0 \qquad \mathbf{R}^T.\,\mathbf{P} = 0 \qquad \mathbf{R}^T.\,\mathbf{R} = \mathbf{I} \tag{3.3}$$

With the above representation, the dynamical equations take the form:

$$\ddot{\mathbf{z}}_s = \mathbf{P}^T.\,\mathbf{F} \qquad \ddot{\mathbf{z}}_r = \mathbf{R}^T.\,\mathbf{F} \tag{3.4}$$

The use of the eigenvectors of the linearized potential for defining the subspace should not be misinterpreted as if the dynamics is linearized. The above equations are exact and the slow and rapid components are coupled nonlinearly through $\mathbf{F}=\mathbf{F}(\mathbf{z_s},\mathbf{z_r})$. Several approximations are possible in solving the coupled nonlinear system above.

- **Decoupled slow component**. This is the zeroeth order approximation and is justified by the high frequency modes having much smaller amplitudes as illustrated in Fig.1a. Within this approximation, only the first equation in (3.4) remains, along with the interpretation that $\mathbf{F} = \mathbf{F}(\mathbf{z_s},\mathbf{0})$:

$$\ddot{\mathbf{z}}_s = \mathbf{F_s}\;(\mathbf{z_s}) \qquad \mathbf{F_s}\,(\mathbf{z_s}) = \mathbf{P}^T.\,\mathbf{F}\,(\mathbf{z_s},\,0) \tag{3.5}$$

Moreover, in view of the coordinate transformation in Eq.(3.2), the slow force formally reads: $\mathbf{F_s} = -dV/d\mathbf{z_s}$. Consequently, the equation for $\mathbf{z_s}$ conserves the energy in a rigorous sense. The approximation can be implemented with a fixed subspace as defined by the eigenvectors at the initial configuration[22]. Alternatively, if the initial subspace ceases to be a good representation as a result of the system having departed considerably from its original configuration, it may be necessary to update the subspace by the use of current Jacobians[23,31]. In switching from a subspace to the next, the two subsequent subspaces are expected to have an imperfect overlap. This point is analogous to the propagation methods in quantum scattering problems[32]. This mismatch will result in an energy drain that needs to be addressed. The subspace equations can be integrated e.g. by a Verlet algorithm[22]. Alternatively, since the local eigenvectors are available, they can be utilised for an exact analytical integration[23,31].

- **Quasi-static rapid component**. For the high frequency components having much smaller amplitudes, a partial linearization in $\mathbf{z_r}$ is a most justifiable approximation. Thus, Eq. (3.4) becomes:

$$\ddot{\mathbf{z}}_s = \mathbf{F_s} - \mathbf{J_{sr}}.\mathbf{z_r} \qquad \ddot{\mathbf{z}}_r = \mathbf{F_r} - \mathbf{J_{rr}}.\mathbf{z_r} \tag{3.6}$$

Above,

Above.

$$\mathbf{F_s} = \mathbf{P}^T . \mathbf{F}(\mathbf{z_s}, 0) \qquad \mathbf{F_r} = \mathbf{R}^T . \mathbf{F}(\mathbf{z_s}, 0)$$
$$\mathbf{J_{sr}} = \mathbf{P}^T . \mathbf{J} . \mathbf{R} \qquad \mathbf{J_{rr}} = \mathbf{R}^T . \mathbf{J} . \mathbf{R} \tag{3.7}$$

The solution of the linear equation for $\mathbf{z_r}$ above consists of a quasi-static component induced by $\mathbf{z_s}$ through $\mathbf{F_r}$ and a rapid component corresponding to self excitations. In the terminology of linear differential equations, the former is the particular solution while the latter is the homogeneous solution. The high frequency oscillations in Fig. 1 originate from the homogeneous solution for $\mathbf{z_r}$. Consequently the neglect of this component of $\mathbf{z_r}$ will filter out the high frequency noise. In a practical manner, dropping the acceleration term achieves this purpose. This idea is the equivalent as in the stroboscopic averaging [33], multiple time scales [34] as well as the static reduction[16-19] methods. The matrix $\mathbf{J_{rr}}$ is diagonal by construction. Furthermore, the high eigenfrequencies remain unchanged even when the molecule has undergone extreme deformations (see Fig.5 below). Therefore, there is sufficient justification for using the values of $\mathbf{J_{rr}}$ at the initial configuration throughout the dynamics. Various calculations indicate that quasi-static strech, which corresponds to the high frequency mode shapes, becomes an important energy reservoir for the system [23]. More specifically, the excess of the energy that can not be stored in torsional form, is distributed to a strech mode temporarily. This energy is later released back into the system with the bond length relaxing to its equilibrium configuration. This is an example where the constrained dynamics lacks the necessary flexibility for storing the excessive localization of energy during the rapid and extreme changes in a rare event. In fact, the energy in the stretch mode is negligible most of the time, expect during the rapid events such as transitions.

- **The rapid component as a random input**. Following the analysis above, the homogeneous solution in Eq (2.10) corresponding to the self excitations for $\mathbf{z_r}$ consist of sine and cosine functions with a large frequency. In following the dynamics at the time scale of the slow motion, the discrete time intervals become as large as many periods of the rapid modes. Consequently the values of these trigonometric functions appear as random when taken at largely spaced time intervals. Therefore, rather than solving for the self excitations of the rapid modes, i.e. the homogeneous part of $\mathbf{z_r}$ exactly, they may be assigned random amplitudes, much in the spirit of the Langevin dynamics[5, 31]. Indeed the Langevin equation is derived through a similar path [6].

4. Results.

The first set of applications with the "Subspace Molecular Dynamics" method are for anharmonic and glassy bulk solids defined by Lennard-Jones potentials[22]. Both solids had 768 degrees of freedom and the major distinction of the glass is the appearence of imaginary frequencies, i.e. temporarily unstable modes. The subspace calculations are carried with a fixed subspace and their dimensions are respectively 22 and 38. It is seen in Fig.2 that the use of as few as 3 to 5 % of the modes yields highly accurate results. It must also be noticed that the linear analysis fails to yield satisfactory results even when all of the modes are kept.

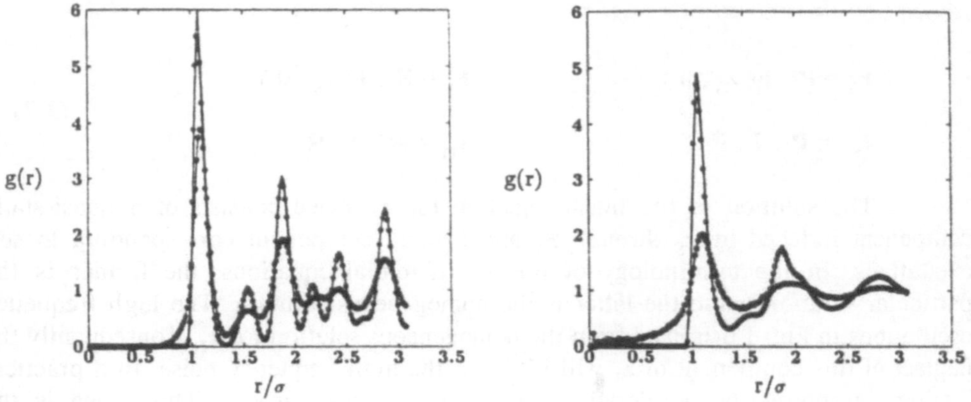

Fig. 2. Comparison of exact and subspace dynamics results for the energy densities [22]. (a) Anharmonic solid.; (b) Glassy solid. The continuous line, the light dots and the heavy dots correspond respectively to the exact, subspace and linear full space calculations.

The second set of applications are for a model polymer which has the stretch, bending and torsional interactions while missing the long range unbinding forces and the hydrogen bonds[23]. The masses in the polymer are taken to be all Carbon atoms and the initial equilibrium configuration is a helix with 32 particles, i. e. 96 degrees of freedom. The snapshots in Fig. 3 show how extreme are the deformations that the system has undergone. This was made possible by designing large impulsive initial velocities solely in the 8th mode of the linearized system at equilibrium. The energy that is initially in the 8th mode is eventually equidistributed to all of the modes, as expected. The computational task undertaken in this example is harder than usual computational problems where the systems are taken to be at thermal equilibrium. Fig.4. Shows the time averaged positions of the particles corresponding to the dynamics shown in the snapshots of Fig.3. Both figures show that the subspace dynamical calculations yield most accurate approximations. In a system at thermal equilibrium, all of the modes having the same energy, the flux between the modes in the energy space is small. A system which is initially at thermal equilibrium while having the same energy as the system shown in Fig. 4 is seen indeed to undergo a much smaller deformation[23]. Finally, Fig.5 shows the time averages and the standard deviations for the eigenvalues of the model system. The model system with 32 particles has 6 degrees of freedom for rigid body motions, 29 torsional angles, 30 bending angles and 31 inter atomic bonds.

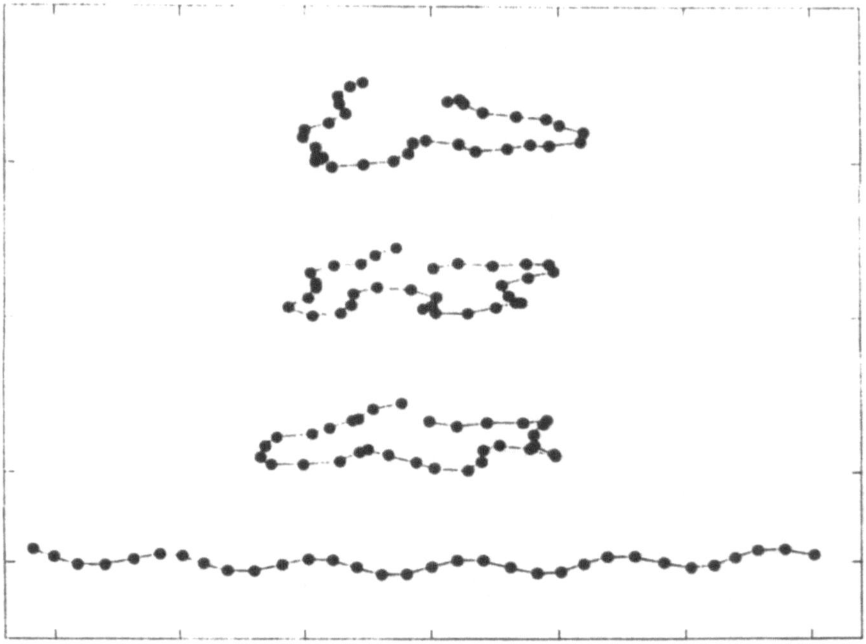

Fig.3. Snap shots of the configuration of the model system[23]. (a) At the initial time; (b), (c) and (d) 2,5 psec, respectively for the exact calculation with 96 degrees of freedom, subspaces with 48, and 24 degrees of freedom.

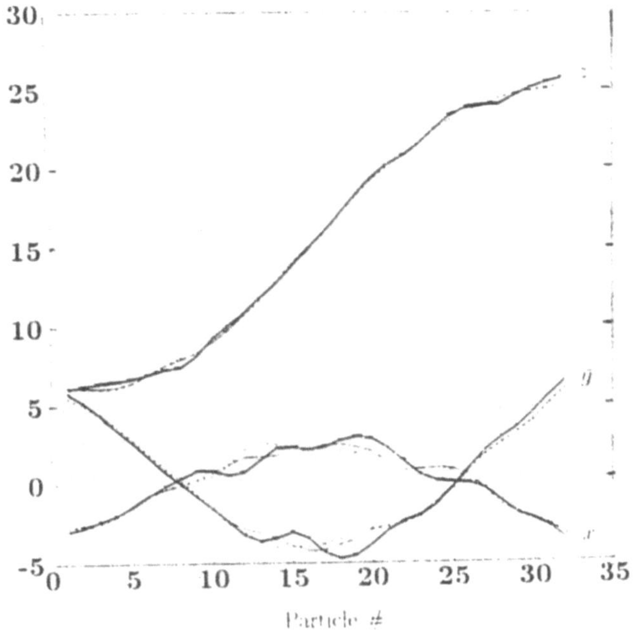

Fig.4. The mean coordinates of the particles as averaged over the duration of 2.5 psec according to the exact calculation with 96 dimension and the subspaces respectively with 48 and 24 dimensions[23].

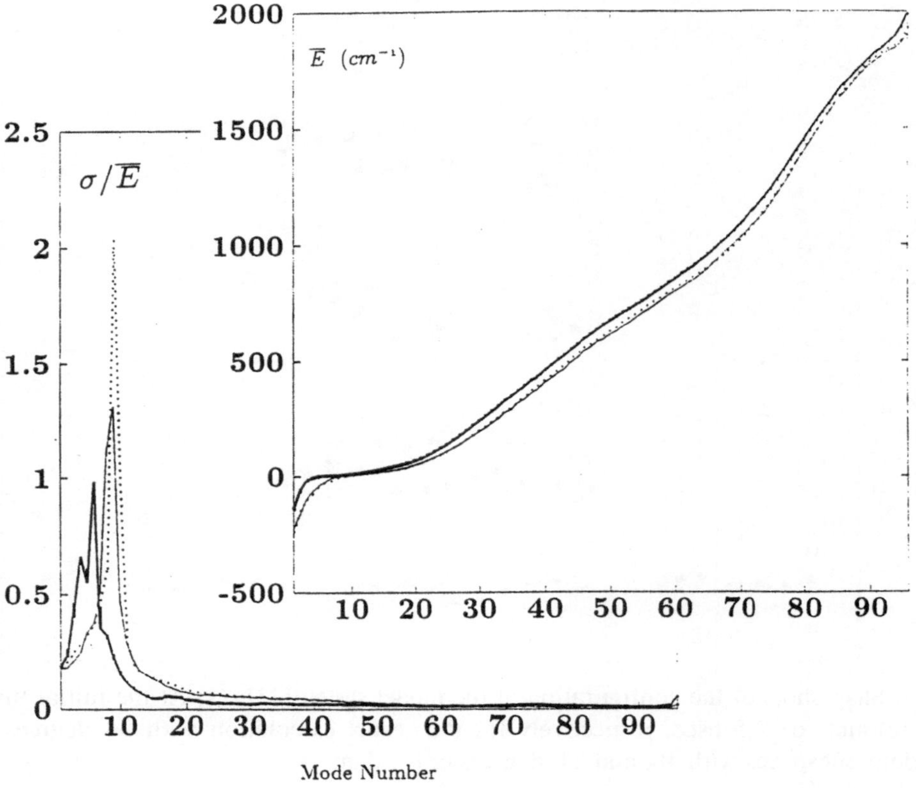

Fig. 5. Time averages of the eigenvalues and the standard deviations in the model system [23].

The dominant torsional modes, seventh to thirtyfifth, are seen to form a very flat plateau followed by shoulder regions. This stratification of the frequency distribution is clearly a product of the vastly different orders of magnitudes in the torsional, bending and the stretching interactions.

5. Conclusion.

The results displayed in Fig. 2 are for bulk solids. These systems have a smooth dispersion relation due to the same type of interaction potential being operations between the near neighbouring particles. In view of the smooth corresponding dispersion relation, there is no natural hyerarchy of frequencies in bulk solids. Consequently, the bulk solid is not expected to be a system suitable to the subspace analysis. In spite of this, the results presented in Fig.3 show that the dynamics can conveniently be described by a relatively small subspace of low frequency modes. Therefore the subspace method is prepared an effective tool for the study of an harmonic phenomena.

The main thrust and motivation for this paper was the polymeric chains. The eigenvalues for the second system, the molecular chain are seen in Fig.5. Three distinct regions with very low frequencies corresponding to the torsion dominated modes, a shoulder region corresponding to the bending dominated modes and finally a high frequency regime corresponding to the stretch modes are observed. The stratified frequency distribution for the polymer chains, makes them more suited to the subspace analysis. Physically, this truncation is happening in a natural way as it takes much higher energies to excite the modes dominated by the bending and stretch interactions. The algorithm summarised here is trying to duplicate nature 's way in the numerical simulations.

References

1. McCammon, A. J. and Harvey, C. S.(1987) Dynamics of Proteins and Nucleic Acids, Cambridge University Press
2. Wilson Jr, E. B. Decius, J. C. and Cross, P.C. (1980) Molecular Vibrations, Dover Pub. NY.
3. Van Gunsteren, W.F., Berendsen, H.J.C., Hermans, J.,Hol, W.G.J. , Postma, J.P.M. (1983) Computer simulation of the dynamics of hydrated protein crystals and its comparison with x-ray data. Proc. Natl. Acad. Sci. U.S.A. 80:4314319.
4. Berendsen,H.J.C., Van Gunsteren, W.F., Zwindermann, H.R.J., Geurtsen, R.G. (1986) Simulations of proteins in water. Ann. NY Acad. Sci. 482:269-286.
5. Allen, M.P. and Tildesley, D.J., (1989) Computer Simulation of Liquids, clarendon, Oxford.
6. Adelman, S.A., (1984) Chemical Reaction Dynamics in Liquid Solution, in Advances in Chemical Physics Vol III ed. by I. Prigogine and S.A. Rice, John Wiley and Sons.
7. Noguti, T. and Go, n. (1983) Dynamics of Native Globular Proteins in Terms of dihedral Angles, Journal of the Physical Society of Japan 52, 3283-3288.
8. Gunsteren, V. and Karplus, M. (1982)The effect of constraints on the dynamics of macromolecules Macromolecules, **15**, 1529-1542.
9. Ryckaert, J.P. (1977), Numerical Integration of the Cartesian Equations of motion of a system with constraints : Molecular Dynamics of n-Alkenes, J. Comp. Phys.**23**, 327.
10. Ryckaert, J.P. (1985), Special geometrical costrains in the molecular dynamics of chain molecules, Molecular Phys., **55**, 549.
11. Turnen, J. Chim H, Lupi V. ,Weiner P., Gallions and Singh Chandra (1992), Researchers Apply Variable reduction Techniques to Molecular simulations, Part1, chem Design Automation News, 7,12,39 and Part2, id, (1993)
12. Go, N., Noguti, T. and Nishikawa, T. (1983) Dynamics of a small globular protein in terms of low frequency vibrational modes, Proc. Natl. acad. Sci. **80**, 3696-3700.
13. Brooks, B. and Karplus, M. (1985) Normal modes for specific motions of macromolecules: Application to the hinge-bending mode of lysozyme, Proc. Natl. Acad. Sci. **82**, 4995-4999.
14. Thacher, H. A. Rabitz, A. Askar, (1990) Discrete-Continuum Hybrid model for Dynamics with Applications:Desorption of Adsorbates and relaxation of Lattice Inclusions, T.J. Chem.Phys. **93**, 4673.
15. Askar, A. (1994) Modelling of Biological Polymers: Discrete and Continuum Mechanics Formulations, NATO ASI Series **H 84, 1-35.**
16. Hao, H.M. and Harvey, C. S. (1992) Analyzing the Normal Mode Dynamcis of

Macromolecules by the Component Synthesis Method, biopolymers **32**, 1393-1405.

17. Hao, H.M. and Scheraga, A.H. (1994) Analyzing the Normal Mode Dynamics of Macromolecules by the Component Synthesis Method : Residue Clustering and Multiple Component Approach, Biopolymers **34**, 321-335.
18. Guyan, R.J. (1965), Amer. Inst. Aero. Astro J.**3**, 380.
19. Ookuma, M. and Nagamatsu, A. (1984) Analysis of Vibration by Component Mode Synthesis Method, Bulletin of JSME **27**, 529-532.
20. Rosenthal, D. (1988), order N Formulationr for equations of motion of multibody systems. Proceedings of the workshop on Multibody Simulation, JPL D-5190, **3**, 1122.
21. Chun, M.H., Turner, D.J. and Frisch, P.H. (1989) Experimental Validation of Order (N) Discos, Paper AAS 89, 457 AAS/AIAA Astrodynamics Specialist Conference.
22. Space, B., Rabitz, H. and Askar, A. (1993) Long time scale molecular dynamics subspace integration method applied to an harmonic crystals and glasses, J. Chem. Phys. **99**, 9070-9079.
23. A. Askar, Space, B. and Rabitz, H. The Subspace Method for Long Time Scale Molecular Dynamics, (1994) J. Phys. Chem. (to appear).
24. Brooks, B. and Karplus, M. (1983) Harmonic dynamics of proteins: Normal modes and fluctuations in bovine pancreatic trypsin inhibitor, Proc. Natl. Acad. Sci. **80**, 6571-6575.
25. Amadei, A., Linssen, B. M. A. and Berendsen, J.C.H., (1993) Essential Dynamics of Proteins, Proteins: Structure, Function and Genetics **17**, 412-425.
26. Levy, M. R. and Karplus, M. (1979) Vibrational Approach to the Dynamics of an α-Helix, Biopolymers **18**, 2465-2495.
27. Levy, M. R., Perahia, D. and Karplus, M. (1981) Molecular dynamics of an α-helical polypeptide: Temperature dependence and deviation from harmonic behavior, Proc. Natl. Acad. Sci. **79**, 1346-1350.
28. Derreumaux, P. and Vergoten, G. (1991) Effect of Urey-Bradley-Shimanouchi Force Field on the harmonic Dynamics of Proteins. Wiley-Liss Inc. 11, 120-132.
29. Levitt, M., Sander, C. and Stern, S. P. (1983) protein Normal-mode dynamics: Trypsin Inhibitor, crambin, ribonuclease and Lysozyme, J. Mol. Biol. **181**,423-447.
30. Teeter, M.M. and Case, (1990) Harmonic anod quasiharmonic Descriptions ou Cramlin, The Journal of Physical Chemistry **94**, 8092-8097.
    Teeter, M.M. and Case, (1990) Harmonic and Quasiharmonic Descriptions of Crambin, The Journal of Ohyscal Chemistry **94,** 8092-8097.
31. Askar, A., Owens, R. and Rabitz, H. (1993) Molecular dynamics with Langevin equation using local harmoncis and Chandrasekhar's convolution, J.Chem. Phys. **99,** 5316-5325.
32. A. Askar, (1993) Finite Element Method for Quantum Scattering, in Numerical Grid Methods and Their Application to Schrödinger's Equation, ed C.Cerjan, Kluwer Academic Publishers, Series C: Mathematical and Physical Sciences, Vol. 412.
33. A. Askar, (1974) A Non-perturbative Approach to Transition Probabilities by the Stroboscopic Method, Phys. Rev. A, **10**, 2395 .
34. Tuckerman, M. and Berne B.J., (1992) Reversible Multiple time scale molecular dynamics, J.Chem. Phys. **97,**1990.

VIBRATIONAL DYNAMICS AT THE ADSORBATE-SUBSTRATE INTERFACE

T.UZER
School of Physics, Georgia Institute of Technology, Atlanta, Georgia 30332-0430, USA

AND

J.T.MUCKERMAN
Chemistry Department, Brookhaven National Laboratory, Upton, New York 11973, USA

1. Introduction

Vibrational relaxation of molecules on surfaces [1] plays the same central role in surface chemistry (microscopic rates, catalytic reactions [2], photochemistry [3], desorption [4]) and surface spectroscopy [5] as does intramolecular energy transfer in isolated-molecule chemistry [6, 7]. While much theoretical attention has been paid to vibrational excitation in molecule-surface collisions [8], an understanding of vibrational decay processes at surfaces has been hampered by the difficulty in unambiguously establishing the nature and time scales of the relaxation processes [9]. The damping of vibrations on surfaces has often been studied by spectral measurements, and there are many different theories relating spectral widths and shifts to the characteristic times and mechanisms of vibrational energy transfer ([9]-[16]). For instance, the insight provided by the line shape of an adsorbate vibrational mode promises to be very powerful [17]. Despite substantial theoretical and experimental efforts in monitoring the decay of population in a set of vibrationally excited, adsorbed oscillators ([9], [13]- [15], [17]), a comprehensive understanding of relaxation time scales has remained elusive. In this context, time resolved measurements offer the ability to measure directly the relaxation rate of vibrationally excited adsorbate ([18] - [21]); and represent the most promising direction in vibrational spectroscopy and

E. Yurtsever (ed.), Frontiers of Chemical Dynamics, 267–290.

reaction dynamics at surfaces ([22] - [24]).

It has been widely anticipated that recent trends towards high resolution and dynamical studies in surface science would elucidate energy flow in adsorbates ([25] - [29]) and surface infrared spectroscopy [9, 16, 24, 31]. But despite giant strides in experiment and theory, the paucity of information concerning time scales and microscopic mechanisms of surface phenomena remains striking. Recent experiments on energy flow in adsorbate-substrate systems have shown that the richness of these energy flow processes is more than a match for that of dynamical processes in the gas phase. It is well recognized that studying such processes demands a microscopic point of view akin to the one in modern isolated molecule dynamics. In particular, recent experimental successes in measuring vibrational energy relaxation in adsorbates provide a powerful incentive to pursue theoretical studies of relaxation [25, 26, 27]. The exciting results of Harris and Levinos [26], for example, show clearly that distinct modes of adsorbates relax differently and furthermore, depending on the complexity of the adsorbate, intramolecular decay processes can compete effectively with energy flow to the substrate's vibrational and electronic degrees of freedom. (In this treatment we avoid relaxation processes involving electronic excitations [29] by choosing materials in which electron-hole processes are negligible.)

Instead, predesorption is used as a case study to provide microscopic detail on energy flow between adsorbates and surfaces [32, 33, 34]. The effect of surface vibrations is simulated using variants of the Generalized Langevin Equation (GLE) ([35] - [39]) which folds the vibrations of the surface into an artificially constructed chain of atoms. GLE is a very useful device by which the temperature of the surface can be included in the calculation while still preserving a microscopic point of view through the chain of fictitious atoms. Consequently, the dynamics calculation at the surface is reduced to an approximate isolated-molecule calculation with some provisions making the "molecule " simulate a heated surface.

First we will discuss our findings [32, 33, 44] on a model that involves a collinear interaction. Although there is increasingly more detailed information about structure and geometry of surfaces [30], this model study is useful for examining the important factors in this problem. Finally, we will consider an adsorbate that can librate [34].

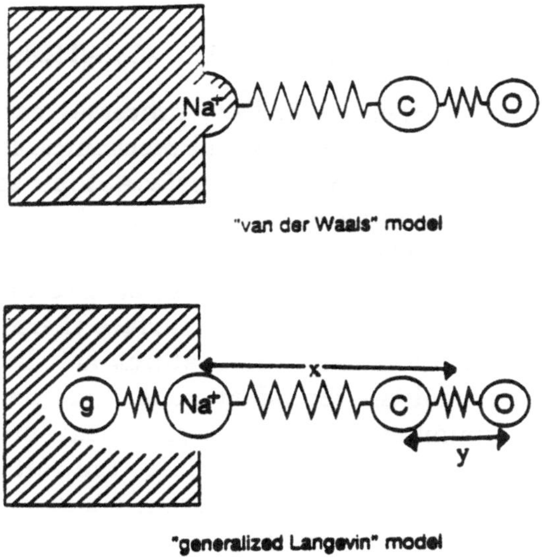

Figure 1. Models used for the predesorption study. In the van der Waals model (abbreviated "vdW"), the surface is assumed to be stationary, whereas in the "Generalised Langevin Model"(abbreviated "gL") the surface vibrations are simulated by the motions of "ghost" atoms. Here one ghost atom is shown; in our simulations we use several.

2. Predesorption

Predesorption ([40, 41, 42], [45], [43], [44]) is the surface counterpart of predissociation: The adsorbate has been given enough energy to break the surface-adsorbate bond and desorb (see Fig.1). However, this energy is not in the reaction coordinate initially, and has to reach it for the reaction to take place. Thus, desorption has to compete with relaxation to other modes of the adsorbate as well as the numerous modes of the substrate and predesorption in a prototypical scenario in which microscopic factors that affect mechanisms and time scales can be investigated. Currently, experimental results exist on desoption of CO excited to its first overtone desorbing from a NaCl surface [40, 41, 42]. In modelling this process, it is essential to treat the CO vibration quantum-mechanically to avoid spurious contributions to the reactivity from the lack of quantization constraint.

3. The model

3.1. GENERALIZED LANGEVIN APPROXIMATIONS

The dynamics of various gas-surface processes can be conveniently and accurately described by using classical mechanics. It is also possible to incorporate the thermal motion of the surface atoms as well as their response to the interaction with the gas particles in this description [46]. Various schemes have been employed for this purpose; from a computational point of view the most successful description is related to the generalized Langevin equation (GLE) representation of the surface thermal motion ([35] - [39]). In this description the solid surface atoms are divided into two groups: the primary zone and the secondary zone (heat bath). The primary zone contains a small number of surface atoms which interact strongly with the adsorbate , while the heat bath simulates the rest of the crystal . As an illustrative example we consider a system which describes the interaction between a gas phase atom (monoatomic adsorbate, denoted by A below) and a solid surface. For simplicity we shall assume a single primary zone atom. The time evolution of the system is described by ([35] - [39])

$$M_A\ddot{R}(t) = -\frac{\partial V(R,X)}{\partial R} \tag{1}$$

$$M\ddot{X}(t) = -KX(t) - \frac{\partial V(R,X)}{\partial R} - \int_0^t \Gamma(t-t')\dot{X}(t') + F(t) \tag{2}$$

where R and X represent the positions of the A atom and the displacement of the primary zone atom from its lattice point, M_A and M are the corresponding masses and $V(R,X)$ is the gas-surface interaction potential. K is the primary zone atom's effective force constant. In equation (2) $\Gamma(t)$ and $F(t)$ are the memory kernel and the random force respectively. These two quantities are related by the second fluctuation-dissipation theorem ([35] - [39])

$$kT < F(0)F(t) >= M\Gamma(t) \tag{3}$$

where T is the surface temperature and $< ... >$ denotes a correlation function.

The exact solution of eqs. (1) and (2) is equivalent to the solution of Avogadro's number of coupled, deterministic Newton's equations corresponding to the gas-crystal system. A number of approximation schemes for solving eqs. (1) and (2) have been reported in the literature , and which were tested for a variety of surface phenomena. The Equivalent Harmonic

Chain Representation (EHCR) [39] and the "ghost" particle approach [48] are the most widely used approximations to the dynamics . These two methods are very similar and they are denoted collectively as the Fictitious Particle Representations (FPR). They both replace eq. (2) by the following set of coupled equations:

$$
\begin{aligned}
\ddot{X}(t) &= -\omega_e^2 X(t) + \omega_{c1}^2 S_1(t) - \frac{1}{M}\frac{\partial V(R,X)}{\partial R} \\
\ddot{S}_1(t) &= -\omega_{e1}^2 S_1(t) + \omega_{c1}^2 X(t) + \omega_{c2}^2 S_2(t) \\
&\vdots \\
\ddot{S}_N(t) &= -\Omega_N^2 S_N(t) + \omega_{cN}^2 S_{N-1}(t) + \beta_{N+1}\dot{S}_N(t) + f_{N+1}(t)
\end{aligned}
\qquad (4)
$$

In writing eqs. (4) we have used the notation of Ref. [39]. Thus the GLE in eq. (2) is replaced by a harmonic chain in which the primary zone atoms are coupled to a chain of N fictitious particles (with mass-weighted positions $S_i(t), i > 1$). The last two terms in the equation (4) are characteristic of the truncated part of the chain. At the limit of eqs. (4) are exactly equivalent to eqs.(2) [39, 48]. The different force constants, coupling coefficients and friction constants are related to the phonon spectral density function of the solid surface [47, 37] . In eqs. (4), is a random force (assumed to have a Gaussian distribution) which is related to the friction constant by

$$< f_i(t) f_i(\tau) >= 3kT\beta_i\delta(t-\tau) \qquad (5)$$

Equation (5) together with equation (4) turns out to be very useful in describing accurately a large variety of surface phenomena [48, 49].

3.2. EXTENSION TO QUANTUM MECHANICS. EHRENFEST COUPLING

Those aspects unique to the quantum mechanical calculations require comment. We partition the (mass-weighted) coordinate space of our model systems into a quantal subspace consisting of x, y and a classical subspace consisting of $z, S_1, \ldots\ldots, S_N$, where x corresponds to the translation of CO relative to the surface, y to the CO stretching vibration, z to the displacement of the surface atom, and S_j to the displacement of the jth fictitious atom in the N-atom equivalent harmonic chain representation of the heat bath (see Fig.1). The time dependent Schrödinger equation for the quantal subsystem is

$$H(x,y,p_x,p_y)\Psi(x,y,t) = i\hbar\frac{\partial\Psi(x,y,t)}{\partial t} \tag{6}$$

where

$$H(x,y,p_x,p_y) = \frac{1}{2}(p_x^2 + p_y^2) + V[x,y,z(t)] \tag{7}$$

and $V(x,y,z)$ is defined as in Ref. [32]. The solution of eq. (6) can be accomplished using the FFT (Fast Fourier transform) and DVR (Discrete Variable Representation) methods [50, 55]. This amounts to solving a set of first order differential equations in time involving the wave function at the points of a two dimensional grid. The classical coordinate(s) are interpereted in the quantal subsystem as time-dependent parameters.

The classical equations of motion for the model surface atom and heat bath chain are

$$\ddot{z}(t) = -\omega_{e0}^2 z(t) + \omega_{c1}^2 S_1(t) - < \frac{\partial V}{\partial z} > \tag{8}$$

$$\begin{aligned}
\ddot{S}_1(t) &= -\omega_{e1}^2 S_1(t) + \omega_{c1}^2 z(t) + \omega_{c2}^2 S_2(t) \\
&\cdot \\
\ddot{S}_2(t) &= -\omega_{e2}^2 S_2(t) + \omega_{c2}^2 z(t) + \omega_{c3}^2 S_3(t) \\
&\cdot \\
&\cdot \\
&\cdot \\
\ddot{S}_N(t) &= -\Omega_N^2 S_N(t) + \omega_{cN}^2 S_{N-1}(t) - \beta_N \dot{S}_N(t) + R_t
\end{aligned} \tag{9}$$

The interaction of the surface atom with the adsorbate molecule is included in the third term of eq. (8). This term is interpreted as a quantal expectation value for the "force operator"

$$- < \frac{\partial V}{\partial z} > = \int_{-\infty}^{\infty}\int_{-\infty}^{\infty} |\Psi(x,y,t)|^2 \left(-\frac{\partial V[u(x,y,z(t))]}{\partial z}\right) dx dy \tag{10}$$

where u is the carbon to surface atom distance, and $\Psi(x,y,t)$ is the solution of the time-dependent Schrödinger equation

$$\left(-\frac{\hbar^2}{2}\frac{\partial^2}{\partial x^2} - \frac{\hbar^2}{2}\frac{\partial^2}{\partial y^2} + V(x,y,z(t))\right)\Psi(x,y,t) = i\hbar\frac{\partial\Psi(x,y,t)}{\partial t} \tag{11}$$

The last two terms in equation (9) are dissipative and stochastic forces. The parameters $\omega_{en}^2, \omega_{cn}^2, \Omega_N^2$ and β_N were discussed in [39].

Equations (8) and (9) constitute a set of second order differential equations in time, with one equation for each of the classical variables. They and Eq. (11) corresponding to the propagation of the quantal subsystem, are allowed to communicate through the semiclassical device which we call "Ehrenfest coupling", and are solved simultaneously and self-consistently. Simply put, Ehrenfest coupling replaces the set of "classical" equations of motion for the coordinates in the classical subsystem by its average over the wave function for the quantal subspace. In the present application, the only effect of this averaging is in the last term of eq. (8). There the expectation value of the force on the surface atom exerted by the adsorbate, i.e. $-<\frac{\partial V}{\partial z}>$, has been averaged over the (x,y) distribution of the wave packet at time t, as in eq.(10) and depends only on the classical coordinate(s).

The justification for this approach lies in the Ehrenfest relations

$$\frac{d<q_j>}{dt} = <\frac{\partial H}{\partial p_j}> \tag{12}$$

$$\frac{d<p_j>}{dt} = -<\frac{\partial H}{\partial q_j}> \tag{13}$$

and the classical-quantum correspondence

$<q_j> \rightarrow q_j, <p_j> \rightarrow p_j$

for the variables q_j in the classical subspace. This approach is a generalization of "classical path" methods [51] and is similar to approaches employed recently by other workers [52, 53]. In the vdW model calculations (see Fig.1), there is no classical subspace, and the quantal results are exact within the context of the model. In the gL model calculations, we expect the Ehrenfest path results to be reliable so long as the frequencies associated with the quantal variables are higher and well separated with respect to those of the classsical variables. This is certainly the case for the NaCl and Si model surfaces (see below).

4. Diagnostics for Desorption and Relaxation

When the adsorbate-surface system is excited excited to the resonance state, it is useful to distinguish four channels for energy transfer that lead either to desorption or relaxation:

predesorption

$$S - CO(v = 1) \xrightarrow{k_p} S + CO(v = 0) \tag{14}$$

relaxation of excited complex adsorbate

$$S - CO(v = 1) \xrightarrow{k_r} S + CO(v = 0) \tag{15}$$

thermal desorption of relaxed adsorbate

$$S - CO(v = 0) \xrightarrow{k_T} S + CO(v = 0) \tag{16}$$

thermal desorption of excited adsorbate

$$S - CO(v = 1) \xrightarrow{k_{T^*}} S + CO(v = 1) \tag{17}$$

Channel (15) is the relaxation process, where energy in the excited high-frequency mode of the resonance state is dissipated into the model surface. Channel (16) is the thermal desorption process wherein thermal energy migrates from the model surface into the CO-surface atom bond of the ground state adsorbate-surface complex and breaks it. Channel (14) is the predesorption process wherein energy transfers from the excited resonance mode directly into the CO-surface atom bond to cause the desorption. In the vdW model, only channel (14) is operative; in the gL model, this pathway has exactly the same interpretation as it does in the vdW model. Channel (17) is another kind of thermal desorption wherein thermal energy transfers from the model surface into the CO-surface bond of the excited surface-adsorbate complex and breaks it before either relaxation or predesorption occurs.

4.1. CLASSICAL SURVIVAL AND RELAXATION CALCULATIONS

In our classical calculations, we compute an ensemble of 1000 trajectories for a specified surface and temperature. The system is initially classified as a $S - CO(v = 1)$ state. The survival probability, the relaxation probability, the ground state desorption probability, and the first excited state desorption probability are defined as the fraction of the ensemble of trajectories identified as $S - CO(v = 1)$, $S - CO(v = 0)$, $S + CO(v = 0)$ and $S - CO(v = 1)$ states at time t, respectively. Desorption is determined by whether the x coordinate is larger than some specified value at which the molecule-surface interaction is negligible. For trajectories with the x coordinate smaller than this specified value, the $S - CO(v = 1)$ and $S - CO(v = 0)$ states are determined by the value of $H(x, y, z)$, defined by

$$H(x, y, p_x, p_y) = T_x + H_{CO} + V_M(x, y, z) \tag{18}$$

If $H(x, y, z)$ is larger than the dissociation energy of the surface-adsorbate complex, the system is classified as $S - CO(v = 1)$ (the surviving initial state), otherwise is considered to have relaxed to $S - CO(v = 0)$. The $S + CO(v = 1)$ and $S + CO(v = 0)$ states are distinguished by whether H_{CO}, also defined in eq. (18), is larger than $\hbar\omega_{CO}$, the boundary between the ground state and the first excited states of the asymptotic CO molecule in a quasiclassical binning approach.

4.2. QUANTAL SURVIVAL AND RELAXATION PROBABILITIES

After determining the quantum initial state by a variational calculation, the quantal survival probability [54] is defined as the modulus squared of the projection of the wave packet onto the initial state

$$P_s(t) = | < \Psi(0)|\Psi(t) > |^2 \tag{19}$$

where $\Psi(0)$ is the initial wavefunction, which is the resonance state, and $\Psi(t)$ is the wavepacket at time t [54]. This definition of the survival probability is also recognized to be the modulus squared of the autocorrelation function of the wavepacket, and is the probability of the initial state surviving to time t.

We define the relaxation probability of the adsorbate on the model surfaces in terms of the projection of the wavepacket onto the bound states of the adsorbate-surface complex,

$$P_r(t) = \sum_{j=0}^{n_b - 1} | < \Phi_{(j)}|\Psi(t) > |^2 \tag{20}$$

where $\Phi_{(j)}(x, y)$ is the wave function for the jth bound state, and the summation is over the n_b bound states. Again, as in the initialization of the wavepacket, the bound states are shifted along the x-axis to follow the motion of the potential well as the surface atom [z(t) coordinate] moves.

4.3. QUANTAL PREDESORPTION AND THERMAL DESORPTION

We monitored predesorption and thermal desorption by computing the outgoing probability flux through a dividing surface this boundary of the grid is far away from the potential well of the adsorbate-surface complex. Thus the probability flux through the dividing surface definitely corresponds to the desorption of the (asymptotic) CO molecule. Let $x = s$ define the dividing surface, then the probability flux through this surface in the positive x direction is

$$J(s,t) = Re\frac{\hbar}{i} < \Psi(s,y,t)|[\frac{\partial\Psi(s,y,t)}{\partial x}]_{x=s} > \tag{21}$$

To determine whether this flux is predesorption or thermal desorption, we consider the state projected flux, i.e. the projection of the flux defined by Eq. 21 onto the basis of eigenfunctions of the asymptotic CO molecule,

$$J_0(s,t) = Re\frac{\hbar}{i}[< \Psi(s,y,t)|\Psi_{(0)}(y) > \times < \Psi_{(0)}(y)|[\frac{\partial\Psi(s,y,t)}{\partial x}]_{x=s} >] \tag{22}$$

$$J_1(s,t) = Re\frac{\hbar}{i}[< \Psi(s,y,t)|\Psi_{(1)}(y) > \times < \Psi_{(1)}(y)|[\frac{\partial\Psi(s,y,t)}{\partial x}]_{x=s} >] \tag{23}$$

where $\Psi_{(j)}(y)$ is the jth eigenfunction of the asymptotic CO molecule. The desorption probabilities for the formation of CO in the ground state and first excited state are then defined respectively, as

$$P_{d0}(t) = \int_0^t J_0(s,t')dt' \tag{24}$$

$$P_{d1}(t) = \int_0^t J_1(s,t')dt' \tag{25}$$

The quantities $P_{d0}(t)$ and $P_{d1}(t)$ do not distinguish directly between predesorption and thermal desorption. The computed values of $P_{d1}(t)$, appropriately shifted in time to allow for the delay time for the appropriate part of the wavepacket to reach the dividing surface, is assumed to correspond to thermal desorption of the excited (initial) state of CO. The computed $P_{d0}(t)$ values have a more complicated interpretation reflecting contributions from both predesorption and relaxed adsorbate molecules. These contributions are distinguished by considering the time dependence of the relaxation probability, in particular its decay at longer times. As in our classical calculations [33], rate constants for the various modes of energy transfer are determined by fitting the integrated rate laws from the assumed mechanism to the computed probabilities.

5. Results

5.1. CLASSICAL CALCULATIONS

In this section we summarize the major results of our classical calculations on the dynamics of desorption and relaxation of vibrationally excited CO on the various model surfaces studied.

In our exploratory work [44] we presented a simple, microscopic collinear model for the predesorption process. We improved the model surfaces and the sampling of initial conditions later, and we have also focused on the competing processes of relaxation and thermal desorption [32, 33]. Despite these improvements, the values of the collinear classical predesorption lifetime for $CO(v = 1)$ on the model NaCl and Si surfaces [32, 33] are about the same as those reported earlier [44].

We have investigated the dependence of the various processes on the Debye frequency of the solid, the length of the heat bath chain, the surface temperature, and the strength of the adsorbate-surface interaction. Of particular interest was the competition between desorption and relaxation. In the present calculations, as the Debye frequency of the model surface is systematically increased, relaxation competes more favorably with predesorption. Thermal desorption is much slower than both at 300 K. Relaxation becomes the dominant channel for energy transfer out of the excited adsorbate when the Debye frequency of the model solid is increased to be commensurate with the adsorbate translational motion. This trend is robust with respect to changes in the length of the heat bath chain. The acceleration of the classical predesorption rate (compared to the van der Waals model) by surface with low Debye frequency can be understood in terms of impulsive collisions between the surface atom and the carbon atom of the adsorbed CO molecule, driven by the high-frequency CO bond vibration (Figs. 2 and 3).

Certain isolated surface properties, on the other hand, were shown to be sensitive to the length of the heat bath chain. These include the magnitude of the fluctuations of the surface temperature from the nominal (long time or ensemble averaged) value, and the persistence of the normal mode

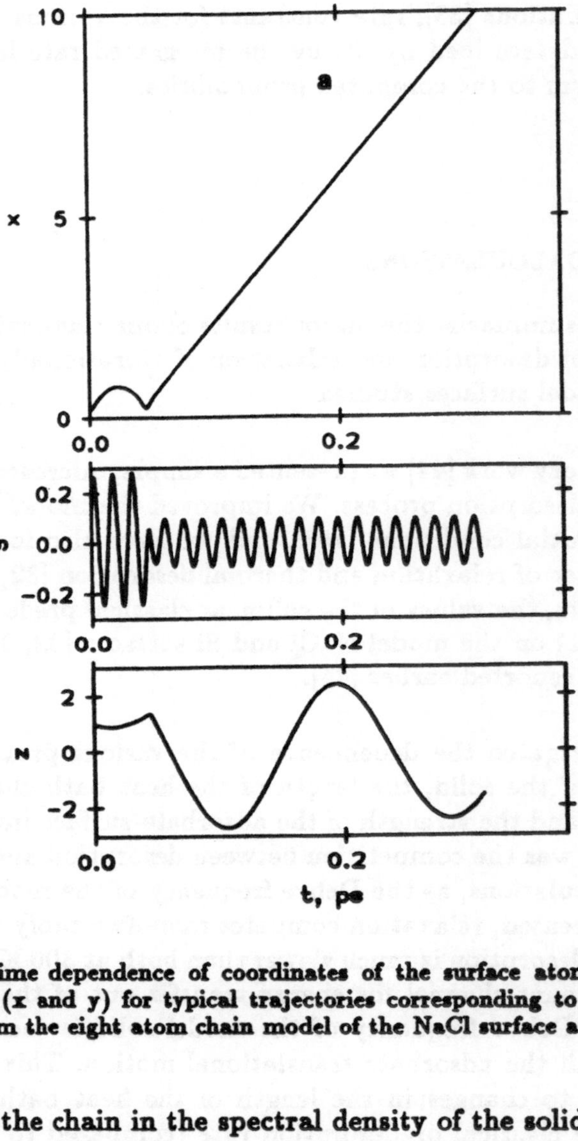

Figure 2. **Time dependence of coordinates of the surface atom (z) and the adsorbate molecule (x and y) for typical trajectories corresponding to the predesorption of $CO(v = 1)$ from the eight atom chain model of the NaCl surface at 300 K.**

structure of the chain in the spectral density of the solid.

5.2. QUANTAL CALCULATIONS

In this section we summarize the major results of our quantal calculations and compare them with the results of our classical calculations reported above. As in our early study [44] of the desorption of vibrationally excited CO from models surfaces, the classical predesorption rate constant determined for the vdW model in the present study (Fig.4) is larger than the

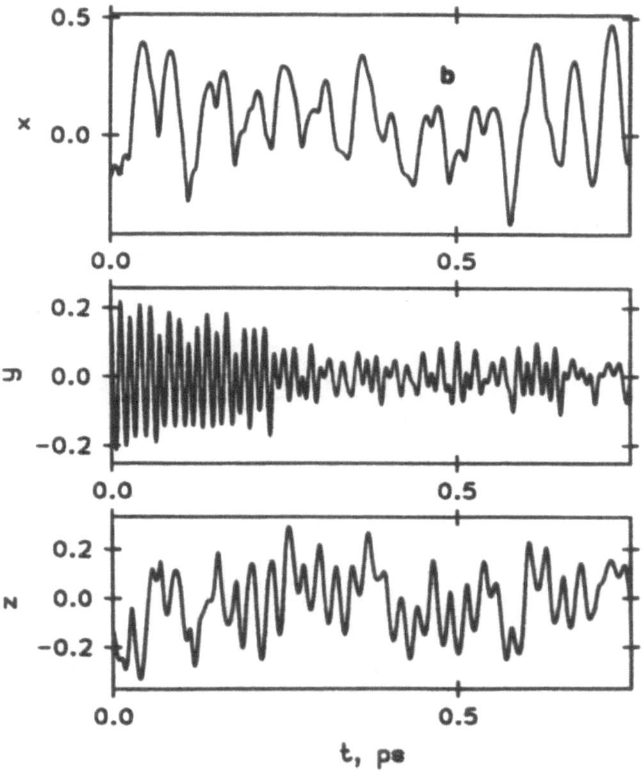

Figure 3. **Time dependence of coordinates of the surface atom (s) and the adsorbate molecule (x and y) for typical trajectories corresponding to the predesorption of $CO(v = 1)$ from the eight atom chain model of the Si-B surface (a Si surface with an artificially high Debye temperature) at 300 K.**

quantal one by a factor of 2, and classical predesorption is accelerated by surface motion in low Debye frequency generalized Langevin (gL) model solids (i.e.NaCl) while such motion has little or no effect on the quantal predesorption rate constant.

For the series of gL model surfaces in the present study, the classical predesorption rate constant decreases monotonically with increasing surface Debye frequency. We attribute these differences between the classical and quantal dynamics to anomalies in the classical mechanics arising from impulsive collisions between the adsorbate and surface atom in the case of

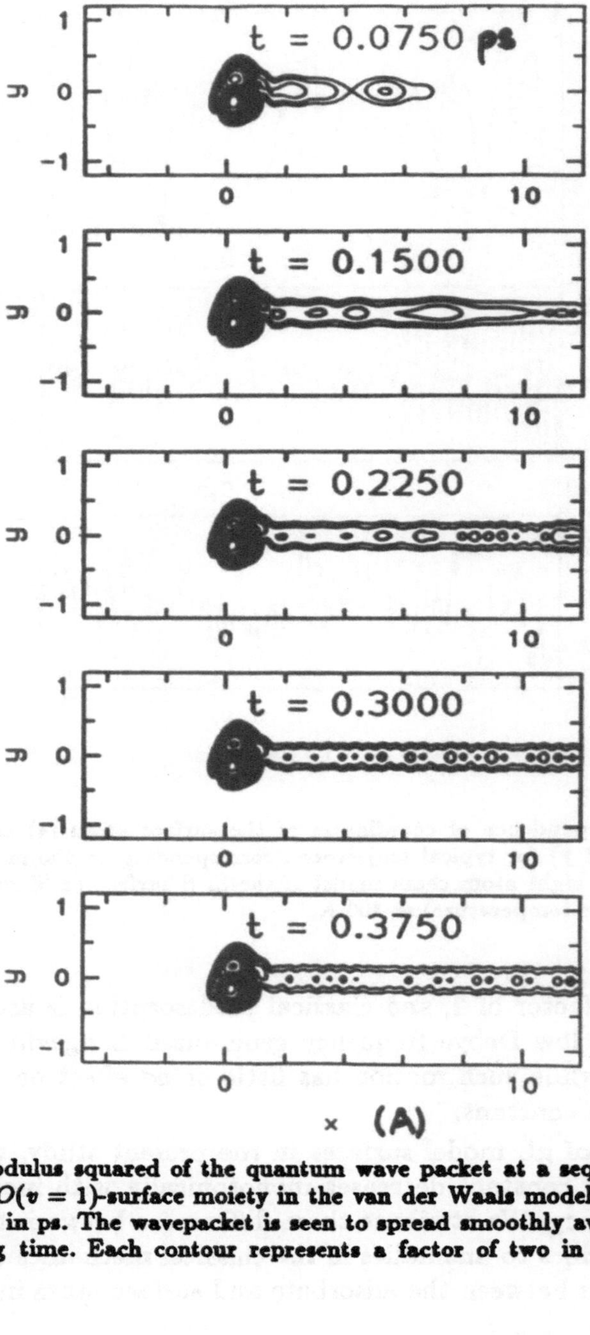

Figure 4. **Modulus squared of the quantum wave packet at a sequence of times after exciting the $CO(v = 1)$-surface moiety in the van der Waals model. Times indicated in each frame are in ps. The wavepacket is seen to spread smoothly away from the surface with increasing time. Each contour represents a factor of two in probability from its neighbors.**

low Debye frequency surfaces, and the lack of a quantization constraint on the amount of energy transferred from the high-frequency CO stretching vibration in all the cases studied.

Apart from the predesorption rate, the most striking feature of the clas-

sical dynamics is the increase of the relaxation rate constant with increasing Debye frequency. Although this result is the "expected" one from the point of view of classical theories of intramolecular vibrational relaxation (IVR) [7], the effect is much less pronounced in the quantal calculations. The classical relaxation rate constants range from over 5 to over 20 times larger than the quantal ones, however, both the classical and the quantal relaxation rate constants increase with increasing surface Debye frequency. Once again, we attribute the larger classical rates to the lack of a quantization constraint on the amount of energy transferred from the CO stretching vibration to the heat bath.

On the other hand, the quantal rate constants for thermal desorption, both of the ground and excited vibrational states of CO, are larger than their classical counterparts, and increase with increasing surface Debye frequency. The classical thermal desorption rate constants are relatively insensitive to the model surface.

The differences in the various rate constants in the classical and quantum calculations lead to an unusual "quantum effect": the trend in the energy flow with increasing surface Debye frequency is in different directions. The flow of energy from the high-frequency CO stretching vibration-either to the reaction coordinate resulting in predesorption, or to the heat bath of the model surface resulting in relaxation- dominates the response of the classical dynamics to increasing surface Debye frequency. By contrast, the energy in the CO vibration tends to remain there in the quantum dynamics: predesorption is relatively slow, and relaxation is quite slow in all cases except "Si-B", which was "invented" to pursue a point related to the classical dynamics. The response of the quantum dynamics to increasing surface Debye frequency is dominated by energy transfer from the model heat bath to the reaction coordinate, resulting in thermal desorption and excited CO adsorbates (Fig. 5).

The Ehrenfest path approach used here in the quantal calculations is most appropriate when the frequencies associated with the quantal coordinates are higher and well separated from those associated with the classical coordinates. This is not the case for the "Si-B"model surface. One must, therefore, interpret the quantal results for that model with some skepticism, but it should be also noted that the quantal results for the Si-B continue a trend (albeit in a somewhat exaggerated fashion) already apparent in the results for systems with lower surface Debye frequency (Fig.6).

The adiabatic following demonstrated in Fig.7 and the slow rate of

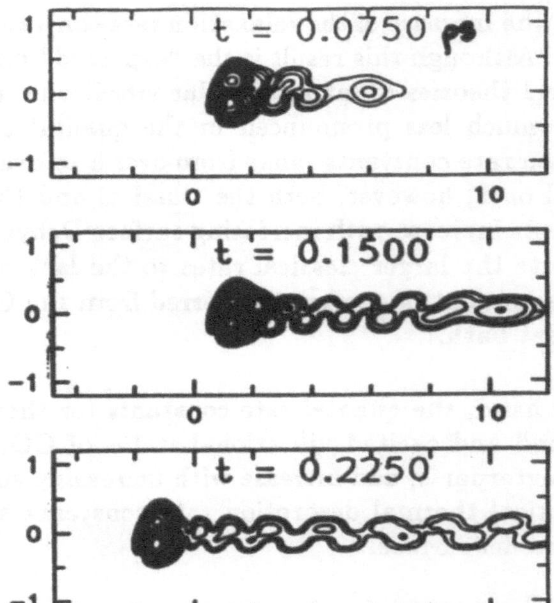

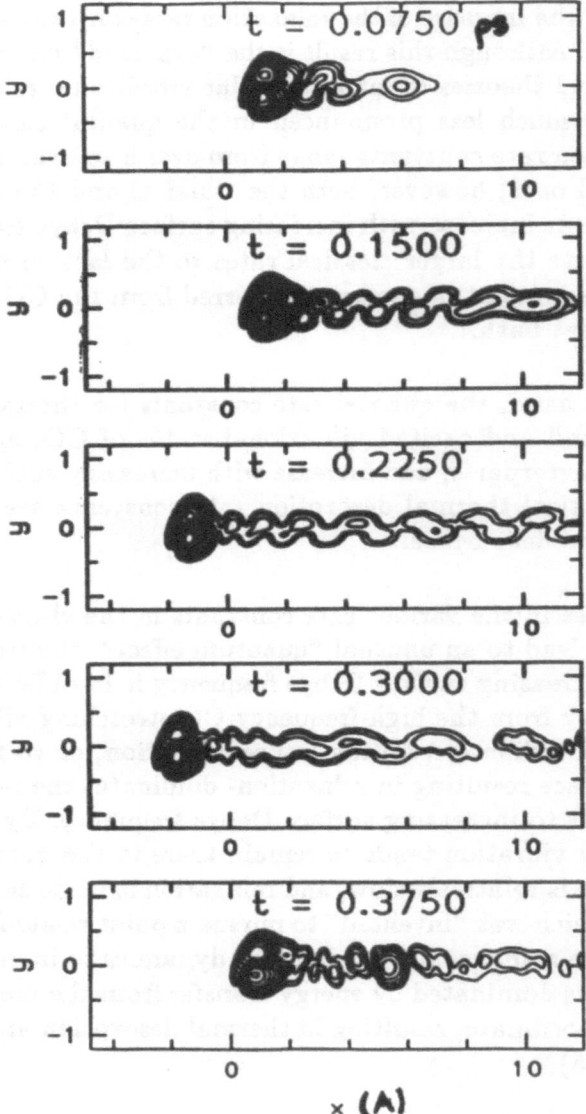

Figure 5. **Modulus squared of the quantum wave packet at a sequence of times after exciting the $CO(v = 1)$-surface moiety in the $N = 7$ gL model NaCl surface. Times indicated in each frame are in ps. The wavepacket is seen to spread somewhat unevenly away from the surface with increasing time.**

energy transfer from the excited adsorbate to the surface (see Table 1) indicate that the assumptions underlying the Ehrenfest path approach are valid for the progression of model surfaces NaCl, Si and Si-A. These results ans trends are summarized in Table 1.

The treatment thus far neglected the effect of libration/rotation on the

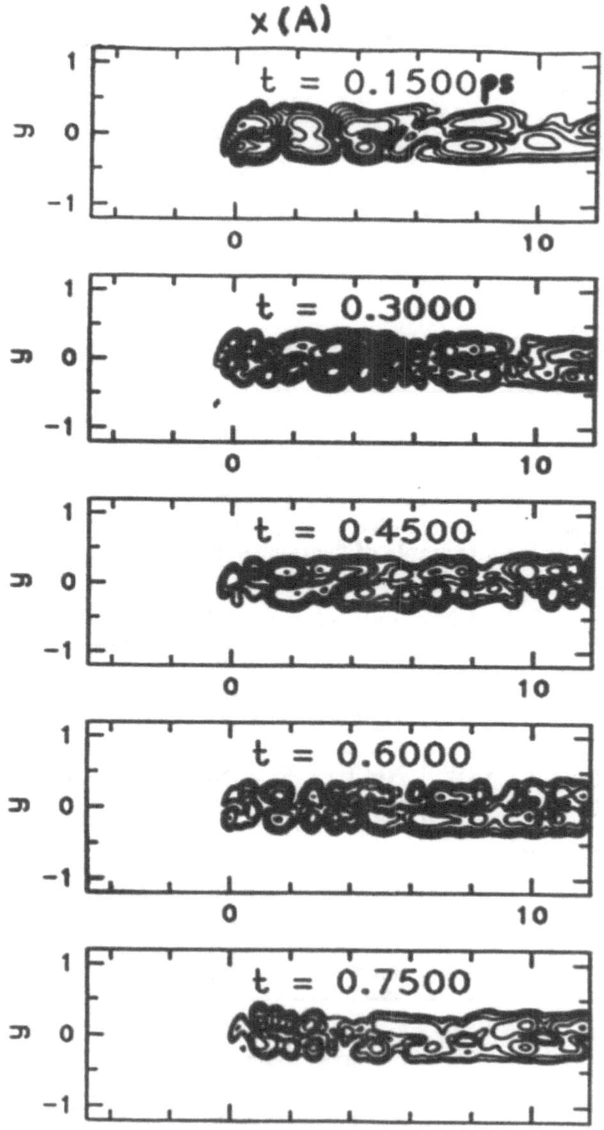

Figure 6. Modulus squared of the quantum wave packet at a sequence of times after exciting the $CO(v = 1)$-surface moiety in the $N = 7$ gL model Si-B surface. Times indicated in each frame are in ps. The wavepacket is seen to spread somewhat unevenly away from the surface with increasing time. "Adiabatic following" (see next figure) has broken down completely.

desorption process. In the next section, we present a three-dimensional model which reflects the true scenario more faithfully.

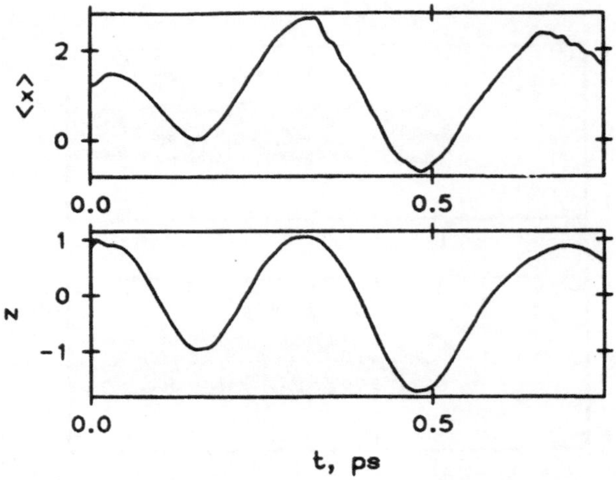

Figure 7. **Coordinate of surface atom s and expectation value of the CO translational coordinate x as a function of time for $CO(v = 1)$ adsorbed on the eight-atom chain gL model NaCl surface. This is an example of "adiabatic following".**

Table 1: Rate constants derived from quantal wavepacket calculations (N=7 gL model surfaces). The Debye temperature increases from left to right.

Quantal Rate Constants

$k_i(ps^{-1})$	vdW	NaCl	Si	"Si-A"	"Si-B"
k_p	0.515	0.506	0.79	0.89	2.45
k_T	—	0.167	0.51	0.6	4.9
k_T^*	—	0.111	0.63	0.64	4.92
k_r	—	0.03	0.04	0.07	0.71

Classical Rate Constants

$k_i(ps^{-1})$	vdW	NaCl	Si	"Si-A"	"Si-B"
k_p	1.22	8.34	6.02	4.72	1.61
k_T	—	0.20	0.1	0.1	0.11
k_T^*	—	0.21	0.11	0.13	0.13
k_r	—	0.14	0.57	1.55	5.36

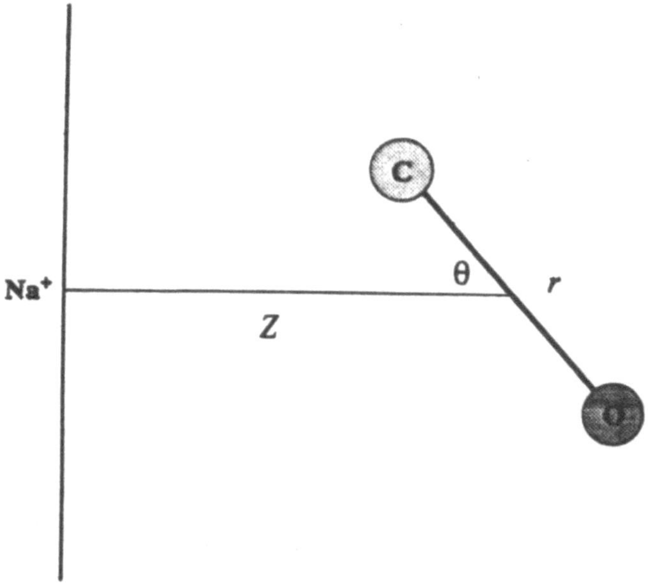

Figure 8. Schematic representation of the model and coordinate system

6. The effect of libration on predesorption

We examined the predesorption of vibrationally excited CO adsorbed on a NaCl(100) surface theoretically using the analytic discrete variable representation (DVR) method in the context of quantum wavepacket propagation [34]. A three dimensional model for the CO-NaCl system was developed which treats the CO stretching , bending (frustrated rotational, or librational), and translational coordinates as dynamical variables (see Fig. 9). We find that exciting the CO stretching mode does not induce significant desorption, while exciting the bending mode can induce discernible desorption. The computed lifetime for the bending (rotational) predesorption is in the nanosecond range. The results of parallel classical calculations are qualitatively consistent with many aspects of the quantum calculation, but for the case of pure rotational predesorption, the classical lifetime is orders of magnitude shorter than the quantal value.

Here we present our results on parallel quantal and classical dynamical studies of the predesorption of CO from a NaCl(100) surface. Computa-

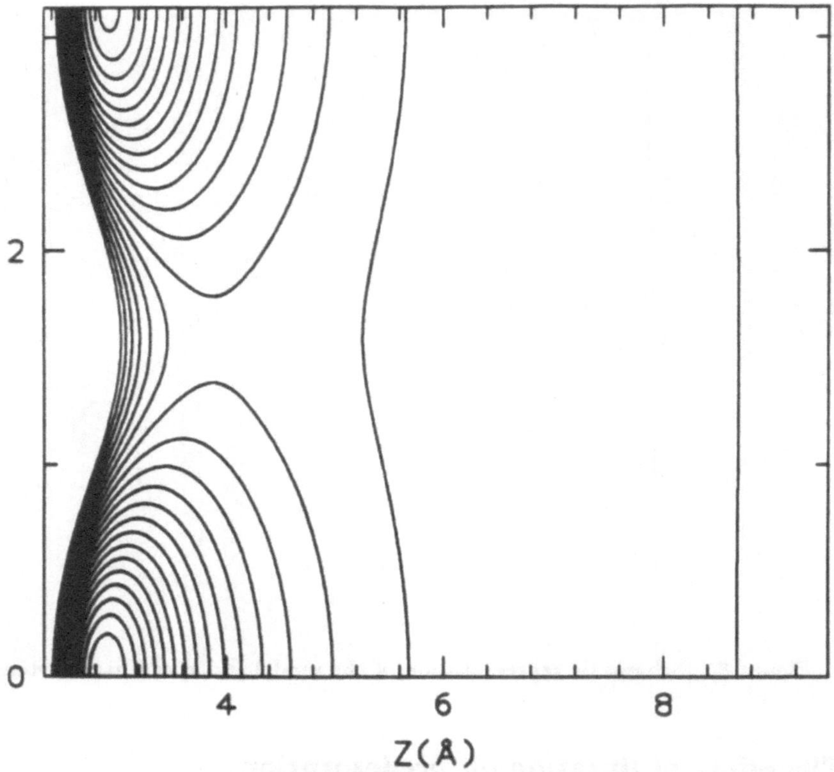

Figure 9. **Contour plot of adsorption potential at $r = r_e$. The contour interval is $100 cm^{-1}$. The (deeper) well at $\theta = 0$ corresponds to the CO oriented normal to the surface with the carbon atom closer to the surface. The well at $\theta = \pi$ corresponds to the normal orientation with the oxygen atom closer to the surface.**

tionally, we have extended techniques of quantal wavepacket propagation by applying analytic discrete variable representations (DVRs) [55] to multidimensional systems. We used DVR basis sets constructed from plane wave functions for the CO translation, harmonic oscillator eigenfunctions for the CO stretching, and Legendre functions for the bending/rotational motion. Compared with the fast Fourier transform method (FFT) used in our previous study of collinear models [32, 33, 44], the DVR basis sets are more natural with respect to the physical problem. Each mode was quantized naturally by using the appropriate DVR basis set. Using the DVR method we were able to propagate the wavepackets for 32.768 ps with very good conservation of norm and energy, and obtained spectral densities with resolutions of 1 cm^{-1}.

We did not find any predesorption induced by exciting the CO stretching mode to its first excited state $n_r = 1$ with the bending and translational modes in their ground states during the time period (32.768 ps) of

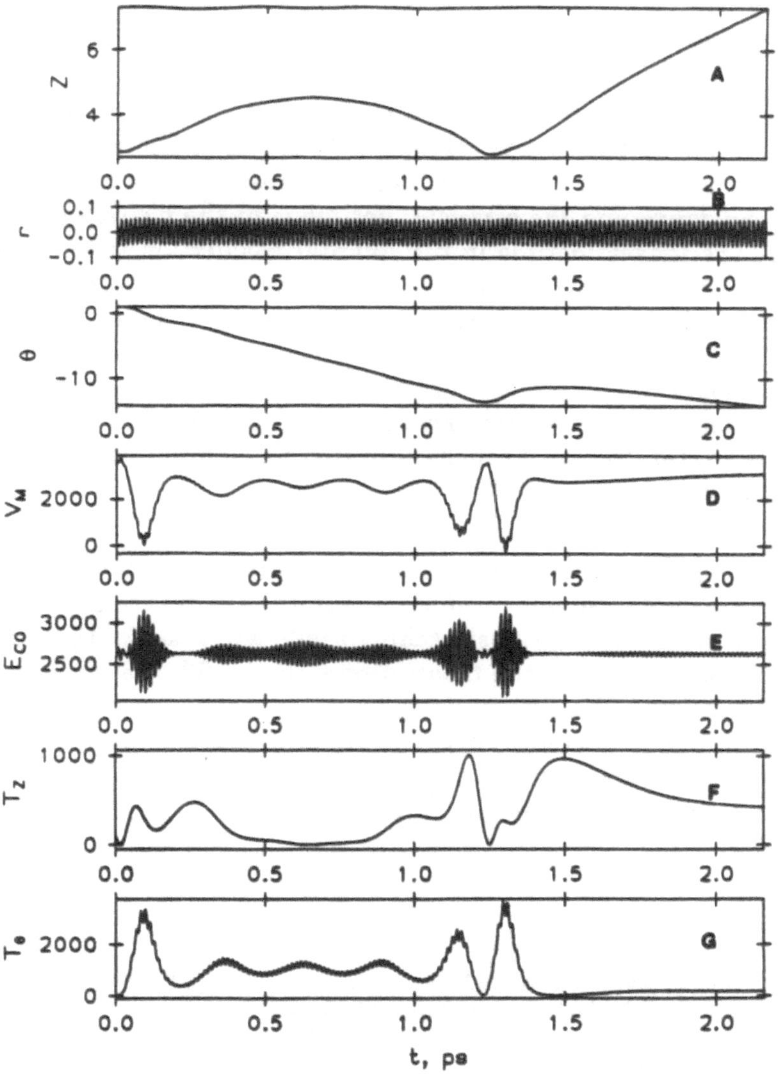

Figure 10. **A typical trajectory for the NaCl-CO system with initially excited bending mode, $n_\theta = 22$: coordinates and terms of energy as a function of time. (a) CO translational coordinate; (b) CO stretching coordinate; (c) bending (rotational) energy; (d) adsorption potential energy; (e) CO stretching energy; (f) CO translational energy; (g) CO bending (rotational) kinetic energy.**

the computation. This result shows the direct coupling between the CO internal stretching mode and the translational mode is very weak, and is in agreement with experimental results [42] that the induced desorption of

NaCl(100)-CO(v=1) is inefficient and the desorption lifetime might be very long (greater than 100 ps).

On the other hand, when the bending mode is excited to the energy range comparable to the energy of $n_r = 1$, we found significant rotational predesorption. The computed quantal predesorption lifetime is in the nanosecond range and the classical lifetime is in the picosecond range. We attribute the longer lifetime in the quantal calculation to the quantization constraint in the energy transfer. It is obvious that the rotational predesorption is due to strong coupling between the bending mode and translational mode. Our results are qualitatively in agreement with the suggestion by Ephraim et al. [43] that the desorption is the analog of rotational predissociation. However, our computed quantal lifetime is much longer than the sub-picosecond range computed by Ephraim et al [43] using the same potential.

If excitation of the CO stretching mode could induce predesorption, the bending mode may play an intermediate role in the energy transfer process, i.e., excitation energy in the CO stretching mode will transfer through the bending mode and then into the translational mode. This is evidenced by our results of exciting the system to $n_r = 1, n_\theta = 10$ state. The lifetime of this desorption is predicted to occur in the microsecond time scale. The drastic acceleration of desorption with libration has been confirmed in the recent calculations of Dzegilenko and Herbst [56].

References

1. Zhdanov,V.P. and Zamaraev,K.I. (1982), *Catal.Rev.Sci.Eng.* 24, 373
2. Tully,J.C. and Cardillo,M.J. (1984), *Science* 233, 445
3. *Surface Studies with Lasers* (1983), Aussenegg,F.R., Leitner, A. and Lippitsch,E. *eds.*, Springer Verlag, Berlin
4. Brivio,G.P. and Grimley,T.B. (1993), *Surf.Sci.Rep.* 17,1
5. Ho,W. (1987), *J.Phys.Chem.* 91, 766
6. Forst,W. (1973), *Theory of Unimolecular Reactions* Academic Press, New York
7. User,T. (1991), *Phys.Rep.* 199, 73.
8. Gadzuk,J.W. (1987), *J.Phys.Chem.* 86, 5196 *and references therein*
9. Gadzuk,J.W. and Lunts,A.C. (1984), *Surf.Sci.* 144 429
10. For a review, see Hoffmann,F.M. (1983), *Surf.Sci.Rep.* 3, 107
11. Chiang,S., Tobin,R.G., Richards,P.L. and Thiel,P.A. (1984), *Phys.Rev.Lett.* 52, 648
12. Trenary,M., Uram,K.J., Bosso,F. and Yates,J.T. (1984), *Surf.Sci.* 146, 269
13. Persson,B.N.J. and Ryberg,R. (1985), *Phys.Rev.Lett.* 54, 2119
14. Ariyasu,J.C., Mills,D.L., Lloyd,K.G. and Hemminger,J.C. (1984), *Phys.Rev.B* 30 507
15. Persson,B.N.J. (1984), *J.Phys.C* 17, 4741

16. Tully,J.C., Chabal,Y.J., Raghavachari, K.,Bowman,J.M. and Lucchese,R.R. (1985), *Phys.Rev.B* **31**
17. Chabal,Y.J (1986), *J.Electron.Spectrosc.Relat.Phenom.* **38**, 159
18. Laubereau,A. and Kaiser,W. (1978) *Rev.Mod.Phys.* **50** 607
19. Cavanagh,R.R., Casassa,M.P., Heilweil,E.J. and Stephenson,J.C. (1987), *J. Vac. Sci.Technol.A* **5**, 469
20. Casassa,M.P., Heilweil,E.J., Stephenson,J.C. and Cavanagh,R.R. (1986), *J. Phys. Chem.* **84**, 2361
21. Cavanagh,R.R., Germer,T.A., Heilweil,E.J. and Stephenson,J.C. (1993), *Faraday Disc.Chem.Soc.* **96**, 235
22. *Ultrafast Phenomena IV*, Auston,D.H. and Eisenthal,K.B. *eds.* (1984), Springer Verlag, New York
23. *Vibrational Spectroscopy of Molecules on Surfaces*, Yates,J.T.Jr. and Madey,T.E. *eds.* (1987), Plenum, New York
24. Gadsuk,J.W. (1987), in *Vibrational Spectroscopy of Molecules on Surfaces*, Yates,J.T.Jr. and Madey,T.E. *eds.* (1987), Plenum , New York pp.49
25. Heilweil,E.J.,Casassa,M.P.,Cavanagh,R.R. and Stephenson,J.C. (1989), *Ann. Rev. Phys. Chem.* **40**, 143 *and references therein*
26. Harris,A.L. and Levinos,N.J. (1989), *J.Phys.Chem.* **90**, 3878
27. Harris,A.L., Kuhnke,M., Jakob,P., Levinos,N.J. and Chabal,Y.J. (1993), *Faraday Disc.Chem.Soc.* **96**, 217
28. Morin,M.,Levinos,N.J., Chabal,Y.J. and Harris,A.L. (1992), *J.Phys.Chem.* **96**, 6203
29. Morin,M.,Levinos,N.J. and Harris,A.L. (1992), *J.Phys.Chem.* **96**, 3950
30. Heidberg,J., Kampshoff,R., Schönekäs,O. and Suhren,M. (1990), *J. Electr. Spectrosc. Relat. Phenom.* **54/55**, 945; Heidberg,J., Kampshoff,R. and Suhren,M. (1991), *J.Chem.Phys.* **95**, 9408
31. Guyot-Siyonnest,P. (1991), *Phys.Rev.Lett.* **66**, 1489; Chabal,Y.J, Dumas,P., Guyot-Siyonnest,P. and Higashi,G.S. (1991), *Surf.Sci.* **242**, 524; *ibid* (1990) *Phys.Rev.Lett.* **64**, 2156; Guyot-Siyonnest,P., Dumas,P. and Chabal,Y.J (1990), *J.Electron.Spectrosc.Relat.Phenom.* **54**, 27
32. Guan,Y., Muckerman, J.T. and User,T. (1990) , *J.Chem.Phys.* **93**, 4383
33. Guan,Y., Muckerman, J.T. and User,T. (1990) , *J.Chem.Phys.* **93**, 4400
34. Guan,Y., Muckerman, J.T. and User,T. (1995), *J.Chem.Phys. to be published*
35. Shugard,M., Tully,J.C. and Nitsan,A. (1977), *J.Chem.Phys.* **66**, 2534
36. Nitsan,A., Shugard,M. and Tully,J.C. (1978), *J.Chem.Phys.* **69** 2525
37. Tully,J.C. (1981), *Acc.Chem.Res.* **14**, 188
38. Adelman,S.A. and Doll,J.D. (1974), *J.Chem.Phys.* **61**, 4242; Doll,J.D.,Myers,E. and Adelman,S.A. (1975), *J.Chem.Phys.* **63**, 4908; Adelman,S.A. and Doll,J.D. (1976), *J.Chem.Phys.* **64**, 2375
39. Adelman,S.A. (1979), *J.Chem.Phys.* 71, 4471; (1981)*ibid* **74**, 4646; (1983) *Adv.Chem.Phys.*, **53**, 61
40. Heidberg,J.,Stein,H.,Riehl,E.,Ssilagyi,Z. and Weiss,H. (1985), *Surf.Sci.* **158**, 553; Heidberg,J.,Stein,H. and Riehl,E. (1985), *Surf.Sci.* **184** L431
41. Richardson,H.H. and Ewing,G.E. (1987),*J.Chem.Phys.* **91**. 5833; Richardson,H.H. and Ewing,G.E. (1987),*J.Electron.Spectry.* **45** 99; Chang,H.-C.,Richardson,H.H. and Ewing,G.E. (1988), *J.Chem.Phys.* **89**, 7561
42. Chang,H.-C.. and Ewing,G.E. (1989), *Chem.Phys.* **139**, 55
43. Ben Ephraim,A.,Folman,M.,Heidberg,J. and Moiseiev,N. (1988), *J.Chem.Phys.* **89**3840
44. Muckerman, J.T. and User,T. (1989), *J.Chem.Phys.* **90**, 1968
45. Heidberg,J., Noseck,U. Suhren,M. and Weiss,H. (1993), *Ber.Bunsenges.Phys.Chem.* **97**, 329
46. Space,B., Rabits,H.A. and Askar,A. (1993), *J.Chem.Phys.* **99** 9070; Askar,A., Owens,R.G. and Rabits,H.A. (1993), *J.Chem.Phys.* **99** 5316; Thacher,T.,

Rabitz,H.A. and Askar,A. (1990), *J.Chem.Phys.* **93**, 4673; Askar,A. and Rabitz,H.A. (1991), *Surf.Sci.* **245**, 411.

47. Tully,J.C. (1980), *J.Chem.Phys.* **73**, 6333; Tully,J.C. (1980), *Adv.Chem.Phys.* **42**, 63 *and references therein*
48. Tully,J.C. (1980), *J.Chem.Phys.* **73**, 1975; Lucchese,R. and Tully,J.C. (1983), *Surf.Sci.* **137**, 570; Tully,J.C. (1985), *J.Vac.Sci.Technol.A* **3**, 1664; Muhlhausen,C.W.,Williams,L.R. and Tully,J.C. (1984) *J.Chem.Phys.* **83**, 2594
49. Zeiri,Y.,Low,J.J. and Goddard,W.A. (1986), *J.Chem.Phys.* **84**, 2408
50. Kosloff,D. and Kosloff,R. (1983), *J.Comp.Phys.* **52**, 351; Kosloff,R. and Kosloff,D. (1983), *J.Chem.Phys.* **79**, 1823; Kosloff,R. (1988), *J.Phys.Chem.* **92**,2087
51. Billing,G.D. (1984), *Comp.Phys.Rep.* **1**, 237
52. Sawada,S., Heather,R.,Jackson,B. and Metiu,H. (1985), *J.Chem.Phys.* **83**, 3009; Heather,R. and Metiu,H. (1985), *Chem.Phys.Lett.* **118**, 558; (1986) *J.Chem.Phys.* **84**, 3250; Jackson,B. and Metiu,H. (1986), *ibid.* **84**, 3535.
53. Kosloff,R. and Cerjan,C. (1984), *J.Chem.Phys.* **81**, 3722; Cerjan,C. and Kosloff,R. (1986), *Phys.Rev.B* **34**, 3832
54. Gray,S.K. (1987), *J.Chem.Phys.* **8**, 2051
55. Muckerman,J.T. (1990), *Chem.Phys.Lett.* **85**, 4594
56. Dzegilenko,F. and Herbst,E. (1994), *J.Chem.Phys.* **100**, 9205

TIME–DEPENDENT NUCLEAR DYNAMICS OF DECAYING STATES

L.S. Cederbaum
Theoretische Chemie, Institut für Physikalische Chemie
Universität Heidelberg
D–6900 Heidelberg, FRG

F. Tarantelli
Dipartimento di Chimica, Università di Perugia
I–06100 Perugia, Italy

The wavepacket dynamics accompanying the excitation to a decaying electronic state and the subsequent decay to final electronic states are discussed. The cross–sections for the excitation and for the production of final states are related to the corresponding wavepackets. The time–dependent formulation adds insight into the process and is amenable to semiclassical approximations and interpretations. It can also be used to compute the gross features of the observed spectra via a spectral moment expansion. An illustrative application demonstrates the usefulness of the expansion.

E. Yurtsever (ed.), Frontiers of Chemical Dynamics, 291–330.

1. Introduction

Most or even all excited electronic states possess a finite lifetime and decay by emitting either photons or other particles, e.g., electrons[1]. The investigation of the excitation process and, in particular, of the emitted particles provide information on the decaying states as well as on the final states following the decay. The observed spectra and cross–sections are greatly influenced by the nuclear dynamics in both the decaying and final states. If the excited state is long lived, the decay process is usually governed by the Franck–Condon principle or by straightforward extensions of it. A shorter lifetime which compares with typical times of the internal degrees of freedom of the system (e.g., vibrations), may give rise to interesting interference effects owing to the overlapping nuclear levels in the decaying electronic state. Further interesting effects are expected if a competition between different decaying channels are present. A vast number of processes involving the decay of states with a lifetime on the latter scale have been investigated of which we mention just a few. Energetically deep levels are preferably populated by photon[2] or electron[3] impact. The decay of those levels is studied by Auger[4] and X–ray emission[5] spectroscopies. In the photoionization cross–sections of molecules the effects of short–lived resonance states have been widely observed[6] for outer valence, inner valence as well as for core electrons. Low lying resonance states constitute a common phenomenon in molecules and markedly influence the electron–molecules scattering cross–sections[7].

In the present work we investigate the dynamics of the general situation shown in Fig. 1. The wavepacket describing the nuclear motion in the target system is excited to a decaying electronic state where it propagates on the corresponding potential energy surface.

During this propagation this wavepacket gradually decays to the final electronic state. A relatively complex wavepacket grows in time on the potential surface of the latter state to which new contributions are continuously added at later times. The knowledge of the wavepacket in the decaying state is needed to compute the cross section for the production of this state. Clearly, the information on the cross–section for the production of the final states of the decay is contained in the wavepacket propagating on the corresponding surface. It is the major task of the present paper to investigate these wavepackets and to relate them to the cross–sections.

A time–dependent formulation of the theory is attempted. The reasons for this choice are twofold. First, a time–dependent picture leads to a better insight of the process itself. This is particularly relevant if several competing degrees of freedom are involved. As an example we mention the competition between dissociation and autoionization following inner–shell excitation of molecules[8]. Second, a time–dependent formulation may also have technical advantages and be the appropriate tool for the computational evaluation of the cross–sections. In the last years efficient time–dependent methods have been developed for propagating wavepackets on potential energy surfaces[9]. Although the present situation indicated in Fig. 1 is more complicated because of the two potential surfaces and two time scales involved, the available techniques will be useful. This applies the more the shorter is the lifetime of the decaying electronic state.

In Section 2 the cross–sections for the production of the decaying and of the final electronic states are discussed within a time–dependent theoretical framework and related to the wavepacket dynamics. In Section 3 the gross features of the observed spectra are discussed in terms of spectral moments and in Sec. 4 a helpful semiclassical theory is presented. In many experiments the structures due to the nuclear dynamics are not resolved. Nevertheless, this dynamics can have strong impact on the gross features of the bands observed in the spectrum. Recently, it has been noticed theoretically[10] that even the energetic positions of the bands can significantly depend on the nuclear dynamics making a reinterpretation of observed spectra necessary. A brief illustrative application to the Auger spectra of CO is presented in Section 5.

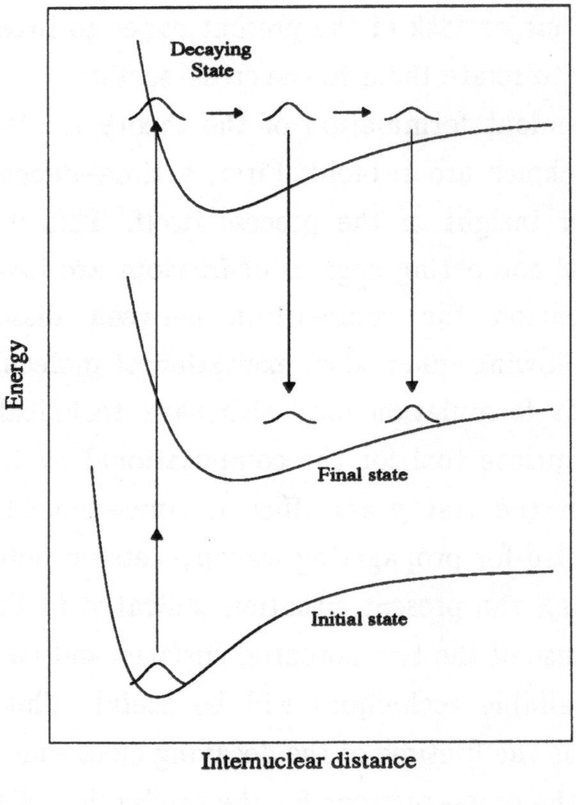

Figure 1. The wave packet is excited from the initial electronic state to the decaying electronic state where it propagates on the potential energy surface of the latter. During this propagation the wave packet continuously decays to the final electronic state. The decayed components propagate on the surface of the final state and may interfere with one another.

Most of the derivations have been omitted. They can be found in Ref. (11). In Ref. (12) one can find numerous additional results which are particularly relevant for polyatomics.

2. The Cross Sections

Let us consider the following basic process. The target system is in the initial electronic state $|i>$ and vibrational state $|n_i>$. By impact with some projectile, e.g., photon, electron, the system is excited (or ionized) to the intermediate state $|d>$. The matrix element for this excitation is V and may depend on the nuclear coordinates of the target and on the difference between the energies of $|d>$ and of the projectile. The state $|d>$ is coupled to a continuum, the coupling matrix element being W, and may thus decay via the emission of a photon or an electron. Clearly, W may also depend on the nuclear coordinates of the system and on the energy of the emitted particle.

The relevant quantity which describes the above general situation is the transition probability per unit time from the initial state to the final state of the decay. This transition probability reads[13]

$$P_{FI} = 2\pi|<n_f|W(E-\mathcal{H}_d)^{-1}V|n_i>|^2\delta(E_I-E_F) \qquad (2.1)$$

where E_I and E_F are the total energies of the initial state plus incoming projectile and of the final state $|f>|n_f>$ plus <u>all</u> the particles present at the end of the process, respectively. It should be noted that the electronic wavefunctions do not appear in expression

(2.1). Their effect is in the transition and decay matrix elements V and W. The energy E is the excess energy which has been absorbed into the system by the impact with the incoming projectile and is available for the decay. If this projectile is a photon and the intermediate state |d> is a neutral state, E is just the photon energy. If |d> is an ionic state, i.e., the primary process is photoionization, then E is the photon energy less the energy of the photoelectron. Analogously, if the projectile is an electron, the excess energy E is simply the energy loss through the impact in case the primary event is excitation and the energy loss less the kinetic energy of the ionized electron in the case of ionization. Unless otherwise stated, we choose the energy of the target system to be the zero of the energy scale.

$\mathcal{H}_d$ is the complex and, in general, energy–dependent Hamiltonian describing the nuclear motion in the intermediate decaying state. It consists of the Hamiltonian H_d for the nuclear motion in the electronic state |d> and terms $\Delta - i\Gamma/2$ due to its interaction with the continuum

$$\mathcal{H}_d = H_d + \Delta - i\Gamma/2 . \tag{2.2}$$

The imaginary part $-\Gamma/2$ of $\mathcal{H}$ represents the decay width and the real part Δ the shift resulting from this interaction. These quantities read

$$\Gamma = 2\pi \sum_k W^* \, \delta(E - E_k - H_f) W \tag{2.3a}$$

$$\Delta = P \sum_k W^* \, (E - E_k - H_f)^{-1} W \tag{2.3b}$$

where the summation is over the continuum states of the particle with energies E_k emitted in the decay and P denotes the Cauchy principle value. H_f is the Hamiltonian describing the nuclear motion in the final electronic states $|f\rangle$. The operator properties of the width and shift Γ and Δ are only of relevance in situations where the decaying state and the final state are energetically very close together. Then non–adiabatic electron–nuclear energy transfer effects may take place in particular in targets with large dipole moments. A prominent example is found[14] in the resonant low–energy electron scattering off $HC\ell$ where interesting threshold peaks have been observed[15]. Apart from such interesting but rare cases, the width Γ and shift Δ can be considered to a good approximation as functions of the nuclear coordinates alone and independent of the energy. The quality of this so–called local approximation has been discussed in the literature[16]. The local approximation is particularly excellent when the decaying state is energetically far from overlapping the final state. In the following we shall restrict ourselves to the local approximation.

It is clear that eq. (2.1) is quite general and the cross sections for numerous experiments can be deduced from it. We may distinguish between two types of experiments. In the first type the primary excitation (or ionization) of the target is detected. Examples are photoelectron spectroscopy where the ionized states can decay, for instance, via Auger decay and electron energy loss spectroscopy involving the excitation of inner–valence or core electrons. The corresponding cross sections are determined from eq. (2.1) by integrating over the momenta of the emitted particle and by summing over the vibrational states $|n_f\rangle$ of the final electronic state. This leads to the cross section

$$\sigma_d(E) \sim \mathrm{Re} \int_0^{\infty} e^{iEt} <n_i|V^* e^{-i\mathcal{H}_d t} V|n_i> dt \tag{2.4}$$

which is nothing but the real part of the Fourier transform of the autocorrelation function. $|\Psi_d(0)> = V|n_i>$ can be viewed as the wavepacket transferred at time t = 0 to the potential energy surface of the intermediate state where it propagates via the Hamiltonian $\mathcal{H}_d$, i.e.,

$$|\Psi_d(t)> = \exp[-i\mathcal{H}_d t]|\Psi_d(0)> \tag{2.5}$$

Consequently,

$$\sigma_d(E) \sim \mathrm{Re} \int_0^{\infty} e^{iEt} <\Psi_d(0)|\Psi_d(t)>dt . \tag{2.6}$$

The proportionality factor contains kinematic quantities and a function of energy arising from the conversion of momentum dk into energy dE_k. The quantity $\sigma_d(E)$ describes the observation of a single vibrational band corresponding to the electronic state |d>. V is actually energy dependent, but since we do not consider overlapping states it can be assumed to be energy independent over the band.

Eq. (2.6) which is well–known for real potential energy surfaces is seen to apply also for decaying states with complex surfaces which is not surprising. In the common Condon approximation the transition matrix element V is treated as a constant instead of a function of the nuclear coordinates and the equations simplify somewhat.

In the second type of experiment one measures the final states or, more precisely, the particle emitted due to the decay of the intermediate state. Examples are Auger spectroscopy, X–ray emission as well as resonant Auger spectroscopy and the corresponding resonant X–ray emission. In the former two examples the system is ionized, for instance by a photon, and one must integrate the transition probability (2.1) over the photoelectron states in order to obtain the cross section for the secondary Auger electron. In the latter two examples the system is excited, for instance by a photon, and one integrates over the bandwidth of the photon source (For details see Ref. (17)). In most experiments the cross section is not separately recorded for each final vibrational level $|n_f\rangle$. The observed spectrum is rather obtained by summing over these levels:

$$\sigma_f(E) \equiv \sum_{n_f} \sigma_f(n_f, E) . \tag{2.7}$$

The cross section (2.4) for observations of the short–lived state does not exhibit interference effects; $\sigma_d(E)$ can essentially be written as an incoherent superposition of Lorentzian lines. In the cross section (2.8) for observations of the final state, on the other hand, there are interference effects: if $\Gamma \geq \omega$ where ω is the typical vibrational spacing in the decaying state, its vibrational levels overlap and coherently contribute to the same final vibrational state $|n_f\rangle$. These interference effects have been discussed in detail and identified in experimental spectra of small molecules[13,18,19].

The cross section $\sigma_f(E)$ can also be formulated in a time–dependent picture in which the interference effects can be unambiguously identified. The result reads

$$\sigma_f(E) \sim \int_0^\infty dT \int_{-T}^{T} dt\, e^{iEt} <n_i|V^* e^{i\mathcal{H}_d^+(T-t)/2} W^*$$
$$e^{iH_f t} W e^{-i\mathcal{H}_d(T+t)/2} V|n_i> \quad (2.8)$$

While for the observation of the decaying state only a single time variable is relevant, see eq. (2.4), we note from eq. (2.8) that two time variables are essential for describing the dynamics leading to the final state. In the former case the time is the propagation time of the initial wavepacket on the potential surface of the decaying state. In the latter case an additional time enters, namely the propagation time on the potential energy surface of the final state.

The interpretation of the spectra $\sigma_d(E)$ where the short–lived state $|d>$ is detected is very straightforward, see eq. (2.6), and there is no need for further discussion. On the other hand, the quantity $\sigma_f(E)$ is less obvious, it exhibits interference phenomena and its time–dependent formulation in eq. (2.8) involves two time variables. In the following we concentrate on the interpretation of $\sigma_f(E)$.

At time $t = 0$ the initial vibrational state $|n_i>$ is excited to the intermediate electronic state and arrives there as $V|n_i>$. This wavepacket propagates via the Hamiltonian $\mathcal{H}_f$ on the potential surface of the decaying state: $|\Psi_d(t)> = \exp[-i\mathcal{H}_d t]V|n_i>$. At every small time interval dt it looses a part which decays to the final state where it propagates via the Hamiltonian H_f. Since Ψ_f is a consequence of the decay, the initial condition is

$$\Psi_f(E,0) = 0 \quad (2.9)$$

and the wavepacket acquires the appearance[11]

$$|\Psi_f(E,t)\rangle = -ie^{-iH_f t} \int_0^t e^{i(E+H_f)t'} W e^{-i\mathcal{H}_d t'} V|n_i\rangle \, dt'. \qquad (2.10)$$

As can be seen from eq. (2.10), the quantity which is propagated by $\exp[-iH_f t]$ is itself an integral over time. This integral represents the contribution which has accumulated on the final surface via the decay up to time t. That strong interferences can take place is near at hand because of the superposition of two contributions each being connected to the motion in a different potential. Due to this superposition, the norm of the wavepacket can vary strongly with E and for short times $t\Gamma \lesssim 1$ also with t. In particular, we notice from eqs. (2.10) and (2.8) that

$$\sigma_f(E) \sim \langle\Psi_f(E,\infty)|\Psi_f(E,\infty)\rangle \qquad (2.11)$$

The norm of the wavepacket at $t \to \infty$ is proportional to the cross section. Note that in actual calculations of the cross section there is no need to compute the full wavepacket Ψ_f: the exponential in front of the time integral in eq. (2.10) does not contribute to the cross section and the integration itself is only needed up to $t \approx 2\Gamma^{-1}$.

Where are the interferences hidden in the time-dependent expression (2.8)? To keep the discussion as simple as possible we first introduce some simplifications which are of interest by themselves. Analogously to the Condon approximation for spectra of the $\sigma_d(E)$ type we substitute the transition and decay matrix elements V and W, which are functions of the target′s nuclear coordinates, by their value at fixed nuclear coordinates. If $|n_i\rangle$ is the target′s ground vibrational state, this will be at the equilibrium geometry of the target. In many cases the decay width will change only slightly with the nuclear coordinates. This is particularly true for all decay processes involving an energetically deep level, e.g., a core level. In these cases the decay width is essentially due to the atom corresponding to the deep level and Γ is essentially a constant. In the following we shall refer to the Condon puls constant Γ approximations as the generalized Condon approximation.

Within the generalized Condon approximation the cross section simplifies and by inserting complete sets of eigenstates between the operators in eq. (2.8) we readily find

$$\sigma_f(E) \sim \frac{\Gamma}{2} \sum_{n_f} \left| \sum_{n_d} \frac{<n_f | n_d> \; <n_d | n_i>}{E-(E_{n_d} - E_{n_f}) + i\Gamma/2} \right|^2 \tag{2.12}$$

which is a well–known result[13,18,19]. If Γ is small compared to the spacings in H_d all interferences are suppressed and σ_f essentially becomes an incoherent superposition of Lorentzian lines. In the limit $\Gamma \rightarrow 0$ these Lorentzians become δ–functions and the interferences have strictly disappeared

$$\sigma_f(E) \sim \sum_{n_f, n_d} |<n_f|n_d><n_d|n_i>|^2 \, \delta(E - E_{n_d} + E_{n_f}) \tag{2.13}$$

Interestingly, putting the integration limits to infinity in the integral over t in eq. (2.8) and letting $\Gamma \rightarrow 0$ after the integrations recovers eq. (2.13). In other words, in the time–dependent formulation the interference effects are due to the finite time intervals over which the Fourier transform must be performed.

Finally, we would like to mention that the time–dependent formulation of the cross sections provides some advantages over the time–independent one. We shall see below that the time–dependent picture amends itself to helpful interpretations of the underlying process. In addition, we shall be able to extract useful informations about the cross section starting with its time–dependent expression. Last but not least, in some cases, for example if dissociative potential surfaces are involved, equations like eq. (2.8) might be simpler to practically compute than the corresponding time–independent equations.

3. Properties of the spectra

3.1. EXACTLY SOLVABLE CASES

There are only very few idealized cases where the spectra can be given explicitly in a closed analytical form. This applies to the spectra $\sigma_d(E)$ where the decaying state is the state under observation and in particular to the much more complicated spectra $\sigma_f(E)$ detecting the final states $|f\rangle$. In the simplest case we can conceive of, all the potential curves involved are harmonic potentials of the same frequency but with different minimum positions and, in addition, the generalized Condon approximation (i.e., constant Γ, W, V) is assumed. The spectrum $\sigma_d(E)$ then consists of a sum of equidistant Lorentz curves weighted by a Poisson distribution[13]. The cross section $\sigma_f(E)$ can also be given[13,18] in this simple case of shifted harmonic oscillators, the resulting expression (which is useful for numerical computations) is, however, still too complicated to allow one to "see" how the spectrum looks like.

In general, the cross sections can be given explicitly whenever the Franck–Condon factors as well as energy differences between the vibrational levels are known explicitly and the generalized Condon approximation applies. If Γ is a function of the nuclear geometry, it can be incorporated into the potential which becomes complex. The energies and the Franck–Condon factors are then complex. In the case of harmonic oscillators, the expansion of Γ up to the quadratic term in the geometry leads to complex harmonic oscillators. The complex harmonic oscillator and its Franck–Condon factors have been discussed in the multi–mode case[20].

Three electronic states, the initial, the decaying and the final states are involved in the process leading to $\sigma_f(E)$. We can expect the spectrum to simplify considerably only if two of the surfaces are parallel and the generalized Condon approximation applies. The situation is particularly simple if the surfaces of the decaying and final states are parallel, i.e.

$$H_d = H_f + E_o \tag{3.1a}$$

where E_o is a constant energy separation. In this case the final wavepacket $\Psi_f(E,t)$ propagates essentially with the same Hamiltonian (up to an energy shift and imaginary decay part) as the decaying wavepacket $\Psi_d(t)$ on the potential surface of the decaying state. The cross section then takes on the appearance of a single Lorentz curve centered at E_o

$$\sigma_f(E) \sim \frac{1}{(E-E_o)^2 + (\Gamma/2)^2} \ . \tag{3.1b}$$

If the potential energy surfaces of the initial and the decaying states are parallel

$$H_d = H_i + E_d \tag{3.2a}$$

the final wavepacket is somewhat more complicated, but again its propagation depends only on H_f. At large times t it takes on the appearance (the energy of the initial vibrational state $|n_i>$ is chosen to be the origin of the energy scale):

$$\Psi_f(E,t) \sim e^{-iH_f t} [E - E_d + H_f + i\Gamma/2]^{-1} |n_i> \tag{3.2b}$$

The cross section can be written as an incoherent superposition of Lorentzians or, equivalently, as the real part of an autocorrelation function

$$\sigma_f(E) \sim \mathrm{Re} \int_0^\infty e^{i(E_d - E)t} <n_i| e^{-i(H_f - i\Gamma/2)t} |n_i> dt \tag{3.2c}$$

By comparing with the general expression (2.6) for $\sigma_d(E)$, we note that, apart from the appropriate definition of the detected energy, $\sigma_f(E)$ just appears as the spectrum of a "decaying" state with the Hamiltonian $H_f - i\Gamma/2$. The true decaying state has transferred its decay width to the final state and is irrelevant otherwise. No essential simplifications arise if the potentials of the initial and of the final state are parallel.

3.2. APPEARANCE OF THE BANDS IN THE SPECTRUM

3.2.1. General Aspects

The cross section $\sigma_f(E)$ for the production of the final state $|f>$ of the decay has been seen to posses a relatively complicated structure. It is thus not surprising that only a few calculations on diatomics are available[13,18,19] within the Condon approximation. In polyatomics several nuclear degrees of freedom are usually of relevance making the accurate determination of $\sigma_f(E)$ prohibitively difficult. Even approximated cross sections are likely to be computed only for a few cases of particular interest. In various kinds of experiments, however, several or even many final states appear and even questions concerning the gross features of the whole spectrum are of relevance, like: at what energies do the corresponding peaks (bands) appear and how large are their widths. Obviously, the gross features of the peaks are those needed to analyze low resolution or unresolvable spectra.

A distribution can be approximated in terms of its moments[21]. The first few moments describe the gross features of the distribution. The k–th moment of $\sigma(E)$ is defined as

$$<E^k> = \int E^k \, \sigma(E) \, dE \tag{3.3}$$

The center of gravity of a band in the spectrum is given by the first moment $<E>$ and the width of the band is related to $<E^2> - <E>^2$.

In the following we discuss the first and second moments of the cross sections $\sigma_d(E)$ and $\sigma_f(E)$ analyzing the decaying and final state, respectively. For the sake of simplicity we assume the generalized Condon approximation and write

$$\sigma_d(E) \sim \frac{1}{2\pi} \operatorname{Re} \int_0^\infty e^{i(E+i\Gamma/2)t} <n_i|e^{-iH_d t}|n_i> dt \tag{3.4}$$

$$\sigma_f(E) \sim \frac{\Gamma}{4\pi} \int_0^{\infty} e^{-\Gamma T/2} \, dT \int_{-T}^{T} e^{iEt} \, \langle n_i | e^{iH_d(T-t)/2} \, e^{iH_f t} \, e^{-iH_d(T+t)/2} | n_i \rangle \, dt \tag{3.5}$$

where the prefactors $1/2\pi$ and $\Gamma/4\pi$ have been chosen such that the zeroth moment of the cross section, i.e., the total integrated intensity of the band, is unity apart from the remaining proportionality factors. Now, the matter of moments of decaying species presents a non–trivial problem. The zeroth and first moments exist and are well defined, but already the second moment diverges as one can easily see by replacing $\sigma(E)$ in eq. (3.3) by a single Lorentzian: the integrand does not vanish for large E. The divergence of the high moments of $\sigma_d(E)$ and $\sigma_f(E)$ originates from the very long Lorentzian tail of the whole band far away from the peak constituting the band. Since this tail does not affect our observations of the band, we can ignore its effect on the observed quantities as, for instance, the full width at half maximum (FWHM) of the band. In the Appendix of Ref. 11 it is shown how this can be done unambiguously. All the moments of the cross sections defined in this way exist and are well defined. For the derivation of the moments we refer to Ref. (11).

The position of the band in the spectrum $\sigma_d(E)$ is simply the initial vibrational state expectation value of H_d:

$$\langle E \rangle_d = \langle n_i | H_d | n_i \rangle \tag{3.6a}$$

The analogous expression in the case of $\sigma_f(E)$ is substantially more involved reflecting the simultaneous dynamics on two potential energy surfaces:

$$<E>_f = \Gamma \int_0^\infty e^{-\Gamma T} <n_i| e^{iH_dT} (H_d - H_f) e^{-iH_dT} |n_i> dT \qquad (3.6b)$$

The influence of the nuclear dynamics on the band position $<E>_f$ is much larger than on $<E>_d$ and gives rise to interesting phenomena.

As mentioned above, the quantity $<E^2> - <E>^2$ is a measure for the width of the band. Assuming that this electronic band takes on the appearance of a Gaussian, then its FWHM is explicitly given by

$$\text{FWHM} = [8\ell n2(<E^2> - <E>^2)]^{1/2} \qquad (3.7)$$

The second moment needed to complete our general considerations reads

$$<E^2>_d = <n_i|(H_d)^2|n_i> + \Gamma^2/8\ell n2 \qquad (3.8a)$$

for the cross section $\sigma_d(E)$ and

$$<E^2>_f = \Gamma \int_0^\infty e^{-\Gamma T} <n_i| e^{iH_dT} (H_d - H_f)^2 e^{-iH_dT} |n_i> dT + \Gamma^2/8\ell n2 \qquad (3.8b)$$

for $\sigma_f(E)$. The origin of the small contribution $\Gamma^2/8\ell n2$ is explained in the Appendix. Γ in this latter contribution may be chosen to include also the experimental resolution of the apparatus.

It should be noted that if the spectral band is highly asymmetric its position $<E>$, i.e., its center of gravity, does not correspond to the maximum of the band. The asymmetry of the band can be computed using the third moment of the spectrum.

3.2.2. Working equations for the moments

In this subsection we derive approximate explicit expressions for the first two moments of the cross sections which can be used in practical calculations. The simpler expressions for $\sigma_d(E)$ are not new[20], but are given for the sake of comparison. We first derive the expressions for the case of a single active nuclear coordinate and give the results for several coordinates at the end of the subsection.

In general, each of the three Hamiltonians H_i, H_d and H_f describing the nuclear dynamics in the initial, decaying and final electronic states has a different potential energy. This severely complicates the calculation of the moments. The starting point of a systematic procedure to derive a useful expansion for the moments is to expand the difference between the potential energy expressions in the final and decaying states and that of the initial state about a reference geometry which we choose to be the equilibrium geometry of the system in the initial state. The merits of such a procedure have been discussed[22] in connection with $\sigma_d(E)$. Since the initial excitation $|i\rangle \rightarrow |d\rangle$ is essentially vertical, it is the potential around the corresponding geometry which contributes and must be well estimated. In our present case of $\sigma_f(E)$ the situation is more intricate. The wavepacket on the decaying intermediate state propagates for a time $\sim 1/\Gamma$ before decaying and the details of a larger portion of the potential energy surfaces become relevant. Nevertheless, the above expansion is still useful in particular if the lifetime $1/\Gamma$ of the decaying state is short.

In the spirit of the above discussion we now write the expansion

$$H_d - H_i = E_d(R_o) - E_i(R_o) + \kappa_d(b + b^+) + \ldots \tag{3.9a}$$

$$H_f - H_i = E_f(R_o) - E_i(R_o) + \kappa_f(b + b^+) + \ldots \tag{3.9b}$$

where b and b^+ are the usual annihilation and creation operators for vibrational quanta[23] in the initial electronic state. $E_d(R_o)$, $E_f(R_o)$ and $E_i(R_o)$ are the electronic energies of the decaying, final and initial states, respectively, at the equilibrium geometry R_o of the target (initial state). The coupling constants are the slopes of the electronic energies at R_o

$$\kappa_d = \frac{1}{\sqrt{2}}\left[\frac{\partial E_d(Q)}{\partial Q}\right]_o \qquad \kappa_f = \frac{1}{\sqrt{2}}\left[\frac{\partial E_f(Q)}{\partial Q}\right]_o \tag{3.10}$$

with respect to the dimensionless[24] normal coordinate $Q = (b + b^+)/\sqrt{2}$. We mention that in a diatomic molecule $Q = \sqrt{\mu\omega}(R-R_o)$ where R is the interatomic distance and μ the reduced mass.

To proceed we use the expansion (3.9) and expand H_i about R_o as well

$$H_i = E_i(R_o) + \omega(b^+ b + 1/2) + \ldots \tag{3.11}$$

and note that the terms not shown explicitly are cubic and of higher order in the operators.

Using only the terms explicitly shown in eqs. (3.9) and (3.11) for the Hamiltonians, the following time–dependent Hamiltonian can quite easily be written in closed–form[11]

$$H_f(T) \equiv e^{iH_dT} H_f e^{-iH_dT}$$

$$H_f(T) = H_f + \beta(1-\cos\omega T) [\frac{2\kappa_d}{\omega} + (b + b^+)] + i\beta \sin\omega T(b - b^+)$$

$$\beta \equiv \kappa_d - \kappa_f \tag{3.12}$$

This time–dependent Hamiltonian describes the nuclear dynamics on a potential surface whose equilibrium geometry and the corresponding energy change periodically in time. Furthermore, a periodical change of momentum also takes place. If higher order terms in the expansion of the Hamiltonians, eqs. (3.9) and (3.11), are included, additional terms will, of course, appear in the expression for $H_f(T)$ which can be evaluated systematically.

Once $H_f(T)$ and the initial vibrational state $|n_i>$ are known, all the moments of the spectrum can be determined. The moments we are usually interested in are those where $|n_i>$ is the vibrational ground state of the target. We choose the corresponding energy to be the zero of our energy scale. Considering only the terms explicitly shown above in the expansion of the Hamiltonians about R_o, the evaluation of the first moment is now very simple. Only the terms without annihilation and creation operators in $H_f(T)$, eq. (3.12), contribute. After performing the integration over T in eq. (3.6b), the first moment reads

$$<E>_f = [E_d(R_o) - E_f(R_o)] - \frac{2\kappa_d\beta}{\omega} [1 - \frac{\Gamma^2}{\Gamma^2+\omega^2}] \tag{3.13a}$$

The first term on the r.h.s. is simply the vertical transition energy and the second term results because of the nuclear dynamics. A discussion of this contribution is given in Sec. IV. Here we just remark that this dynamically induced contribution vanishes if the decaying state has zero lifetime $\Gamma \to \infty$, i.e., the nuclei have no time to move before the decay takes place.

The first moment of the spectrum $\sigma_d(E)$ measuring the decaying state is simply given by

$$<E>_d = E_d(R_o) - E_i(R_o) \tag{3.13b}$$

In words, the center of the band in the σ_d spectrum is just the vertical transition energy and is not affected by the nuclear dynamics. We mention here that the dynamically induced contribution in eq. (3.3a) for the σ_f spectrum can be large. Our numerical calculations indicate that it can amount even up to a few eV!

Using the closed–form expression (3.12) for $H_f(T)$, the second moment also follows easily. The expectation value in eq. (3.8b) is evaluated most straightforwardly if $[H_d - H_f(T)]$ is first applied to $|n_i>$ and subsequently the self–scalar–product is performed. Introducing the abbreviation $C_f^2 = <E^2>_f - <E>_f^2 - \Gamma^2/8\ln 2$ we obtain

$$C_f^2 = \beta^2 + \frac{2\kappa_d^2\beta^2}{\omega^2}\left[1 + \frac{\Gamma^2}{\Gamma^2+4\omega^2} - 2\left[\frac{\Gamma^2}{\Gamma^2+\omega^2}\right]^2\right] \tag{3.14a}$$

If we neglect the motion of the wavepacket, i.e., if dynamic effects are not present, only the first term β^2 on the r.h.s. of the above equations contributes. The width of the band would then be FWHM $= [\beta^2 8\ln 2 + \Gamma^2]^{1/2}$ as can be seen from $\Gamma \to \infty$ in eqs. (3.14a) and (3.7). The dynamically induced contribution to the second moment, given by the second term on the r.h.s. of eq. (3.14a), is always positive and can be large and even exceed the static contribution β^2

as long as the lifetime of the decaying state is not very short.

In the case of the $\sigma_d(E)$ spectrum the analogous quantity C_d^2 takes on the following simple appearance

$$C_d^2 = \kappa_d^2 \tag{3.14b}$$

giving rise to a band with the width FWHM $= [\kappa_d^2\, 8\ln 2 + \Gamma^2]^{1/2}$. Clearly no dynamical contributions of the wavepacket's motion to the width are present. The width of the spectrum is just the result of the fact that the initial wavepacket itself possesses a width. The projection of the initial wavepacket on the potential curve of the decaying state leads to the observed width of the band.

The expression derived above for the position and width of a band in a spectrum are amenable to ab initio computations. To be calculated are the electronic energies at a fixed geometry as well as the slope of the corresponding potential curves at this geometry. We would like to restress here that the above expressions are the leading terms of a systematic expansion. The corrections are also amenable to ab initio computations.

For polyatomic molecules the dynamically induced shifts and widths are of particular interest because of the non–additivity of the contributions of the individual nuclear degrees of freedom. A thorough discussion can be found in Refs. (11) and (12).

4. Semiclassical Approach

At time T = 0 the wavepacket is vertically excited from the initial state to the potential surface of the decaying state where it propagates and decays to the final state. At any time T > 0 the center of the wavepacket is at a normal coordinate position Q(T) and the momentanuous energy of the particle ejected via the decay is

$$\Delta(Q(T)) = E_d(Q(T)) - E_f(Q(T)) \tag{4.1}$$

as indicated in figure 2. Since the wavepacket is not located at a single point but describes a distribution of positions, the correct energy is obtained by averaging over this distribution. The wavepacket decays continuously and hence we have to average over the decay probability as well. Our energy which is the band position, therefore, reads

$$<E>_f = \Gamma \int_0^\infty dTe^{-\Gamma T} \int_{-\infty}^\infty dq |\Psi(q,T)|^2 \Delta(q) \tag{4.2}$$

where Ψ is the wavepacket (compare with eq. (3.6b) and notice that $\Psi_d(T) = \Psi(T)\exp(-\Gamma T/2)$).

To simplify the expression we note that

$$Q(T) = \int |\Psi(q,T)|^2 q \, dq$$

and approximate $|\Psi|^2$ semiclassically by neglecting the spreading of the wavepacket with time which should be an excellent approximation in most cases of interest, i.e., $\Psi(q,T)$ is replaced by $\Psi(q-Q(T))$. Assuming the initial wavepacket to be the vibrational ground state of the target, our final result takes on the appearance

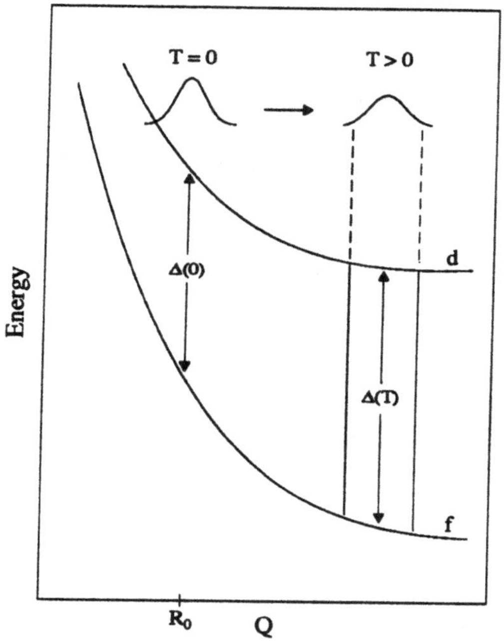

Figure 2. At time T = 0 the wave packet starts to propagate on the potential energy surface of the decaying state d. A decay to the final state f at T = 0 will lead to the observation of a peak at the vertical energy $\Delta(O)$. A decay at T > 0 will lead to a peak at $\Delta(T)$. In reality, of course, the decay is continuous.

$$<E_{s.c.}>_f = \frac{\Gamma}{\sqrt{\pi}} \int_0^{\infty} dT\, e^{-\Gamma T} \int_{-\infty}^{\infty} dq\, e^{-q^2}\, \Delta(Q(T)+q)\,. \qquad (4.3)$$

In complete analogy the second moment can be expressed by the semiclassical formula

$$<E^2_{s.c.}>_f = \frac{\Gamma}{\sqrt{\pi}} \int_0^{\infty} dT\, e^{-\Gamma T} \int_{-\infty}^{\infty} dq\, e^{-q^2}\, [\Delta Q(T) + q)]^2\,. \qquad (4.4)$$

The quantity yet undetermined is $Q = Q(T)$ which can be compute classically. Following Ehrenfest's theorem this is expected to be an excellent approximation for short times T and is even exact for harmonic potentials. Newton's equation of motion reads

$$\frac{d^2Q}{dt^2} = -\omega \frac{\partial E_d}{\partial Q} \tag{4.5}$$

where $E_d(Q)$ is the potential surface of the decaying state and ω appears because the normal coordinate is dimensionless[24].

Let us discuss an explicit example where the potential curves of the decaying, final and initial states read

$$\begin{aligned} E_d(Q) &= \frac{1}{2}\omega Q^2 + \sqrt{2}\,\kappa_d\, Q + E_d(R_o) \\ E_f(Q) &= \frac{1}{2}\omega Q^2 + \sqrt{2}\,\kappa_f\, Q + E_f(R_o) \\ E_i(q) &= \frac{1}{2}\omega Q^2 \end{aligned} \tag{4.6}$$

The equation of motion (4.5) under the initial condition for vertical excitation ($Q(0) = 0$, $dQ(0)/dt = 0$) has the solution

$$Q(T) = -\frac{\sqrt{2}\kappa_d}{\omega}(1-\cos\omega T). \tag{4.7a}$$

To evaluate the integrals in (4.3) we need $\Delta(Q(T))$ which follows from (4.6):

$$\Delta(Q(T)) = [E_d(R_o) - E_f(R_o)] + \sqrt{2}(\kappa_d - \kappa_f)Q(T)\,. \tag{4.7b}$$

The integrals are readily calculated and we find

$$<E_{s.c.}>_f = \Delta(0) - \frac{2\kappa_d(\kappa_d - k_f)}{\omega}(1 - \frac{\Gamma^2}{\Gamma^2+\omega^2}) \qquad (4.8a)$$

which is identical with the result (3.13a) of the full quantum mechanical calculation.

Inserting (4.6) into (4.4) we straightforwardly find

$$<E^2_{s.c.}>_f - <E_{s.c.}>_f^2 = <E^2>_f - <E>_f^2 - \Gamma^2/8\ell n2 \qquad (4.8b)$$

Up to a usually irrelevant contribution the semiclassical width of the band is identical with the quantum result of the preceding section. Because the potential $E_d(Q)$ is harmonic, Q(T) is exact. At first sight, it seems surprising that the spreading of the wavepacket neglected in our calculation does not contribute at all. However, the wavepacket at T = 0 corresponds to an eigenstate of $E_i(Q)$ which according to eq. (4.6) has the same frequency as $E_d(Q)$ on which it propagates and hence no spreading occurs (see, e.g., Ref. (25)). If $E_i(Q)$ and $E_d(Q)$ are still harmonic but with different frequencies, spreading of wavepacket with time takes place. In this case the spreading contributes even to the position of the band. We compared $<E_{s.c.}>_f$ with $<E>_f$ and identified the term due to spreading[11]. This term will usually be small. We mention that the effect of spreading can be incorporated into the basic semiclassical equations (4.3) and (4.4) if the curves are harmonic: one just has to replace $\exp(-q^2)$ by $\exp(-\alpha(T)q^2)$ where $\alpha(T)$ can be given explicitly[25].

The above semiclassical expressions (4.3) and (4.4) are useful in practice also when the potentials $E_d(Q)$ and $E_f(Q)$ are anharmonic. Although the classical equation of motion for Q(T) is not strictly valid in this case, the time scale of the decay is usually sufficiently short to ensure satisfactory accuracy. An advantage of the semiclassical theory or modifications of it is that they can be applied at relatively low cost to polyatomic systems with many degrees of freedom.

Let us now turn to the more involved issue of deriving a semiclassical approximation to the cross section $\sigma_f(E)$. Starting from the basic equation (3.5) we notice that it can be rewritten to give

$$\sigma_f(E) \sim \frac{\Gamma}{2\pi} \int_0^\infty e^{-\Gamma T} dT \int_{-2T}^{2T} e^{iEt} <\Psi(T)| e^{-iH_d t/2} e^{iH_f t} e^{-iH_d t/2} |\Psi(T)> dt \tag{4.9}$$

One readily notices that the product of the three exponential operators in (4.9) has the form of a symmetric split operator. For short times t we may write

$$e^{-iH_d t/2} e^{iH_f t} e^{-iH_d t/2} = e^{-i\Delta(q)t} + 0(t^3) \tag{4.10}$$

where $\Delta(q) = E_d(q) - E_f(q)$ is just the difference between the two potential energy surfaces of the decaying and final states. Using (4.10), our approximation to the cross section takes on the following simple appearance

$$\sigma_f(E) \sim \frac{\Gamma}{2\pi} \int_0^\infty e^{-\Gamma T} dT \int_{-2T}^{2T} e^{iEt} dt \int dq\, e^{-i\Delta(q)t} |\Psi(q,T)|^2 \tag{4.11}$$

which is amenable to a simiclassical treatment. Replacing $|\Psi|^2$ by a semiclassical expression provides a semiclassical expression for the cross section. The moments of $\sigma_f(E)$ in eq. (4.11) can be given in closed form:

$$<E_{sc}^{k}>_f = \Gamma \int_0^{\infty} e^{-\Gamma T}\, dT \int dq\, |\Psi(q,T)|^2\, [\Delta(q)]^k \qquad (4.12)$$

They concide with the moments discussed above.

In (4.11) The integration over t is easily performed

$$\sigma_f(E) \sim \frac{\Gamma}{\pi} \int e^{-\Gamma T}\, dT \int \frac{\sin(2Tf)}{f}\, |\Psi(q,T)|^2 dq \qquad (4.13)$$

where $f = E-\Delta(q)$ is the deviation of the actual energy E of the emitted electron from the vertical difference between the involved potential energy surfaces at the nuclear configuration q. We have carried out numerical calculations on $\sigma_f(E)$ using (4.13) and compared them to the exact results obtained[27] via propagation of the wave packet Ψ_f and eq. (2.11). The exact results will be dicussed elsewhere.[17]

In Fig. 3 the cross section $\sigma_f(E)$ is shown for three examples as a function of energy of the emitted electron. For convenience we have chosen all states involved, i.e., the initial, the decaying and the final state, to posses a harmonic energy curve. Each curve $E_i(R)$, $E_d(R)$ and $E_f(R)$ has its own frequency ω_o, ω_d and ω_f and its own equilibrium geometry R_o, R_d and R_f, respectively. We work with dimensionless quantities, where ω_o is the unit of energy and R is identical to the dimensionless normal coordinate Q introduced above for the initial electronic state. Each curve then reads

$$E_s(R) = \frac{1}{2}\, \omega_s^2 (R-R_s)^2 + E_s(R_s) \qquad (4.14)$$

where s = o,d or f.

Figs. 3a and 3b have been obtained using the same imput data except of the decay width Γ which is twice as large for the first curve. The wave packet is vertically excited at R_o from the ground electronic state to the decaying state. The classical Auger energy $\Delta(R)$ at R_o corresponds approximately to the maximum fo the band in Fig. 3a. The wave packet starts to propagate to $R > R_o$ values and the Auger energy E grows for the present set of parameters. Because of the quadratic dependence of $\Delta = E_d(R) - E_f(R)$ on R, there is a sharp classical maximum for the Auger energy either at $R_{max} = (\omega_d^2 R_d - \omega_f^2 R_f)/(\omega_d^2 - \omega_f^2)$ or at the classical turning point on the decaying curve $R_{turn} = R_d + 2[E_d(R_o) - E_d(R_d)]/\omega_d^2$. At values of R larger than R_{max} the classical Auger energy falls off again. For the data of Fig. 3a and Fig. 3b, R_{max} is slightly larger than R_{turn} and is hence not reached by the classical nuclear motion. At $R_{max}(\approx R_{turn})$ the Auger energy $\Delta = 30\ 2/3$ corresponds to the sharp drop of $\sigma_f(E)$ at its high energy side. The difference between the curves in Fig. 3a and Fig. 3b is due to Γ. Because of the large value of Γ in Fig. 3a, the wave packet on the decaying state has already decayed substantially when arriving at R_{turn} in contrast to the situation in Fig. 3b where an intense peak developes in the vicinity of $\Delta(R_{turn}) = 30.5$. In the full quantum calculations a similar intense peak is observed which mainly corresponds to the $0 \rightarrow 0$ vibrational transition between the decaying and the final electronic states.

If $\omega_d = \omega_f$, there is a linear relationship between the classical Auger energy $\Delta(R)$ and R. Correspondingly, if R_d and R_f are very different, one expects to see two peaks in the cross section. An intensive peak related to the vertical transition at R_o and a less intense peak related to the classical Auger energy at R_{turn}. The intensity ratio of these two peaks depends on the value of Γ: the

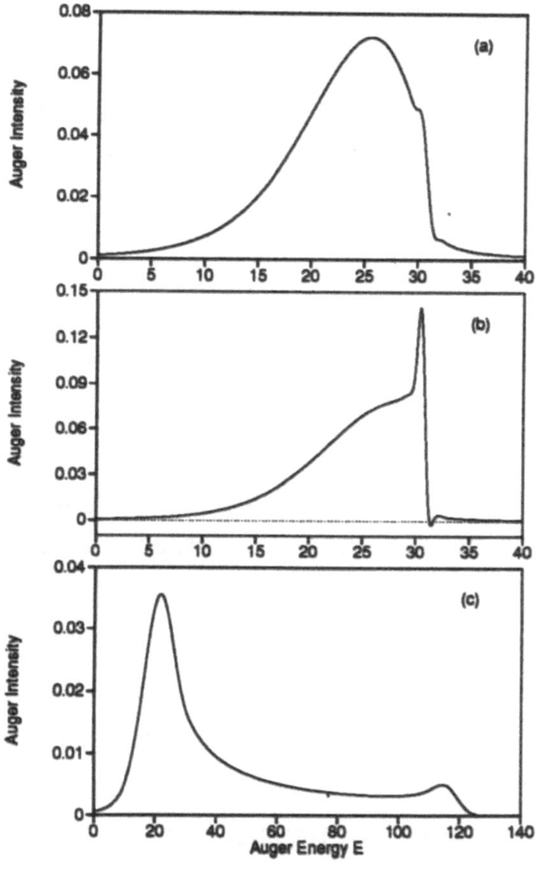

Figure 3. The cross section $\sigma_f(E)$ as a function of the Auger energy E of the emitted electron computed in the semiclassical approximation. The corresponding exact quantum results are available.[27] The parameters used are: (a) $\omega_d=1$, $\omega_f = 2$, $R_d = 1$, $R_f = 2$, $R_o = 0$, $E_d(R_d) = 30$, $E_f(R_f) = 0$, $\Gamma = 2$. (b) as in (a) except of $\Gamma = 1$. (c) $\omega_d = \omega_f = 1$, $R_d = 10$, $R_f = 15$, $R_o = 0$, $E_d(R_d) = 80$, $E_f(R_f) = 0$, $\Gamma = 1$.

larger Γ, the smaller the second peak. Fig. 3c shows a typical example.

The exact quantum results exhibit a similar overall behavior to the semiclassical ones. One finds, however, typical shifts of the peaks due to quantum effects and, for $\Gamma \lesssim \omega_f$, vibrational and interference fine structures which have been smoothed out by the semiclassical ansatz.[17,27]

5. Illustrative Application of Spectral Moments

To illustrate the practicability of spectral moments we apply the theory of the preceding section to the Auger spectra of carbon monoxide which have been discussed many times in the literature[19,26], one due to the creation of a C1s hole and the other following the ejection of a O1s electron. Although the same final dicationic states of CO^{++} are populated by the decay, the bands observed in the two spectra appear at different energies because of dynamical effects. This has been discussed for the first time recently[10]. The computed positions of the bands corresponding to the first states of CO^{++} are depicted in Fig. 3. They are calculated using the expression for the first moment given in eq. (3.13a). The ab initio data for the vertical transition energies as well as for the coupling constants κ_d and κ_f are taken from ref. (10) where all details can be found. We just mention that the latter two constants are calculated as the slopes of the energy curves of the decaying and final states, respectively, at the equilibrium geometry of CO in its ground state (see eq. (3.10)). During the lifetime of the core hole states the nuclei perform a fraction $\omega/(2\pi\Gamma) \approx 0.4$ of a vibrational period.

The vertical transition energies are also shown in Fig. 4. It is

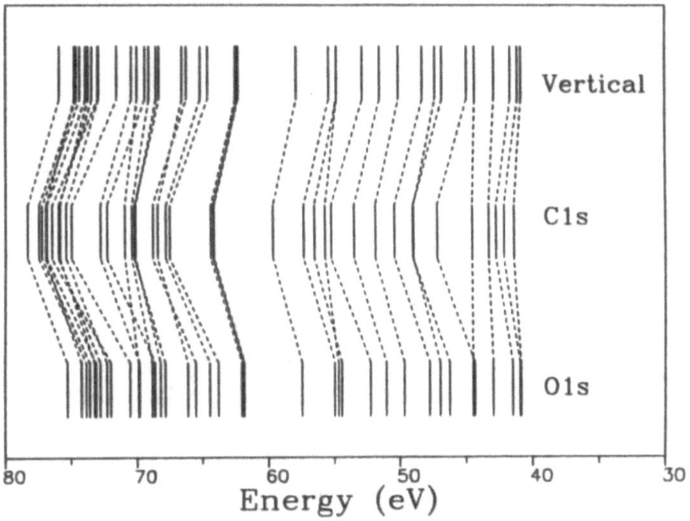

Figure 4. Band positions corresponding to the first dicationic states of CO. Shown are the vertical transition energies and the band positions in the C1s and O1s Auger spectra corrected for dynamical effects.

clearly seen that the dynamical shifts of the band positions can be substantial. For the Auger C1s spectrum many of the shifts exceed 2 eV. The shifts can also be quite different for different states in the same spectrum. The smallest is 0.2 eV for the $^1\Sigma^+$ state at 44.4 eV and the largest reaches 3.5 eV for the $^1\Delta$ state at 73 eV. In general,we can expect the ordering of the states to change due to dynamical effects. This change of ordering is also encountered in CO (see Fig. 4) and is expected to prevail in polyatomic molecules where

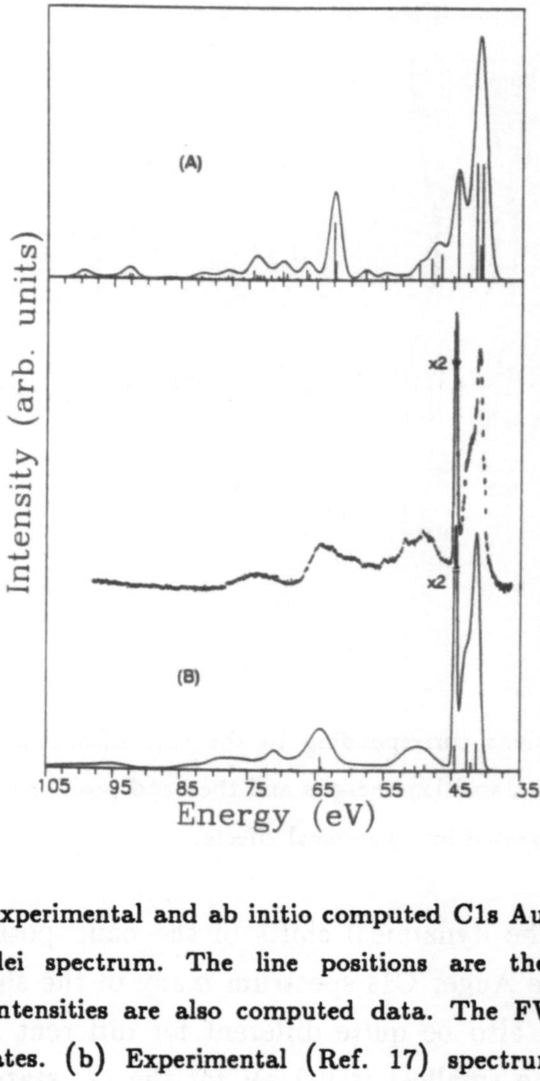

Figure 5. The experimental and ab initio computed C1s Auger spectrum of CO. (a) Fixed—nuclei spectrum. The line positions are the vertical transition energies. The intensities are also computed data. The FWHM is constant (2 eV) for all states. (b) Experimental (Ref. 17) spectrum compared to the spectrum computed using line positions and widths corrected for dynamical effects. The figure is taken from Ref 10.

several active nuclear modes participate in the dynamics. Furthermore, the shifts will usually be different for the same final states in different spectra. In our example of CO the dynamical effects increase the energies of the band positions in the C1s Auger spectrum whereas they decrease the energies of most bands in the O1s spectrum. Obviously, an assignment of the spectrum performed by equating the energies of a final state in all observed spectra, as often carried out in the literature, easily leads to incorrect conclusions.

The widths of the bands in both spectra have also been computed using the second moment. These widths vary strongly over the states and spectra. In CO they range from 0.1 eV to over 7 eV (!). This underlines again the difficulties one encounters in discussing such spectra on the grounds of experimental data alone. In Fig. 5 we show the experimental[19] C1s Auger spectrum and compare it with two computed spectra using ab initio calculated[10] vertical transition energies and coupling constants. One spectrum is based on vertical transition energies and the second is calculated incorporating the shift and broadening due to dynamical effects using the formulas of Section 3 derived from the time–dependent formulation. The substantial difference between these two computed spectra and the eye catching agreement of the latter with experiment demonstrates the relevance of dynamical effects and the usefulness of spectral moments.

6. Summary

The motions of the wavepackets propagating on the potential surface of a decaying electronic state and on the surface of a final electronic state are discussed. The latter wavepacket is found to depend on the energy of the particle emitted by the decay. Its dynamics is relatively complex and interference effects can take place. The cross–sections for the production of the decaying state and of the final state can be related to the corresponding wavepackets. Explicit expressions for these cross–sections in a time–dependent framework have been given. The latter cross–section σ_f is represented as an integral over two time variables. One time variable is needed for the description of the decay, the other is related to the propagation on the final state's potential surface. The interference effects known to appear in σ_f can be identified in the time–dependent picture used.

Simplified interpretation of the decay process is one advantage of a wavepacket dynamical study; computational effort is another, in particular, if the lifetime of the decaying state is short. In the present case the time–dependent picture is, in addition, suitable for deriving the spectral moments of the cross–sections. Spectral moments are commonly used to approximate a distribution and can be used here to describe the gross features of the cross–sections, like the energetic positions and widths of electronic bands observed in the spectrum under consideration, e.g., autoionization spectrum. However, relevant moments of cross–sections involving exponentially decaying states diverge. The time–dependent formalism enabled us to identify and to unambiguously eliminate the divergencies and determine well–defined moments. Approximate explicit expressions for the position and width of a band are given in terms of properties of the potential energy surfaces of the decaying and final states. An illustrative application to the Auger spectra of carbon monoxide is briefly discussed demonstrating the usefulness of spectral moments. It is shown that even the ordering of the electronic bands in the spectra can change due to the dynamics.

The time–dependent formulation is amenable to semiclassical approximations which also add insight into the decay process and its dynamics. The semiclassical approximations are of practical relevance in particular for polyatomic molecules. Propagation of the wavepackets in the framework of the present formulation are currently being computed and will be used to interpret the dynamics of the decay process.

Acknowledgements

Enlightening discussions with H.–D. Meyer are gratefully acknowledged.

References

1. Lim, E.C.(1974) *Excited States*, Academic Press, New York
2. Ma, Y., Sette, F., Meigs, G., Modesti, S. and Chen, C.T. (1989) Phys. Rev. Lett. 63, 2044; Rabus, H. et al. (1990) Phys. Scr. T 31, 131; Gadea, F.X. et al. (1991) Phys. Rev. Lett. 66, 883
3. Hitchcok, A.P. and Brion, C.E. (1977) J. Electr. Spectr. 10, 317; Tronc, M., King, G.C. and Read, F.H. (1979) J. Phys. B 12, 137
4. Thompson, M., Baker, M.D., Christie, A. and Tyson, J.F. (1985) *Auger Electron Spectroscopy,* Wiley, New York; Correira, N., Flores–Riveros, A., Ågren, H., Helenelund, K., Asplund, L. and Gelius, U. (1985) J. Chem. Phys. 83, 2035; Svenson, S. and Karlsson, L. (1992) Phys. Scr. T 41, 132; Plummer, E.W., Chen, C.T., Ford, W.K., Eberhardt, W., Messmer, R.P. and Freund, H.–J. (1985) Surf. Sci. 158, 58
5. Nordgren, J., Selander. L., Pettersson, L., Nordling, C., Siegbahn, K. and Ågren, H. (1982) J. Chem. Phys. 76, 3928; Deslattes, R.D. (1986) Aust. J. Phys. 39, 845
6. Southworth, S.H., Parr, A.C., Hardis, J.E. and Dehmer, J.L. (1987) J. Chem. Phys. 87, 5125; Dehmer, J.L., Dill, D. and Parr, A.C. (1983) *Photophysics and Photochemistry in the Vacuum Ultraviolet*, eds. S. McGlynn, G. Findley and R. Huebner, Reidel Publ., Dordrecht; Ferrett, T.A., Piancastelli, M.N., Lindle, D. W., Heimann, P.A., Medhurst, L.J., Liu, S.H. and Shirley, D.A. (1987) Chem. Phys. Lett. 134, 146

7. Schulz, G.J. (1973) Rev. Mod. Phys. 45, 423; Borrow, P.D., Michejda, J.A. and Jordan, K.D. (1987) J. Chem. Phys. 86, 9; Domcke, W. (1991) Phys. Rep. 208, 97

8. Morin, P. and Nenner, I. (1986) Phys. Rev. Lett. 56, 1913; Phys. Scr. T 17, 171

9. Le Forestier, C., Bisseling, R., Verjan, C., Feit, M.D., Friesner, R., Guldberg, A., Hammerich, A., Jolicard, G., Karrlein, W., Meyer, H.–D., Lipkin, N., Roncero, O. and Kosloff, R. (1991) J. Comp. Phys. 94, 59; Schinke, R. and Engel, V. (1990) J. Chem. Phys. 93, 3252; Kulander, K.C., Cerjan, C. and Orel, A.E. (1991) ibid. 94, 2571; Manthe, U. and Köppel, H. (1991) Chem. Phys. Lett. 178, 36; all articles in *Time–dependent Quantum Molecular Dynamics: Theory and Experiment* (1992), eds. L. Lathouwers et.al., Plenum Press, New York

10. Cederbaum, L.S., Campos, P., Tarantelli, F. and Sgamellotti, A. (1991) J. Chem. Phys. 95, 6634

11. Cederbaum, L.S. and Tarantelli, F. (1993) J. Chem. Phys. 98, 9691

12. Cederbaum, L.S. and Tarantelli, F. (1993) J. Chem. Phys. 99, 5871

13. Kaspar, F., Domcke, W. and Cederbaum, L.S. (1979) Chem. Phys. 44, 33

14. Domcke. W. and Cederbaum, L.S. (1981) J. Phys. B 14, 149

15. Rohr, K. and Linder, F. (1975) J. Phys. B 8, L200

16. Bienek, R.J. (1980) J. Phys. B 13, 4405; Cederbaum, L.S. and Domcke, W. (1981) ibid. 14, 4665

17. Pahl, E., Meyer, H.–D. and Cederbaum, L.S. to be published

18. Gel′mukhanov, F.K., Mazalov, L.N. and Kondratenko, A.V. (1977) Chem. Phys. Lett. 46, 133
19. Correira, N., Flores–Riveros, A., Ågren, H., Helenelund, K., Asplund, L. and Gelius, U. (1985) J. Chem. Phys. 83, 2035; Cesar, A., Ågren, H. and Carravetta, V. (1989) Phys. A 40, 187
20. Domcke, W. and Cederbaum, L.S. (1977) Phys. Rev. A 16, 1465
21. Cramer, H. (1946) *Mathematical Methods of Statistics* Princeton University Press, Princeton
22. Cederbaum, L.S. and Domcke, W. (1977) Adv. Chem. Phys. 36, 205
23. Davydov, A.S., (1965) *Quantum Mechanics*, Pergamon Press, New York
24. Dimensionless normal coordinates are obtained from the usual normal coordinates by dividing the latter by the square root of the corresponding frequency
25. Meyer, H.D. (1981) Chem. Phys. 61, 365
26. Carrol, T.X. and Thomas, T.D. (1987) J. Chem. Phys. 86, 5221; Kelber, J.A., Jennison, D.R. and Rye, R.R. (1981) ibid. 75, 652; Ungier, L. and Thomas, T.D. (1985) ibid. 82, 3146; Moddeman, W.E., Carlson, T.A., Krause, M.O., Pullen, B.P., Bull, W.E. and Schweizer, G.K. (1971) ibid. 55, 2317
27. Pahl, E. (1993) Diplom Theses, University of Heidelberg

Collisional Energy Transfer; New Light on an Old Problem

A.J.McCAFFERY,

School of Molecular Sciences

University of Sussex, Brighton BN19QJ, U.K.

1 Introduction

Energy transfer collisions in molecules play a central role in many areas of chemistry and physics and the detailed understanding of these relatively simple interactions could lead to greater insight into the more complex processes that involve bond breaking and making. However, it would be wrong to suppose that our level of understanding of the simplest (molecular) collision process is satisfactory. The contrast to photon-induced quantum state changes, is stark. Here it is possible to predict energy levels and transition probabilities with high precision once some relatively basic information on the molecule is established. In *collisionally* induced quantum state changes, no such predictability is available. Theory is not sufficiently transparent that even simple 'rules of thumb' concerning energy transfer may be extracted. This unsatisfactory state of affairs may not be inherent in the collisional problem and may be more a function of our chosen way of addressing the issue.

There is also something of a watershed in experimental approaches to collisional quantum state change. An important goal in experimentation over several decades has been the development of techniques that are sensitive to the potential and may be used directly to obtain quantitative data on the PE surface. In this regard the determination of the state-to-state differential scattering cross-section (dcs) is known to be the experiment most directly relateable to the PE surface since the least averaging of events is involved. In effect the complete disposal of kinetic and potential energy

E. Yurtsever (ed.), Frontiers of Chemical Dynamics, 331–347.

is determined before and after collision together with *directions* of relative velocity vectors.

Such an experiment however is rarely performed. Very few collisional processes have been studied with full angle, velocity and quantum state resolution. The reason is that the experiment is difficult, time-consuming and fearsomely expensive. The standard approach starts with crossed molecular beams to give velocity selection followed by two or more lasers for state selection and/or detection. The experimenter makes a committment to a system often many years before any results are forthcoming.

In this contribution alternative approaches to both of these issues is presented. Although these are new and only just about to appear in the literature, the title of this Advanced Study Institute, *Frontiers* of Chemical Dynamics, indicates that this material is perhaps more appropriate than reviews of past work. First is a description of a wholly spectroscopic method through which the state-to-state dcs may be obtained. Secondly an alternative theoretical description of rotational energy transfer is given.

2 Spectroscopic Determination of the State-to-State DCS

2.1 THEORY

Spectroscopy is uniquely suited to the study of quantum state populations and changes. If we could develop an effective *spectroscopic* technique for selection and detection of relative velocities, a process normally associated with molecular beam techniques, then we have a flexible, (relatively) inexpensive and more readily accessible technique for obtaining the high quality experimental data needed to calculate potential energy surfaces. Wholly spectroscopic experiments may be done in cells or in bulbs and the expense and complexity of molecular beams may be avoided. Although lasers are inevitably involved which generally are not cheap, technological advances continue to improve range, linewidth and performance of these devices.

The method adopted to determine the state-to-state dcs spectroscopically is that of velocity selected double resonance (VSDR). This technique is described fully in recent publications [1, 2]. An outline is now given with fuller details in the publications cited. The basis of the method is active use of the Doppler shift, a well known spectroscopic phenomenon often regarded as an obstacle to high resolution spectroscopy, to select and detect molecu-

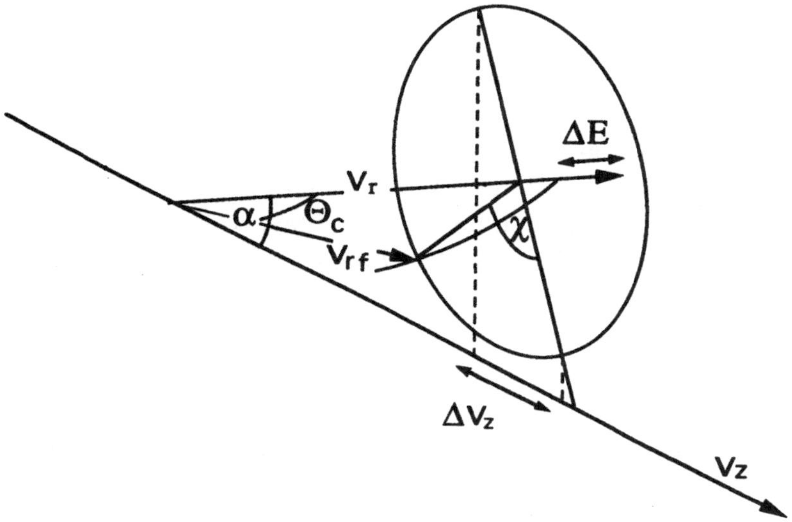

Figure 1: Newton diagram showing velocity vector (v_r) selected by the pump laser and the 3D distribution of velocities after a collision. In plane (Θ) and out-of plane scattering angles are shown together with the velocity component distribution

lar velocities. This latter function is performed using lasers with linewidths narrower than the Doppler width.

A narrow line tuneable laser selects molecular velocity v_{mz} and from this known function it is possible to calculate the probability distribution of *relative* velocity (i.e.) direction and magnitude from the relation

$$P(\alpha, v_r \mid v_{mz}) = v_r^2 exp(\frac{-v_r^2 sin^2\alpha}{2(s_a + s_m^2)} - \frac{(v_z - v_r cos\alpha)^2}{2s_a^2}) \quad (1)$$

This is the basis of the spectroscopic velocity selection (and detection) process in which laser detuning is converted into relative collision velocity of target molecule(m) and colliding atom (a). The effect of a collision on a selected velocity is shown in the Newton diagram (Fig.1).

Represented here is an inelastic collision (excitation) resulting in a shortening of the relative velocity vector and scattering by in-plane angle Θ_c and out-of-plane angle χ. This results in a spread of velocities with projections on the detuning axis shown in the figure. The broadening of the spectral line is directly related to the in-plane scattering angle (that out-of-plane is considered to be isotropic).

The experiment is a double resonance technique in which a probe laser determines the width of a transition from a level populated by collisions from that pumped from the ground state using narrow line laser excitation. Two single frequency tuneable dye lasers are used for this work each having linewidths of $1MHz$. The collision system chosen for this work was Li_2-rare gas for spectroscopic convenience and because the mass ratio (with Xe) gives excellent velocity selection in Eqn 1. The Doppler width is $8GHz$ for Li_2 in this experiment and thus excellent velocity selection is obtained together with a wide range of initial velocities so that dependence on initial velocity may be determined.

Final velocity component distribution is a complex function of several probability densities. These are incorporated in the expression for the VSDR lineshape;

$$P(v_{mzf}|v_{mz}) = \int\int\int\int\int P(v_{mz}|v_z)P(\Theta_c|v_r)P(v_r,\alpha|v_{mz})P(\chi) \times sin\alpha sin\Theta_c d\chi d\Theta_c d\alpha dv_z dv_r \tag{2}$$

The δ functions representing the important kinematic constraints have been omitted from this expression.

The unknown function in Eqn.2 is the dcs function $P(\Theta_c|v_r)$. In this work we have used an assumed form of the dcs with adjustable parameters (width and most probable scattering angle) and have varied these in a series of nested integrations to obtain the best fit to experimental lineshape data. This is a computer-intensive approach and an alternative has been developed relating width of VSDR line directly to most probable scattering angle. This is now described in more detail.

Figure 2 is a vector diagram representing the most-probable scattering vectors from an initial value of v_{rp}. New vector length after collision will be $(v_r^2 - 2\Delta E/\mu)^{1/2}$. The projections of this vector on the detuning axis determine the VSDR lineshape but the line*width* is determined by three factors; the size of the circle that forms the base of the scattering cone, its angle of inclination to v_z and the length of v_{zf}. The projection of the extremes of this circle on the detuning axis will be related to the VSDR linewidth.

A simple geometric relation connects the linewidth Δv_z to the most

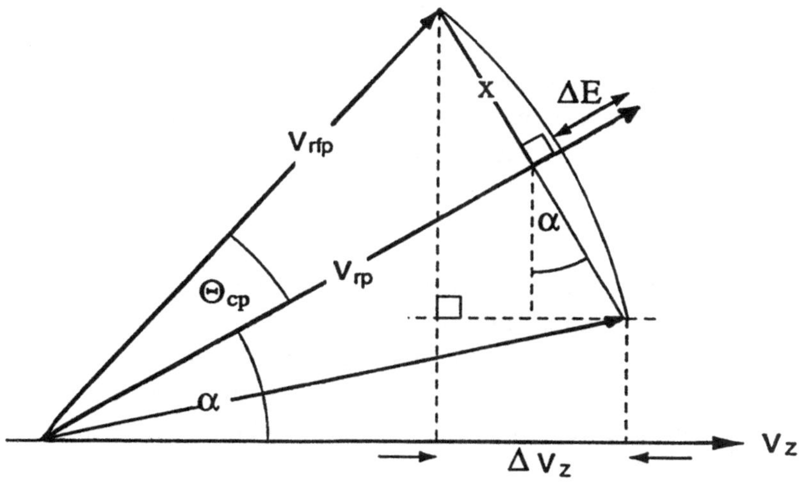

Figure 2: 2-dimensional c.o.m. Newton diagram showing the relationship between velocity component distribution and double resonance linewidth

probable scattering angle Θ_{cp}

$$sin\Theta_{cp} = \frac{\Delta v_z}{2(v_r^2 + 2\Delta E/\mu)^{1/2}[1 - (v_z/v_r)^2]^{1/2}} \tag{3}$$

This simple relation between linewidth and scattering angle may become an important tool in spectroscopic methods of velocity distribution analysis. A preliminary account of the method has appeared in J.Phys.B [3]

There are a number of points of note worth making regarding the VSDR method. There are several clear advantages over the molecular beams method of velocity selection/detection. First of these is the direct relation between laboratory and collision frame that characterises the VSDR method. All quantities are referred to a common quantisation direction (the detuning axis). This contrasts the problems that traditionally surround the relating of lab and collision frame events in molecular beam methods. A further gain is in signal strength in a manner similar to the Fellgett advantage of FT methods due to summing contributions from many azimuthal angles. The ability to select widely over a range of relative velocities is a great advantage and we have measured the velocity dependendent state-to-state dcs here *for the first time.* A five-fold range of relative velocities in the input channel is available in the Doppler selection method described here for $Li_2 - Xe$. Finally there is the ease with which the method may be made into a three or four vector correlation experiment using polarisation techniques in pump and probe lasers. We have demonstrated this in two recent publications with the first four vector correlation molecular dynamics experiments [4, 5].

2.2 RESULTS

In developing a completely new method of obtaining velocity and angular variables in a scattering experiment it is important to check that the experiment does indeed measure those quantities that theory predicts. We demonstrate below that the results obtained are of unprecedented sensitivity and precision yielding the best rotationally inelastic state-to-state dcs data yet obtained. In addition we have made a comprehensive study of the velocity dependence of the state-to-state dcs, this also for the first time, yielding cross-sections differential both in angle and in velocity.

The systems investigated in detail are the lithium dimers $^7Li_2$, $^6Li_2$ [7] and each in its $A^1\Sigma_u^+$ state. Some results are summarised in figure 3. We discuss these shortly after first describing how one might expect VSDR lineshapes to be relateable to angular variables and means of verifying the technique. Two tests were devised, the first examining the change in linewidth *for a given inelastic process* as v_r, i.e. laser detuning, is changed. This changes the angle of inclination of the scattering cone to the detuning axis and should lead to *narrowing* as v_r increases. The results show that this was indeed observed. In a related test, the variation of linewidth was measured for fixed v_r as inelasticity (and hence Θ_c) increased. Simple physical arguments would predict a steady increase in linewidth and this indeed was the behaviour observed.

A final test is of course comparison with other data on state-to-state inelastic transfer. Only one data set exists that has some element of comparability, that by Bergmann and colleagues on Na_2-rare gas collisions [8]. We note however that comparability is limited since the Bergmann experiments were carried out on Na_2 in its ground electronic state the repulsive potential of which is considerably less anisotropic than that of Li_2 in its first excited state with any of the rare gases (there is more than a factor of 2 in the maximum anisotropy difference with Ne,0.7Å for Na_2 and 2.1Å for Li_2). Scattering angles would be expected to be smaller in the case of Li_2 and this indeed is found but quantitative comparability is not possible.

Our data is more precise than that of Bergmann and in addition we have

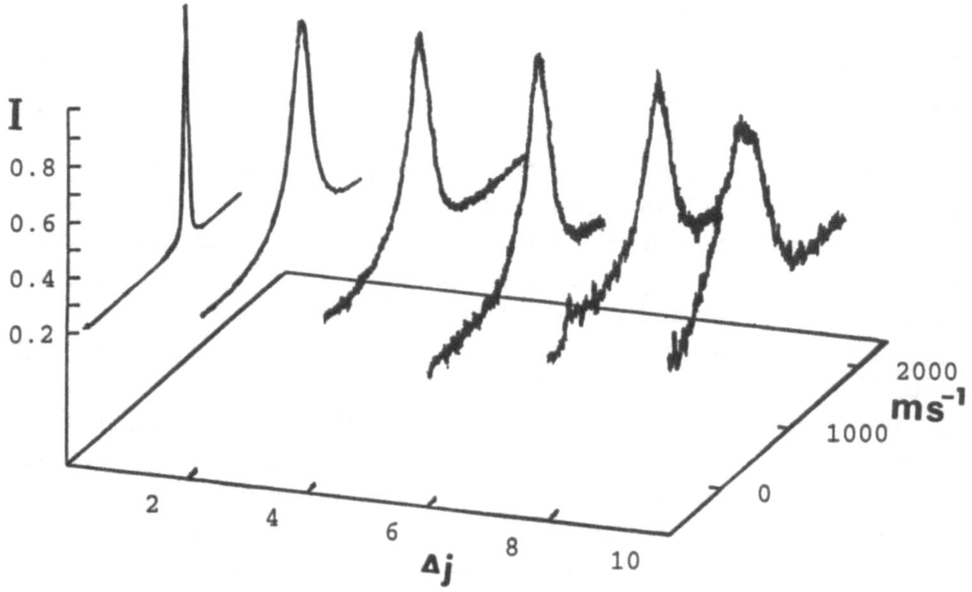

Figure 3: Normalised plots of double resonance lineshapes for rotationally inelastic collisions for $v_z = 900ms^{-1}$

measured velocity dependence. A key test is our observation of a difference between linehapes (and the dcs) for upwards and downwards Δj jumps. This has not been observed before experimentally though the processes are clearly different (for the same value of j) due to change in the energy threshold for upwards and downwards processes.

The figure 3 shows results on the $^7Li_2 - Xe$ system. Analysis of results with other rare gases is in progress together with those on the isotopomers. The main trends are that the rotational rainbow angle increases with increasing Δj and that scattering angle decreases with increased relative velocity. Both these trends have been measured quantitatively. These features are readily explained in the classically impulsive limit using a hard ellipse model. Here the hard wall of the potential is represented by an elliptical energy contour giving predominantly in-plane scattering. Conversion of linear to angular momentum is via a torque-arm or effective impact parameter b_n the maximum value of which is related to the difference of semi-major and semi-minor axes of the ellipse. Full discussion of the model is contained in a recent publication [7]. In this we use the simple arguments put forward above together with a realistic distribution of velocities to relate observed dcs values to the topological features of the potential.

2.3 CONCLUSIONS

We have demonstrated the validity of this *spectroscopic* method of determining the state-to-state differential scattering cross-section. The results obtained are of higher quality, in terms of precision, range and velocity dependence, than previously achieved. The method is clearly of considerable value, may be used widely and is capable of extension to the three and four vector correlation level. For the first time, we have recorded state-to-state cross-sections that are differential both in angle and velocity.

The technique is flexible and adaptable to the study of ground and excited states. It possesses elements of the Fellgett advantage that markedly improves signal strength and the lab - molecular frame transformation that causes much difficulty in conventional beam experiments is no longer a problem since all quantities are relateable to the unique quantisation axis. State-to-state dcs values have been measured as a function of relative velocity and of rotational inelasticity. A particularly significant observation is the difference seen (for the first time) between upwards and downwards Δj transitions. For rotational transitions a difference is anticipated though not previously observed and it highlights the power of the VSDR technique.

3 An Angular Momentum View of Rotational Transfer

3.1 BACKGROUND

In a recent publication we surveyed a number of experiments in the area of collision-induced rotational transfer (RT) and concluded that the magnitude of angular momentum change plays a crucial role in the outcome of a collisional event. Rotational transfer (RT) is of great significance since it is usually the most efficient of the gas phase excitation and relaxation mechanisms and hence plays a major role in chemical and physical processes. It is also significant as the simplest phenomenon in molecular collision physics and full understanding of this might presage deeper insights into the more complex collisions that result in chemical reaction. It is a paradox that despite the presence of a full quantum theory for RT together with numerous approximate methods based on this, there is little in the way of insight into the physics of the process, let alone predictive power. The overwhelming majority of experiments are 'interpreted' by fitting data to empirical fitting 'laws' that have no basis in the physics of the interaction and no relationship to the intermolecular potential.

A key experiment in the development of an angular momentum (AM) based model for RT was based on a study of state-to-state RT rates in asymmetric rotor-atom collisions [9]. Symmetric and asymmetric rotors are important test molecules since energy change and angular momentum change are unrelated, unlike the case of the diatomic molecule where the two are very hard to disentangle. We found in the case of $NH_2 - H$ collisions that the rate constant for state to state RT was related to the amount of AM transferred in the form of an inverse exponential gap law. No relationship to amount of energy transferred could be established (except of course that of overall energy conservation).

This and other observations from experiments in our and other labs over a period of years led to a re-examination of the atom-diatom collision problem. The probability of RT (rate constant or cross-section) appears to be controlled by the amount of AM generated in the collision. In this case it may be simpler to calculate directly the *probability* of conversion of linear to angular momentum in the collision rather than the conventional approach in which (in effect) the probability of converting kinetic to potential energy is calculated. The evidence on which the assertions above are based and the theoretical development of an angular momentum model are described in two recent publications [10, 11].

The basis of the theoretical expression is the well known relationship,

$$l = \mu v_r b \tag{4}$$

In this equation the orbital angular momentum l which becomes the transferred AM Δj, the relative velocity v_r and impact parameter b have their characteristic probability densities. The first is discrete and the remainder are in effect continuous. The densities for v_r and b are convoluted together in a simple form to yield the probability density of Δj;

$$P(\Delta j) = \mu \int P(v_{rel})dv_{rel} \int P(b)db\delta(E' - E)\delta(J' - J) \tag{5}$$

The delta functions have the effect of imposing energy and AM conservation on all combinations of the random variables. This equation is the basis of the calculation of RT probabilities. Clearly it is considerably simpler to operate than the close-coupled equations of scattering theory.

The AM theory was shown to be a considerable simplification over conventional approaches and in addition contains a number of significant insights into the physics of the process. For example the ubiquitous exponential-like fall of RT probabilities with ΔE of with Δj, which has no simple origin in the rigorous or approximate theories, has a natural explanation in the AM model. It originates in the exponential-like fall of the repulsive intermolecular potential (the repulsive *anisotropy* in particular). The model accounts for the remarkable conservation of the (lab. and collision frame) m_j quantum number and contains the basis of a simple (and potentially parameter-free) inversion routine to extract the repulsive anisotropy from data.

3.2 PREDICTIVE RELATIONS

Two further advances have followed from the establishment of the basic ideas embodied in the AM theory, both of which could have significance for the general field of collision physics. These are now briefly discussed. The first of these relates to RT in diatomic molecules. It would be very valuable to have some *predictive* capability for collisional RT based on accessible molecular and collision condition properties. No such ability is found in current formulations even at the level of simple rules of thumb. Partly this is the result of lack of transparency in the theory. The AM approach on the other hand reveals the physics of the RT process very readily and we have utilised this to develop what potentially is a predictive relationship for RT in atom-diatom collisions [12].

The AM model is readily cast into the form of a predictive or fitting relation which may be compared with known data sets. This differs from earlier approaches to this topic in that the basis of the relation is in the simple physics of linear to angular mometum conversion via the anisotropy of the intermolecular potential and that the parameters have genuine physical significance in terms of the model. The basic fitting relation is given by

$$P(j_f|j_i)dj_f = C\int_0^{b_n^{max}} P(l|b_n)P(b_n)\delta(|E-E'|)\delta(|j_i-j_f|)b_n db_n dj_f \qquad (6)$$

In this equation the probability distribution of relative velocity will generally be known, that for effective impact parameter $P(b_n)$ was obtained empirically by fitting the expression to known data sets. There are three parameters each with physical significance. First is the maximum anisotropy b_n^{max}. We identify this with the anisotropy of the zero contour of the potential. Second is γ, related to the steepness of the repulsive potential wall. Finally the parameter C converts the probability density into a bimolecular rate constant.

Fits to data using this expression are very good as shown in figure 4. This is a difficult task in view of the distinctive shape of the semilog. plot of most RT data sets. However, the values of the parameters are most revealing. The value of b_n^{max} that occurs with greatest frequency is one half the bond length of the diatomic. The steepness parameter γ varies over only a narrow range around 2.0 and Marks [14] has shown that C is readily calculated from simple data on the repulsive potential. We note that the half bond length (HBL) value emerges in what we term the AM constrained limit [12] in which the full rotor arm is needed to generate the RT available within energy conservation. In the converse energy constrained region the maximum HBL value is not explored due to energy restrictions.

Thus in the AM constrained case, the AM model for RT genuine predictive capability exists regarding unknown RT. Even under less favourable conditions, simple rules of thumb are available and we see for example that the principal requirement for extensive RT is the presence in the diatomic of a long bond. This is just one of the improvements in transparency that the AM model brings. It may soon be possible to predict RT in a manner comparable to the way in which rotational state energies are now computed and we feel that this is a valuable contribution to the understanding of fundamental collision physics.

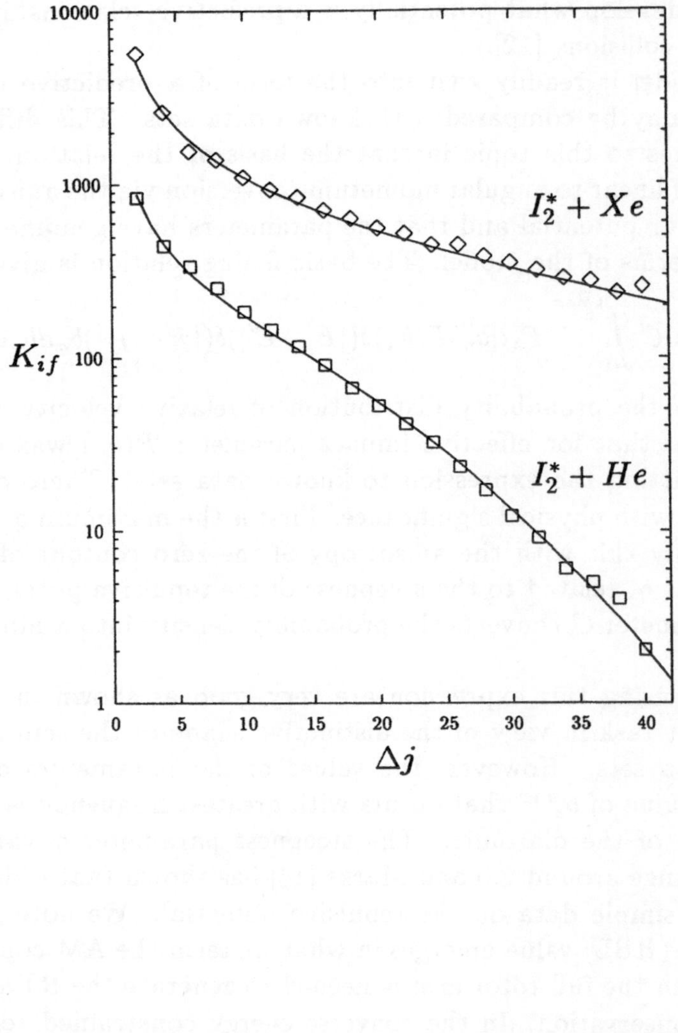

Figure 4: Semilog plot of RT rate constants for I_2 with He and Xe together with fits using the AM law

3.3 THE A.M. SPHERE

The second development from our basic AM model concerns the much more complex issue of RT in polyatomic molecules. Here, energy gap relations cannot begin to tackle this problem due to the strong dependence on component of AM in the molecule frame arising from rotations about more than one inertial axis. Our approach to this is also new and is based on earlier work in which AM gap in the molecular frame is a key parameter [9]. We now recognise that in the molecular frame, the AM vector that represents each JK (or Jk_ak_c) state is not fixed but precesses in a trajectory that may be calculated classically for each Jk_ak_c vector and for each molecule.

This begins with the rotational Hamiltonian

$$H = AJ_a^2 + BJ_b^2 + CJ_c^2 \tag{7}$$

and on expressing this in terms of the spherical polar coordinates J, θ, ϕ, the classical trajectory of each Jk_ak_c state vector is obtained from those values of θ, ϕ that fulfil the conditions

$$E(\theta, \phi) = E(Jk_ak_c) \tag{8}$$

The set of trajectories of a single J level form an *angular momentum sphere* of radius $\sqrt{J(J+1)}$ and are shown in figure 5. Related spheres may be drawn for other values of J and are shown overlaid in figure 6. The AM spheres represent the dynamical motion of vectors in AM space and the calculation of AM *gaps* may be made in this representation. We have used the AM sphere representation to fit data on $NH_2 - H$ collisions obtained in our laboratory. The results are very promising and further details are given in a forthcoming publication [13]. Several models for calculating AM gaps from the sphere representation are proposed all of which provide a good basis for fitting RT rate constants. Averaging of the dynamical motion of the AM trajectory and of the vibrational motion of the molecule appear to be important components of successful use of the AM sphere representation.

3.4 CONCLUSIONS

This is the first time the AM sphere has been used as the basis of an approach to molecular dynamics problems. In view of the success of the AM model in diatomics, our findings in the case of NH_2 and the growing realisation that energy gap relations are incapable of dealing with energy transfer in polyatomics, the model proposed has great potential. In addition the AM

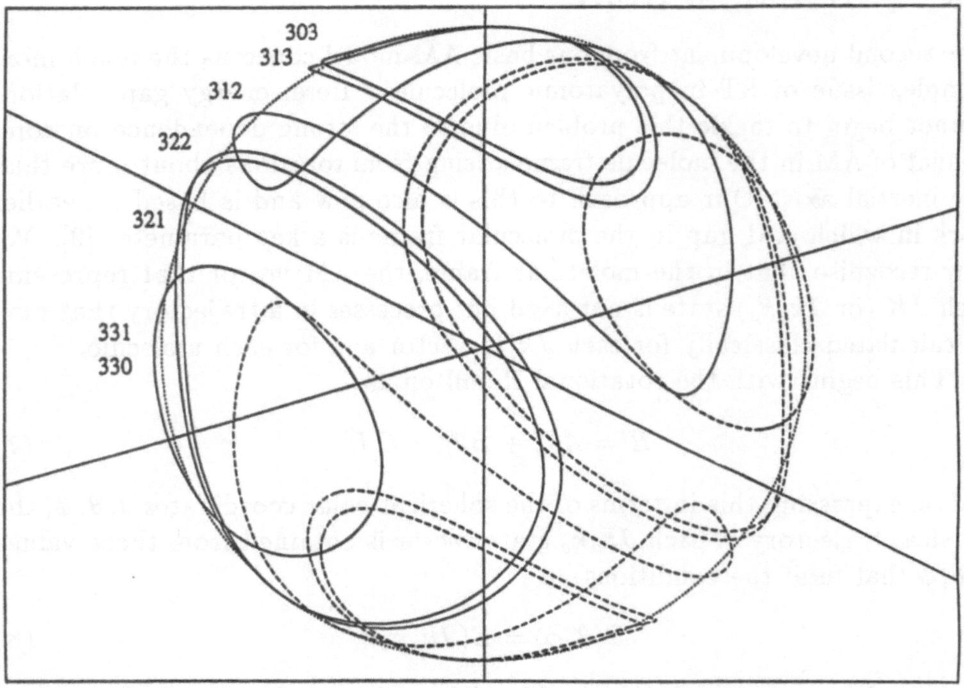

Figure 5: The angular momentum sphere for the $J = 3$ level of NH_2

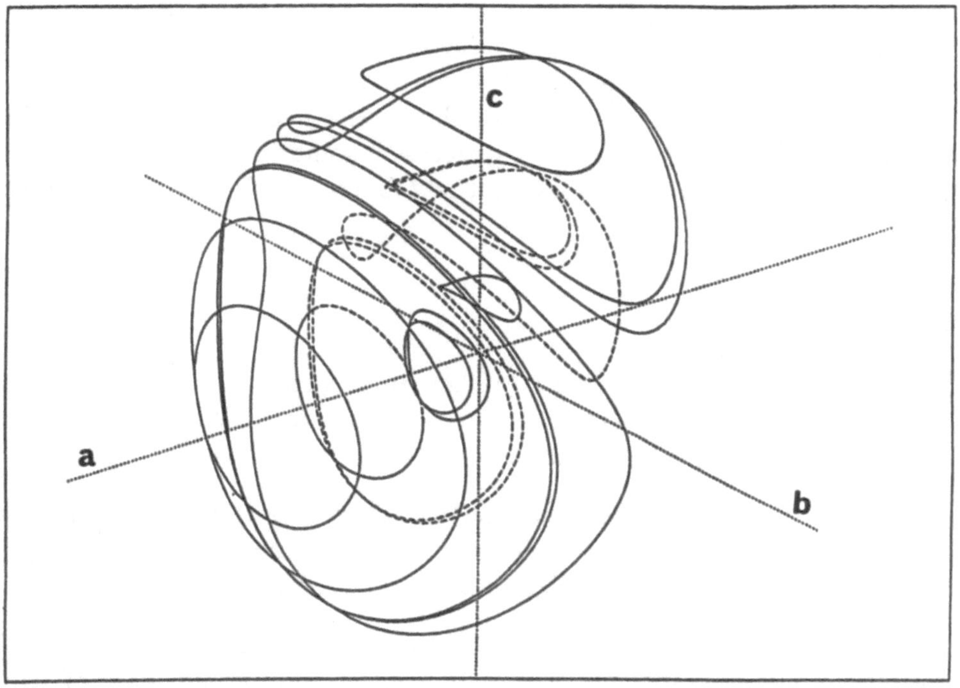

Figure 6: Angular momentum (half) spheres for the J = 1,3 and 5 rotational levels for NH_2

sphere may be a useful basis for consideration of vibration-rotation coupling and the origins of ivr in polyatomics. The experimental data base on which the model rests is still very slender and an important priority would be to expand this data base. This work has shown the need for more extensive studies of the fundamental collision processes, sometimes neglected in the haste to apply reaction dynamics to the full complexity of the chemical reaction.

References

[1] K.L.Reid and A.J.McCaffery (1992) Spectroscopic determination of the state-to-state differential cross-section for inelastic collisions, *J.Chem.Phys.* **96**, 5789-5796.

[2] A.J.McCaffery, K.L.Reid and B.J.Whitaker (1988) Velocity selective double resonance, a novel technique for determining differential scattering cross-sections, *Phys.Rev.Letters* **61**, 2085-2087.

[3] A.J.McCaffery, J.P.Richardson, R.J.Wilson and M.J.Wynn (1993) A vector model of double resonance linewidths; a direct estimate of scattering angle, *J.Phys.B* **26**, L705-709.

[4] T.L.D.Collins, A.J.McCaffery and M.J.Wynn (1991) Two colour sub-Doppler circular dichroism, a four vector correlation molecular dynamics experiment. *Phys.Rev.Letters* **66**, 137-139.

[5] T.L.D.Collins, A.J.McCaffery and M.J.Wynn (1991) Sub-Doppler circular dichroism, a four vector correlation experiment. *Far. Disc. Chem.Soc.*, **91**, 91-106.

[6] S.Kasahara and H.Kato private communication.

[7] T.L.D.Collins, A.J.McCaffery, J.P.Richardson, R.J.Wilson and M.J.Wynn (1995) Velocity dependent state-to-state differential cross-sections for rotational transfer in Li_2-Xe using velocity selected double resonance *J.Chem.Phys.* to be published.

[8] K.Bergmann, U.Hefter and J.Witt (1980) State to state differential cross sections for rotationally inelastic scattering of Na_2 by He. *J.Chem.Phys.* **72**, 4777-4790.

[9] Z.T.AlWahabi, C.G.Harkin, A.J.McCaffery and B.J.Whitaker (1989) Stereochemical influences in atom-triatomic collisions. *J.Chem.Soc. Far. Trans.*, **85**, 1003-1016.

[10] A.J.McCaffery, Z.T.AlWahabi, M.A.Osborne and C.J.Williams (1993) Rotational transfer, an angular momentum model. *J.Chem.Phys.*, **98**, 4586-4602.

[11] A.J.McCaffery and Z.T.AlWahabi (1991) Mechanism of rotational transfer *Phys. Rev.A* **43** 611-614.

[12] M.A.Osborne and A.J.McCaffery (1994) A fitting law for rotational transfer rates, an angular momentum model with predictive power. *J.Chem.Phys.*, **101** 5604-5614

[13] Z.T.AlWahabi, N.A.Besley, A.J.McCaffery, M.A.Osborne and Z.T.Rawi, (1994) Dynamical angular momentum models for rotational transfer in polyatomic molecules *J.Chem.Phys.*, submitted.

[14] A.Marks (1994) A multiple hard-ellipsoid model for rotationally inelastic collisions. *J.Chem.Soc.Far.Trans.*, **90** 2857-2863.

Angular Momentum Projection in Quantum Molecular Dynamics

L. Lathouwers
Departement Wiskunde-Informatica
Universitair Centrum Antwerpen,
University of Antwerp,
Groenenborgerlaan 171, B2020 Antwerpen, Belgium

Abstract

We investigate the possibilities of a new scheme in time dependent quantum molecular dynamics. It is based on the use of angular momentum projection as an alternative to frame transformation to account for the conservation of total angular momentum. The new methodology applies irrespective of the number of atoms involved and its computational cost scales favourably with particle number.

1 INTRODUCTION

Molecular collision dynamics at the quantum level has evolved from the colinear collision of three masses to calculations of state to state cross sections of triatomic systems in three dimensions. The number three seems to show a strange affection for the field ! What are the reasons that seemingly prevent quantum molecular dynamics to boldly go beyond its present limitations ?

Difficulties arise from both operators that make up the total Hamiltonian of the system viz. the kinetic energy and potential energy operators. Let us first consider the potential energy since it is the basic input to the theory. For an N particle system it is a scalar function that depends on $3N - 6$ degrees of freedom. Depending on the nature of the molecules involved it is most conveniently described in terms of a set of generalized coordinates appropriate for the dynamics of the system. Here a first fundamental problem arises. Electronic structure computer codes will provide an approximate potential energy surface at various points of a cartesian frame for the nuclear positions. Therefore one needs to fit the numerical potential energy surface to an analytical form expressed in a set

E. Yurtsever (ed.), Frontiers of Chemical Dynamics, 349–356.

of internal coordinates. It has been found that the results of quantum molecular dynamics calculations are very sensitive to the potential energy surface and therefore also to the fiting procedure used.

The kinetic energy is the sum of the Laplace operators for each of the atomic nuclei in the system. Its simplest representation is in the cartesian coordinates of every particle. However, since the potential energy is rotationally and translationally invariant, total linear and angular momentum are conserved quantities. Whereas the separation of centre-of-mass motion does not pose any problem, a proper treatment of angular momentum conservation is far from trivial. One has to supplement the $3N - 6$ internal coordinates with a specific choice of three coordinates which determine the orientation of the N-particle configuration, or equivalently, one has to specify three Euler angles relating the laboratory frame to a rotating frame. In general the transformation between the centre-of-mass vector $\vec{X}$, the Euler angles $\Omega = (\phi, \theta, \gamma)$ and the internal coordinates $q = (q_1, q_2, ..., q_{3N-6})$ reads symbolically

$$\vec{X} = \vec{X}(R) \quad , \quad \Omega = \Omega(R) \quad , \quad q = q(R) \tag{1}$$

where R stands for all cartesian components $R_k(i)$ ($i = 1, ..., N$ and $k = x, y, z$) of the particle position vectors $\vec{R}(i)$. In order for (1) to be an admissible coordinate transformation it should have an inverse which we can write as

$$R_k(i) = R_k(i)(\vec{X}, \Omega, q) \tag{2}$$

The standard approach is to transform the Hamiltonian for the nuclear motion on a potential energy surface

$$\mathbf{H} = -\sum_{ik} \frac{1}{2M(i)} \frac{\partial^2}{\partial R_k(i)^2} + V(R) \tag{3}$$

to the coordinates (1). This can be done in a generic way (i.e. without specifying more than the above about the transformation (1)) by deriving the momenta conjugate to the variables $(\vec{X}, \Omega, q)$. In spectroscopic problems, however, one resorts to the Watson Hamiltonian which is derived using the Eckart conditions for a semi-rigid molecule. The derivation is tedious but the result standard. For triatomic bound state or scattering problems several other coordinate systems have been used (Jacobi coordinates, hyperspherical coordinates, interatomic distances, etc.) according to an intuitive feeling about the dynamics and knowledge of the potential. However as one goes to larger systems the relation between the choice of coordinates and a limited knowledge of the potential becomes less obvious.

What is more important, the transformation of (3) to the chosen coordinates has to be done analytically. Even with the help of present day symbolic manipulation programs this remains a formidable task with a complex outcome of the form

$$\mathbf{H} = \mathbf{H}_0(q, \frac{\partial}{\partial q}) + \mathbf{A}_k(q, \frac{\partial}{\partial q})\,\mathbf{J}_k + \frac{1}{2} I_{kl}^{-1}(q)\,\mathbf{J}_k\mathbf{J}_l \tag{4}$$

see e.g. reference [1].

We now turn to the true evolution of a given initial state and total angular momentum, i.e.,

$$\mid \Psi(t)\,\rangle \;=\; \exp\left(-\imath \mathbf{H}_J\; t\right) \mid \Psi(0)\,\rangle \tag{5}$$

where $\mathbf{H}_J$ is the projection of the full Hamiltonian on the subspace of total angular momentum J. One obtains $\mathbf{H}_J$ from (4) by setting up this operator in the basis of Wigner D-functions

$$D^J_{KM}(\Omega) = \exp\left(-\imath K\phi\right)\, d^J_{MK}(\theta)\, \exp\left(-\imath L\gamma\right) \tag{6}$$

The resulting $2J+1$ dimensional matrix contains scalar, first and second order partial differential operators as its elements. One can now proceed in essentially two ways: expand and represent the initial state and $\mathbf{H}_J$, respectively, in a basis set or map both quantities on a grid in q-space. In the former case one has to evaluate matrix elements of the above mentioned operator in the chosen basis. One needs to accomodate both potential energy and kinetic energy terms which may become cumbersome or impossible if one retains the full kinetic energy and uses and accurate analytical representation of the potential. In the second case one can chose between Fast Fourier Transform (FFT) techniques or discrete variable representation (DVR) to evaluate the action of the kinetic energy operator. A recent generic DVR introduced by Colbert and Miller [2] is very helpful in this context. They showed that the grid point representation of the kinetic energy on an arbitrary interval for a uniform grid and associated particle-in-a-box eigenfunctions can be calculated anlytically. From these formulae several specific cases of coordinate ranges can be derived. For a triatomic system using Jacobi coordinates and $J = 0$ the kinetic energy differntial operators sufficiently decouple, yielding a sparse matrix representation. However, even with the above tool at hand the full kinetic energy matrix for four or more atoms and $J \neq 0$ will be of dimension $(2J + 1)\, n^{3N-6}$, and even though the relative sparsity of a grid representation will help, calculations become prohibitively large if $N \geq 4$.

From the above expose it is clear that methods based on the transformed version (4) of the Hamiltonian are cumbersome to apply to systems with four atoms and will most probably become unfeasible for five or more particles. It is also apparent that any alternative methodology should be numerically insensitive to the choice of coordinate system to represent the potential and its analytic formand should preferably be generic and effective (in both storage and number of multiplications) in the use of a time evolution algoritm. In the following we outline such a scheme based on the grid representation of the full kinetic energy in cartesian coordinates and the use of angular momentum projection operators.

We first observe that all molecular properties such as spectra, cross sections, etc., can be obtained by Fouriertransforming the appropriate time-correlation function. For specified total angular momentum J the vibration-rotation spectra require the calculation of the autocorrelation function

$$C_J(t) = \langle \Psi(0) \mid \exp(-\imath \mathbf{H}_J\, t) \mid \Psi(0) \rangle \tag{7}$$

while for other prorerties the appropriate operator has to be inserted in the matrix element. We now consider the rotations about the center of mass which constitute the rotation group of the system. The corresponding rotation operators on the systems wave functions are

$$\mathbf{R}(\Omega) = \exp(-\imath\phi \mathbf{J}_z)\ \exp(-\imath\theta \mathbf{J}_y)\ \exp(-\imath\gamma \mathbf{J}_z) \tag{8}$$

while the associated irreducible representations are the Wigner functions defined earlier. From the group elements and the irreducible representations one can construct the group projection operators

$$\mathbf{P}^J_{MK} = \frac{2J+1}{8\pi^2} \int d\Omega\ D^J_{MK}(\Omega)^*\ \mathbf{R}(\Omega) \tag{9}$$

Their properties have been listed and proven elsewhere [3]. The ones we need in the present context are

$$\sum_J \mathbf{P}_J = 1 \quad \textit{with} \quad \mathbf{P}_J = \sum_K \mathbf{P}^J_{KK} \tag{10}$$

i.e., one can construct the projector onto the space of total angular momentum J as the K sum of the operators $\mathbf{P}^J_{KK}$. The operators $\mathbf{P}_J$ being true projection operators are idempotent and hermitian and, because the Hamiltonian is rotationally invariant, they commute with $\mathbf{H}$

$$\mathbf{P}^2_J = \mathbf{P}_J \quad , \quad \mathbf{P}^\dagger_J = \mathbf{P}_J \quad \textit{and} \quad [\mathbf{H}, \mathbf{P}_J] \tag{11}$$

Since the operator $\mathbf{H}_J$ appearing in (7) is really $\mathbf{P}_J\,\mathbf{H}\,\mathbf{P}_J$, one can rewrite the formula for the autocorrelation function as

$$C_J(t) \;=\; \langle \Psi(0) \mid \mathbf{P}_J \exp(-\imath \mathbf{H}\, t) \mid \Psi(0) \rangle + \langle \Psi(0) \mid 1 - \mathbf{P}_J \mid \Psi(0) \rangle \tag{12}$$

where the second term removes the non-J contributions that are present in the initial condition. This expression is the key to the new method. Indeed it shows that evaluating $C_J(t)$ can be regarded as a three stage process: the projection of the initial state onto the subspace J, the time evolution of the initial state under the total Hamiltonian and the calculation of the overlap of the projected and time evolved states.

2 OUTLINE OF THE ANGULAR MOMENTUM PROJECTION METHOD

The present description does not aim at presenting a fully worked out and detailed computational scheme. Rather it intends to demonstrate the feasibility of combining the three steps outlined above. As a result of this we will be able to estimate the computational effort involved and to compare this estimate with a similar one for the efficiency of the frame transformation method.

2.1 ANGULAR MOMENTUM PROJECTION

The initial step in frame transformation theory, the derivation of the Hamiltonian in a rotating frame, is replaced by the projection of the initial state onto a subspace of total angular momentum J. This, according to (10), amounts to summing up the action of the operators P^J_{KK} on the state $\Psi(0)$. Explicitly this boils down to evaluating the three dimensional integrals

$$\int_0^{2\pi}\int_{-1}^{+1}\int_0^{2\pi} d\phi\, d\cos\theta\, d\gamma\ \exp(\imath K\phi)\, d^J_{KK}(\theta)\, \exp(\imath K\gamma)\ \mid \Psi(R_\alpha(\phi,\theta,\gamma)\,|\,0)\,\rangle \quad (13)$$

where $R_\alpha(\phi,\theta,\gamma)$ denotes the grid obtained by subjecting all particle position vectors to the inverse of the orthogonal tranformation perform the integrations in (13) effectively,keeping in mind the subsequent evaluation of matrix elements, it has been found advantageous to introduce an alternative parametrisation of the rotation group in terms of the angles $\epsilon = (\phi+\gamma)/2$, $\delta = (\phi-\gamma)/2$ and θ. Using these angles one obtains from (13)

$$\int_{-\pi}^{+\pi}\int_{-1}^{+1}\int_{-\pi/2}^{+\pi/2} d\delta\, d\cos\theta\, d\epsilon\ \exp(\imath 2K\epsilon)\, d^J_{KK}(\theta)\, \exp(\imath K\gamma)\ \mid \Psi(R_\alpha(\delta,\theta,\epsilon)\,|\,0)\,\rangle \quad (14)$$

The parametrisation in terms of δ, θ and ϵ has the advantage that small rotations correspond to small ϵ and θ but arbitrary δ. For bound state problems one can therefore restrict the integrations in (14) to small ϵ and θ since over several vibrational periods the system will not rotate significantly such that all overlaps with macroscopically rotated states will essentially be zero.

2.2 TIME EVOLUTION

Since angular momentum conservation has been taken care of through the action of $\mathbf{P}_J$, the evolution operator in (12) is the one corresponding to the full Hamiltonian. In the center of mass frame the kinetic energy part of $\mathbf{H}$ remains fully seperable in cartesian coordinates. Therefore one can use the representation on the real line with grid spacing Δx for each of the $3N-3$ directions. The abovementioned generic formula for this case gives

$$T_{ij} = \frac{1}{2m(\Delta x)^2}(-)^{i-j}\begin{cases}\pi^2/3 & i=j\\ 2/(i-j)^2 & i\neq j\end{cases} \tag{15}$$

with grid points labelled as $x_i = i \Delta x$, $i = 0, \pm1, \pm2, \ldots$ the grid spacing being the only parameter. When applied to each of cartesian components of Jacobi vectors the full kinetic energy matrix conists of $3N-3$ blocks of type (15). The grid spacing in every direction can be chosen in a massweighted way such that $\Delta x = \sqrt{M(i)}\, \Delta R_k(i)$.

The potential energy is of course diagonal on the grid. If we denote a grid point by R_α we can write

$$V_{\alpha\beta} = V(q(R_\alpha))\,\delta_{\alpha\beta} \tag{16}$$

i.e., in order to have the potential energy values on the cartesian grid one merely needs the expression of the internal cooordinates in terms of the cartesian ones. For channel selection in scattering problems, hypespherical coordinates for four or more particle systems will be usefull [4].

Formula (15) corresponds to an infinite grid. Any real calculation must of course be made on a finite grid. One may first chose to restrict the number of grid points in each massweighted direction to say n. However one can go further and discard those points where the wavefunction is essentially zero, i.e., those regions that are energetically inaccessible. This can be accomplished by eliminating those points for which the potential energy is greater than some safe cutoff value. Convergence can be checked by increasing V_{cutoff} until results are stable against any further increase [2].

Having obtained the grid representation of the full Hamiltonian one can evaluate its action on the initial state mapped onto the grid. Selecting the time evolution scheme of ones choice (Chebyshev, Lanczos, second or higher order differencing schemes, ...) one can propagate the initial state to time t.

2.3 CORRELATION FUNCTION EVALUATION

The two previous sections outlined the evaluation of the the time evolved and projected states on the chosen grid. The overlap of these two functions gives us the time dependent part of the autocorrelation function. This integral is readily evaluated by summing over the grid points. At this time it is important to compare the two forms, (7) and (12) of $C_J(t)$.

First we reemphasize that doing the time evolution with the full Hamiltonian instead of the projected one eliminates the analytical derivation of the latter. Secondly it is clear that in the present scheme the angular momentum projection computational effort is negligable compared to the time evolution part since it has to be performed only once. In the traditional approach however the time evolution has to be performed for every J seperately which implies that the total computational effort will scale with the total number of angular momenta.

Let us try to estimate the number of floating point operations in every time step in both methods. In the projection method for N nuclei a grid in Jacobi coordinates consists of $p = n^{3N-3}$ points if we take n points per cartesian dimension. Due to its block diagonal structure the action of the kinetic energy matrix involves $(3N-3)\, n^2\, n^{3N-4}$ multiplications. But since $p = n^{(3N-3)}$ and therefore $(3N-3) = \ln p / \ln n$, the number of floating point operations is $(n/\ln n)\, p \ln p$. This scaling law is due to the seperability of the kinetic energy operator in cartesian coordinates implying a growing sparsity with increasing number of particles. A similar estimate for the frame tranformation approach leads to $J(J+1)(3N-3)\,(p\ln p)/(n \ln n)$ where $p = n^{3n-6}$ and the prefactor is due to angular momentum part of the Hamiltonian and the appearance of cross terms of internal coordinate partial derivatives. For $N = 4, J = 10$ and $n = 10$ this gives comparable numbers for the two methods while for more than 4 nuclei the projection method is less computationally intensive.

3 CONCLUSIONS

We have presented a new approach to quantum molecular dynamics that is capable to go beyond three particle systems in a generic way. This is accomplished using angular momentum projection operators and the representation of the kinetic energy on a cartesian grid. The computational effort scales favorably with increasing number of particles and total angular momenta.

References

[1] H. Herold. *J. Phys. G: Nucl. Phys.*, **5**:341, 1979.

[2] D. T. Colbert and W. H. Miller. *J. Chem. Phys.*, **96**:1982, 1992.

[3] L. Lathouwers and E. Deumens. *J. Phys. A: Math. Gen.*, **15**:2785, 1982.

[4] Y. Öhrn and J. Linderberg. *Mol. Phys.*, **49**:53, 1983.

THREE PARTICLE SYSTEMS AND HYPERSPHERICAL HARMONICS

JUERGEN HINZE AND ALEXANDER ALIJAH
Fakultät für Chemie, Universität Bielefeld
D-33615 Bielefeld, Germany

AND

L. WOLNIEWICZ
Institute of Physics, Nicholas Copernicus Univ.
Torun, Poland

1. Introduction

The general quantum mechanical description of three particle systems is a challenge, in particular if higher values of the total angular momentum, J, are required and large amplitude internal motions are strongly coupled, being also coupled to the overall rotation. For $J = 0$ only $3N - 6$, here three internal degrees of freedom need to be considered. However, for $J \neq 0$, with strong coupling of the internal motions with the overall rotation, it is necessary to consider all $3N - 3$, here six degrees of freedom explicitly and simultaneously. Such a fully general quantum mechanical description of three particle systems would be desirable:

i. for the bound states of a variety of three particle Coulomb systems, i.e. H^-, He, H_2^+ etc;
ii. for the rotation vibrational states of triatomics, in particular if large amplitude motions are expected with strong Coriolis coupling; and
iii. for the characterisation of the reactive scattering of an atom with a diatomic molecule, i.e. the simplest, elementary chemical reaction.

More complex chemical reactions, with many more degrees of freedom in the reactive system could be reduced to the elementary reaction by folding the additional degrees of freedom approximately into those of the fragments, "atom" or "diatomic".

E. Yurtsever (ed.), Frontiers of Chemical Dynamics, 357–369.

2. Hyperspherical Coordinates

Different types of hyperspherical coordinates, with one distance coordinate and five angular coordinates,[1, 2, 3] have been used in the past for the description of three particle problems.[4, 5, 6, 7, 8, 9] Of these different coordinate systems we will focus here on the so called "democratic" hyperspherical coordinates introduced by Whitten and Smith,[10] modified and detailed by Johnson.[11, 12, 13] The advantage of this coordinate choice is that the three Euler angles, specifying the orientation of the system fixed coordinates relative to the laboratory frame, appear explicitly, correct for any instantaneous arrangement of the three particles, thus permitting the use of the Wigner functions to characterize the overall rotation. Furthermore, with this coordinate choice, the three possible arrangement channels are treated equally, providing the highest symmetry for the description of the system.

The particular coordinate choice is based on mass normalized Jacobi coordinates, defined as follows. With $\mathbf{x}_i$ and m_i, for $i = 1, 2, 3$, the coordinates and masses of the three particles, the coordinate for the centre of mass is

$$\mathbf{X} = \sum_{i=1}^{3} m_i \mathbf{x}_i / M \ , \tag{1}$$

with $M = \sum_{i=1}^{3} m_i$ the total mass. The three pairs of mass scaled Jacobi vectors are obtained as

$$\mathbf{r}^k = (\mathbf{x}_j - \mathbf{x}_i)/d_k \tag{2}$$

and

$$\mathbf{R}^k = d_k \left[\mathbf{x}_k - (m_i \mathbf{x}_i + m_j \mathbf{x}_j)/(m_i + m_j)\right] \ , \tag{3}$$

with the mass scaling parameters

$$d_k^2 = (1 - m_k/M) m_k/\mu \ , \tag{4}$$

and the reduced mass is given by

$$\mu^2 = m_1 m_2 m_3 / M \ . \tag{5}$$

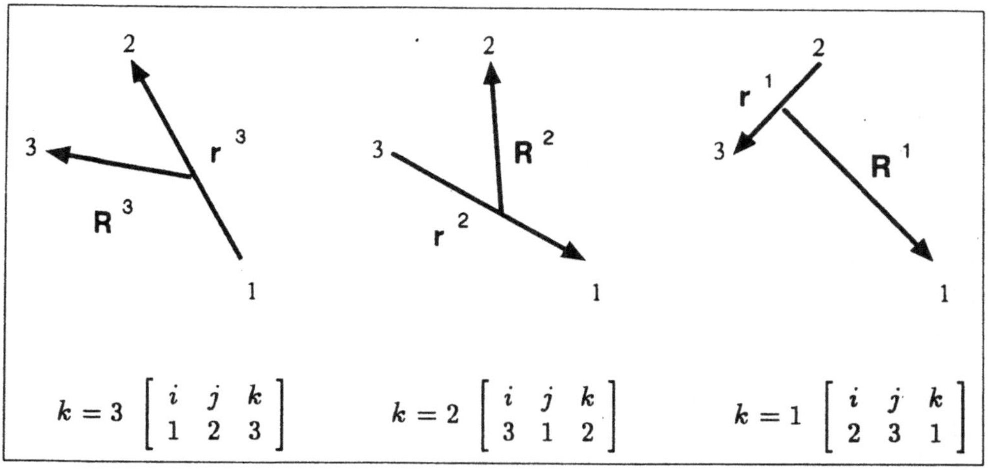

Figure 1. The three equivalent Jacobi systems

The three equivalent Jacobi systems which result by letting (i, j, k) be the cyclic permutations of $(1, 2, 3)$ are displayed in Fig. 1. These three Jacobi systems are related to each other via a kinematic rotation

$$\begin{pmatrix} \mathbf{r}^i \\ \mathbf{R}^i \end{pmatrix} = \begin{pmatrix} \cos\beta_{ij} & \sin\beta_{ij} \\ -\sin\beta_{ij} & \cos\beta_{ij} \end{pmatrix} \begin{pmatrix} \mathbf{r}^j \\ \mathbf{R}^j \end{pmatrix} , \tag{6}$$

with

$$\tan\beta_{ij} = -m_k/\mu \; ; \qquad \beta_{ij} = -\beta - ji \; . \tag{7}$$

The six dimensional Schrödinger equation in these coordinates is

$$\left[-\frac{\hbar^2}{2\mu} \left(\nabla_{\mathbf{r}}^2 + \nabla_{\mathbf{R}}^2 \right) + V(\mathbf{r}, \mathbf{R}) - E \right] \Psi(\mathbf{r}, \mathbf{R}) = 0 \; . \tag{8}$$

In terms of these Jacobi coordinates, the hyperspherical coordinates are defined by choosing the Euler angles α, β, γ such that the body fixed frame is given with the z-axis perpendicular to the plane defined by the three particles, i.e.

$$\mathbf{z} || \mathbf{A} = \mathbf{r} \times \mathbf{R}/2 \; , \tag{9}$$

and the $\mathbf{x}$ and $\mathbf{y}$ axis defined such that they form a right-handed system along the other instantaneous principle moments of inertia, defined such that

$$I_y = \mu[r_y^2 + R_y^2] \geq I_x = \mu[r_y^2 + R_x^2] \; . \tag{10}$$

Now the internal coordinates are defined as

$$\rho^2 = |\, \mathbf{r} \,|^2 + |\, \mathbf{R} \,|^2 \; , \tag{11}$$

and

$$\cos\theta = 2 \,|\, \mathbf{A} \,|\, /\rho^2 = u \; . \tag{12}$$

The only coordinate, which depends on the reference system chosen, is the internal ϕ-angle defined as

$$\cos\phi^k = 2(\mathbf{r}^k \cdot \mathbf{R}^k)/(\rho^2 \sin\theta) \ , \tag{13}$$

where the reference system, i.e. the k chosen specifies just the origin for the angle ϕ. The ranges for the internal coordinates are $0 \leq \rho \leq \infty$, $0 \leq \theta \leq \pi/2$ or $1 \geq u \geq 0$ and $0 \leq \phi \leq 4\pi$. Thus for fixed ρ the "northern" hemisphere of a globe is covered twice, since the longitudinal angle ϕ ranges through 4π. The "north-pole" corresponding to $\theta = 0$ or $u = 1$ represents an equilatteral triangle, while the "equator" with $\theta = \pi/2$ or $u = 0$ corresponds to the collinear arrangement of the three particles.

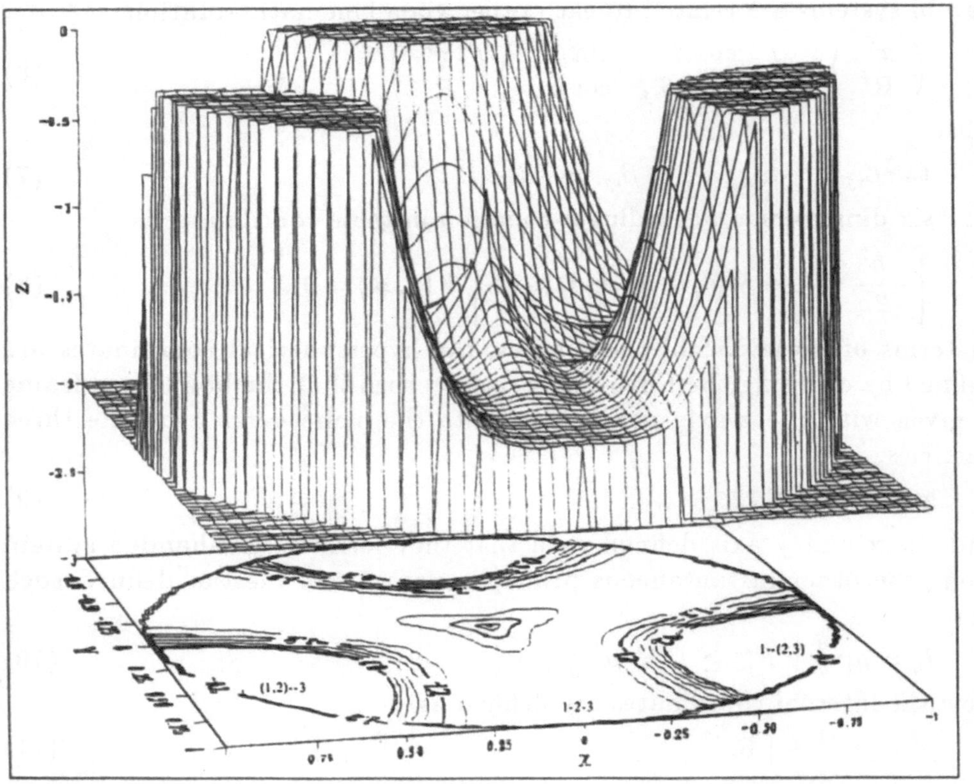

Figure 2. The potential of H_3 for $\rho = 3$.

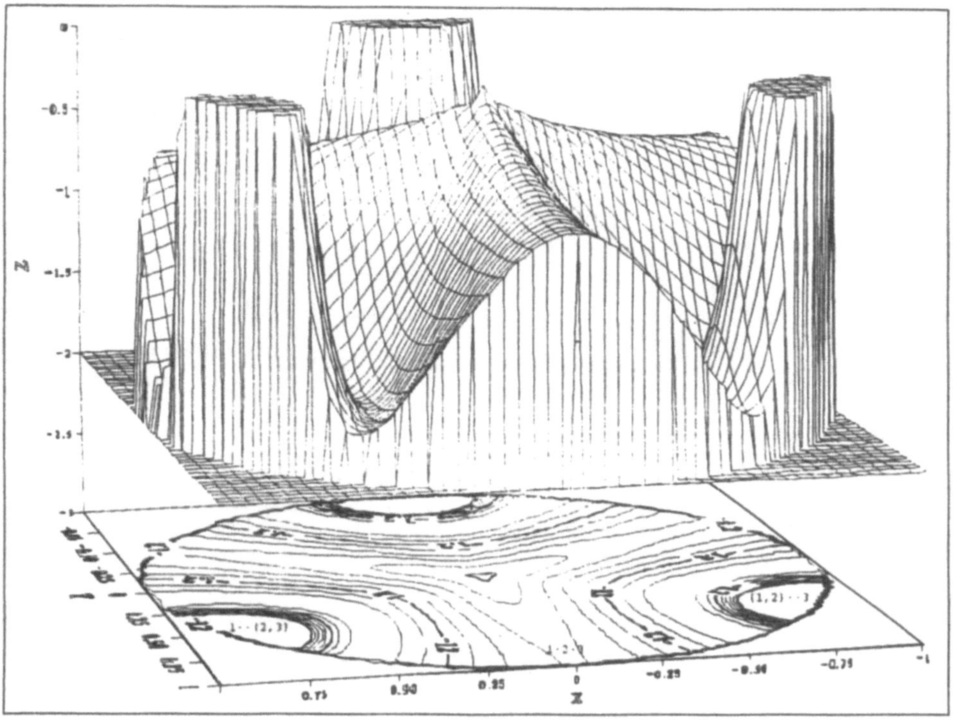

Figure 3. The potential of H_3 for $\rho = 6$.

To get an impression of the appearance of different types of potentials in these hyperspherical coordinates, we have displayed in Figs. 2 through 6 a few examples, each for a fixed value of ρ. In Fig. 2 the potential for H_3 is displayed for $\rho = 3$, roughly the reaction distance. Here the left singularity is marked $(1,2) - -3$ to indicate that the hydrogens 1 and 2 are coincident, while hydrogen 3 is removed. Moving from left to right along the front part of the picture, corresponds to the motion of hydrogen 2 away from 1 and towards hydrogen 3. In the middle, the symmetric collinear arrangement, the transition state is marked as $1 - 2 - 3$, and on the right the sigularity is $1 - -(2,3)$. In Fig. 3 the same potential is displayed for $\rho = 6$, where the narrow entrance or exit channels for the reaction appear clearly. In Fig. 4 the potential for the bound system H_3^+ is presented with $\rho = 2$, corresponding roughly to the equilibrium distance. In Fig. 5 an extreme light-light-heavy example is given with the Coulomb potential for He, and in Fig. 6 the other extreme, heavy-heavy-light is presented with the Coulomb potential for H_2^+. It is seen that for this last case the hyperspherical coordinates discussed here will certainly be poor, since the two proton-electron singularities are very close to each other, leaving very little space for the electronic part of the wave function.

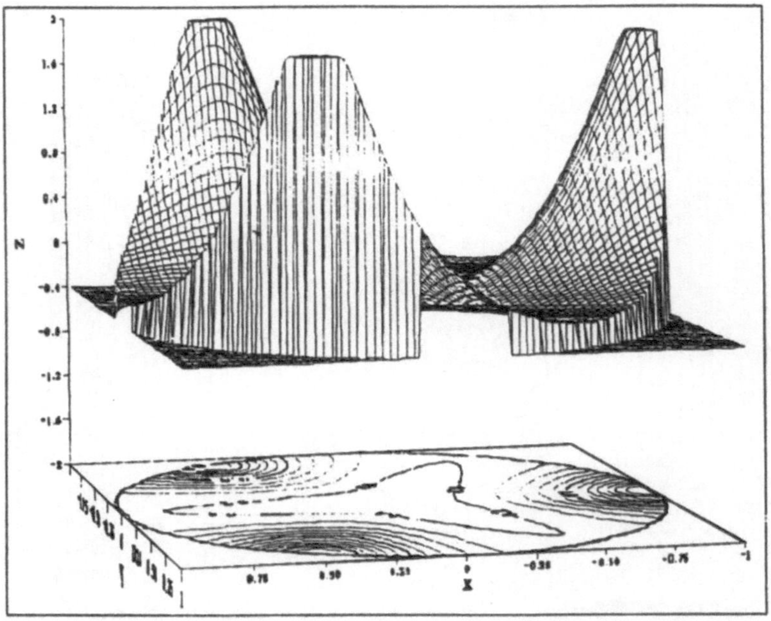

Figure 4. The potential of H_3^+ for $\rho = 2$.

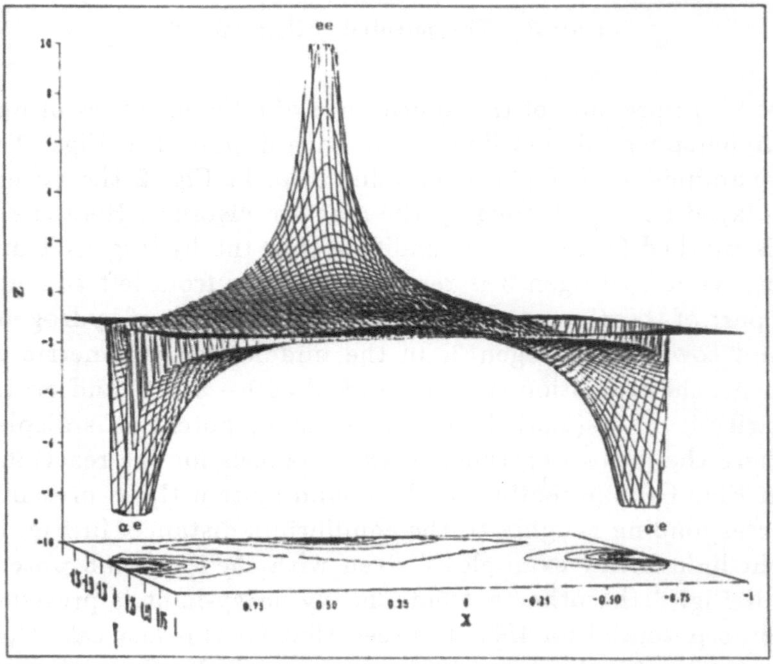

Figure 5. The potential of He for $\rho = 2$.

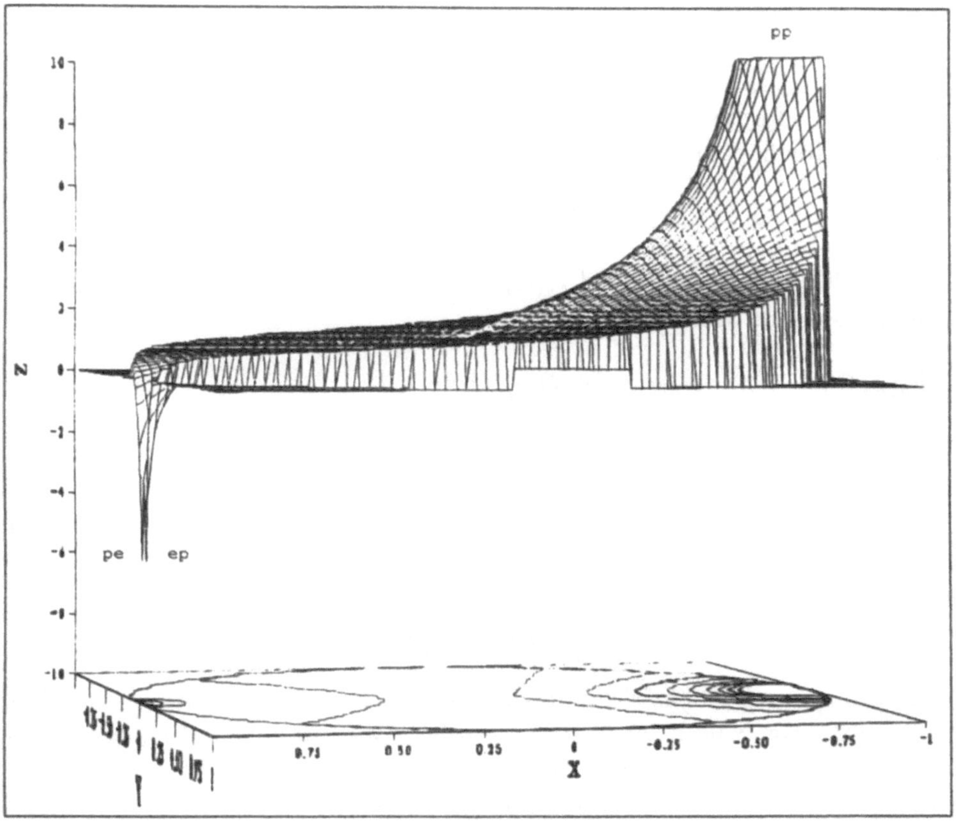

Figure 6. The potential of H_2^+ for $\rho = 2$.

3. Method for Solving the Schrödinger Equation

The Hamiltonian in the hyperspherical coordinates is

$$H = -\frac{1}{2\mu\rho^5}\frac{\partial}{\partial\rho}\rho^5\frac{\partial}{\partial\rho} + \frac{\Lambda^2}{2\mu\rho^2} + V(\rho,\theta,\phi)\,, \tag{14}$$

with the grand angular momentum operator defined as

$$\begin{aligned}\Lambda^2 \;=\; & -\frac{4}{u}\frac{\partial}{\partial u}u(1-u^2)\frac{\partial}{\partial u} - \frac{1}{(1-u^2)}\left[4\frac{\partial^2}{\partial\phi^2} - J_z^2 - 4iJ_z u\frac{\partial}{\partial\phi}\right] \\ & +\frac{2}{u^2}\left[J_x^2 + J_y^2 + \sqrt{1-u^2}(J_x^2 - J_y^2)\right]\,,\end{aligned} \tag{15}$$

with $u = \cos\theta$.

In order to solve the Schrödinger equation, using the six-demensional Hamiltonian, eq.(14), an ansatz for an adiabatic coordinate separation can be used of the type

$$\Psi_A(\rho,\Omega) = \sum_q \Phi_q(\rho;\Omega)\Theta_{qA}(\rho)/\rho^{5/2} , \tag{16}$$

where Ω is to represent the set of five angles, i.e. $\Omega \equiv \theta, \phi, \alpha, \beta, \gamma$ and the $\Phi_q(\rho,\Omega)$ are for fixed ρ solutions of

$$\left[\frac{\Lambda^2}{2\mu\rho^2} + V(\rho,\theta,\phi) - \epsilon_q(\rho)\right] \Phi_q(\rho,\Omega) = 0 . \tag{17}$$

Using the Wigner functions $D^J_{NM}(\alpha,\beta,\gamma)$ to characterise the overall rotation of the system with the ansatz

$$\Phi_q(\rho;\Omega) = \sum_{N=-J}^{J} G_{qN}(\rho;\theta\phi)D^J_{NM}(\alpha\beta\gamma) \tag{18}$$

reduces the eq. (17), in five angles to a set of coupled equations in two angular coordinates of the type

$$\sum_{N'} \{T_{NN'} + \delta_{NN'}[V(\rho;\theta\phi) - \epsilon(\rho)]\} G_{qN'}(\rho;\theta\phi) = 0 , \tag{19}$$

with

$$T_{NN'} = \langle D^J_{NM}(\alpha\beta\gamma) \mid \frac{\Lambda^2}{2\mu\rho^2} \mid D^J_{N'M}(\alpha\beta\gamma)\rangle_{\alpha\beta\gamma} . \tag{20}$$

The coupling in eq. (19) is through the kinetic energy operator only, i.e. the Coriolis coupling, which leads to J coupled two-dimensional equations to be solved for each value of ρ. This becomes prohibitive for larger values of the total angular momentum J, since the boundary conditions for $\theta = 0$ and $\theta = \pi/2$ are by no means simple. Thus it is advantageous, in particular if higher J terms are desired, to use an expansion of Φ_q in terms of the eigenfunctions $\Psi_i(\Omega)$ of the grand angular momentum operator Λ^2, the hyperspherical harmonics, i.e.

$$\Phi_q(\rho;\Omega) = \sum_i \Psi_i(\Omega)c_{iq}(\rho) \tag{21}$$

With this, eq. (17) reduces to a simple algebraic eigenvalue equation of the form

$$\sum_j \left\{V_{ij}(\rho) - \delta_{ij}\left[\frac{K_i(K_i+4)}{2\mu\rho^2} + \epsilon(\rho)\right]\right\} c_{jq}(\rho) = 0 . \tag{22}$$

This requires that we know the hyperspherical harmonics, which are the eigenfunctions of the grand angular momentum operator, i.e. satisfy

$$\left[\Lambda^2 - K_i(K_i+4)\right] \Psi_i(\theta\phi\alpha\beta\gamma) = 0 . \tag{23}$$

These funcions have been discussed in detail recently,[14, 15] and they can be written as

$$\Psi_i(\Omega) = \Psi^{JM}_{K\nu s}(\Omega) = e^{i\nu\phi/2} \sum_{N=-J}^{J} F^{JM}_{K\nu s,N}(u)D^J_{NM}(\alpha,\beta,\gamma) , \tag{24}$$

with $u = \cos\theta$ and the F's finite polynomials of the form

$$F_{K\nu s,N}(u) = (1+u)^{(\nu+N)/4}(1-u)^{|\nu-N|/4} \sum_{k=s}^{k_{max}} b_k^{K\nu s} u^k \;, \tag{25}$$

An algorithm to compute the coefficients b has been presented.[15]

Using the hyperspherical harmonics as basis functions to obtain the angular solutions to eq. (17) for fixed values of ρ has several advantages:

i. There is no need to solve coupled equations for higher total angular momentum J, since this coupling, the Coriolis coupling, through the grand angular momentum operator has been taken care of analytically;
ii. The boundary conditions, singularities in the kinetic energy operator, for the collinear as well as for the equilateral triangle arrangement are treated exactly;
iii. the symmetry requirements in case of identical particles, including the spin symmetry, can be handled easily.

There are, however, also disadvantages in the use of the hyperspherical harmonics as basis functions:

i. If the potential on the semi-sphere has deep structures, leading to rather localised wave functions, the global hyperspherical basis functions are expected to perform poorly;
ii. the same will be true if cusps are expected in the wave functions, as in the case of Coulomb potentials.

These problems can be overcome by the use of contracted basis functions and the introduction of additional weighting functions.[16, 17]

With the solutions of eq. (17) at hand on a fine grid of ρ, the resulting coupled one dimensional differential equations

$$\sum_q \{T_{pq}(\rho) + \delta_{pq}\left[\varepsilon(\rho) - E_A\right]\}\,\Theta_{qA}(\rho) = 0 \tag{26}$$

can be integrated numerically. To be sure, there is room for improvement in the algorithms for the integration of such equations.

4. Results

We have used the procedure outlined above for the description of the reactive collision of H with H_2,[16] where we employed the R-matrix propagation method[18] to integrate the eqs. (26) for various total energies. Some of the results are presented in Figs. 7 and 8. To obtain these it was in addition necessary to match the R-matrix results obtained in hyperspherical coordinates to the asymptotic channel functions in Jacobi coordinates. We will not detail this procedure here. Suffice it to state here, that these computations became quite elaborate, since calculations with J up to 30

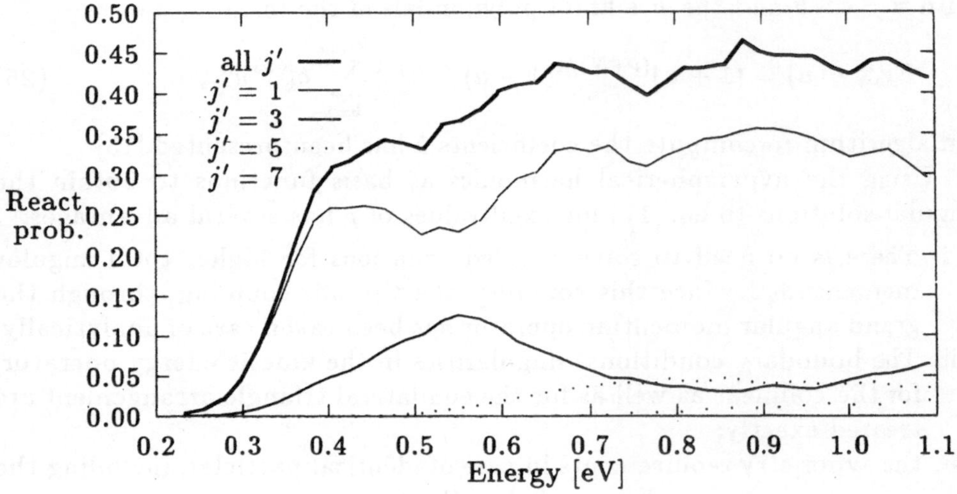

Figure 7. **p-H/o-H reaction probability, $v = 0, j = 0 \rightarrow v' = 0$, $J = 1$**

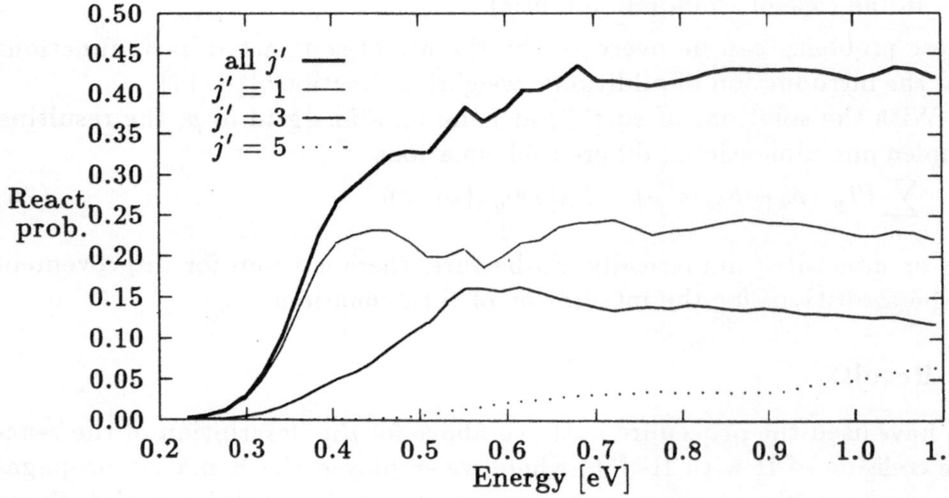

Figure 8. **p-H/o-H reaction probability, $v = 0, j = 0 \rightarrow v' = 0$, all j', $J = 3$**

are necessary to obtain converged results and the number of physically allowed channel increases rapidly with J and the total energy.

We have also computed the rotation vibrational states of H_3^+[17, 19] and its isotopomers[20] with the procedure described. As the potential surfaces[21, 22] used in these calculations are already excellent, the results obtained agree exceptionally well with all the observed transitions in these systems. A comparison of the computed and observed transition energies is presented in Figs. 9 and 10. The computed term energies can become, due to their accuracy, quite useful for the assignment and identification of additional rotation-vibrational transitions. The remaining discrepancies could be reduced significantly with the addition of the diagonal adiabatic correction to the potential.[23]

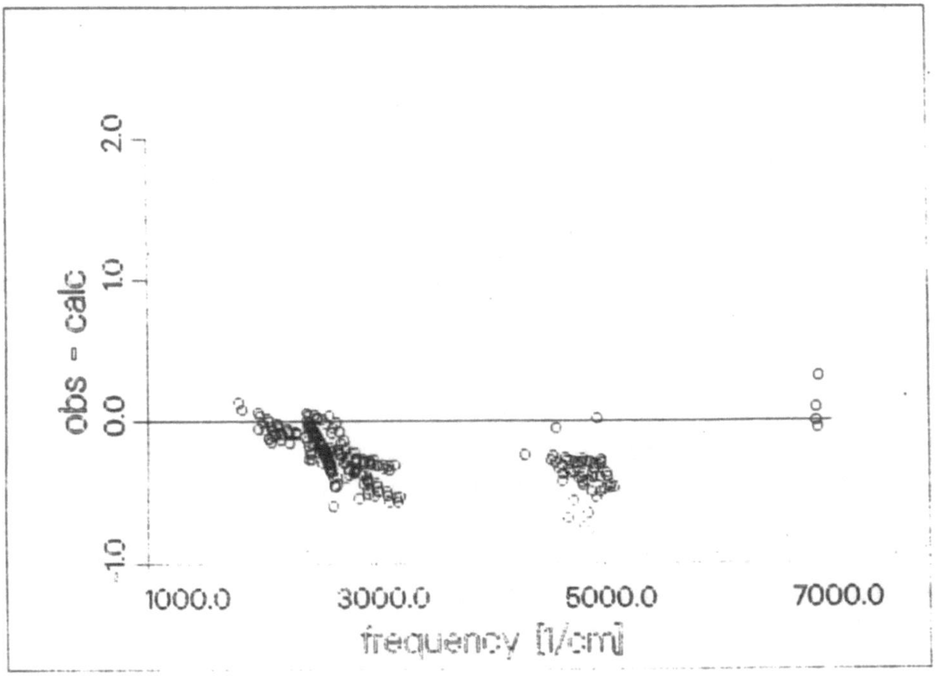

Figure 9. Differences between observed and calculated frequencies for H_3^+

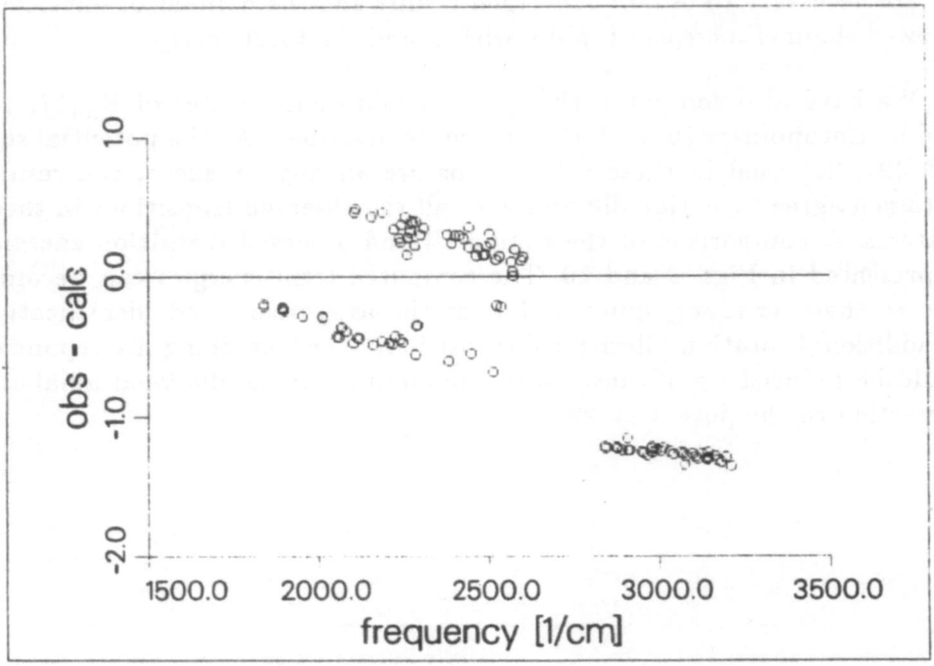

Figure 10. Differences between observed and calculated frequencies for H_2D^+

For the characterisation of rotation vibrational spectra of triatomics with large amplitude motion and expected strong Coriolis coupling the procedures developed are very promising indeed, and we plan to employ them for the description of other triatomic systems, provided good potential surfaces can be obtained.

5. Acknowledgement

The authors are grateful for the able assistance of Udo Welz and acknowledge the support of the Deutsche Forschungsgemeinschaft, Sonderforschungsbereich 216 "Polarisation und Korrelation in atomaren Stoßkomplexen" and of the "Fonds der chemischen Industrie". This work was also supported in part by the Polish government through a KBN grant no. 2 0399 91 to LW.

References

1. in *Higher Transcendental Functions*, Vol. 2 of *Bateman Manuscript Project*, edited by A. Erdélyi (McGraw-Hill, New York, 1953), Chap. 11, p. 232.
2. P. M. Morse and H. Feshbach, in *Methods of Theoretical Physics* (McGraw-Hill, New York, 1953), Vol. 2, p. 1730.

3. F. T. Smith, Phys. Rev. **120**, 1058 (1960).
4. A. Kupperman and P. G. Hipes, J. Chem. Phys. **84**, 5962 (1986).
5. P. G. Hipes and A. Kupperman, Chem. Phys. Lett. **133**, 1 (1987).
6. R. T. Pack and G. A. Parker, J. Chem. Phys. **87**, 3888 (1987).
7. J. G. Frey and B. J. Howard, Chem. Phys. **99**, 415 (1985).
8. R. M. Whitnell and J. C. Light, J. Chem. Phys. **89**, 3674 (1988).
9. J. M. Huts and S. Jain, J. Chem. Phys. **91**, 4197 (1989).
10. R. C. Whitten and F. T. Smith, J. Math. Phys. **9**, 1103 (1968).
11. B. R. Johnson, J. Chem. Phys. **73**, 5051 (1980).
12. B. R. Johnson, J. Chem. Phys. **79**, 1906 (1983).
13. B. R. Johnson, J. Chem. Phys. **79**, 1916 (1983).
14. M. I. Mukhtarova and V. D. Efros, J. Phys. A **19**, 1589 (1986).
15. L. Wolniewicz, J. Chem. Phys. **90**, 371 (1988).
16. L. Wolniewicz, J. Hinze, and A. Alijah, J. Chem. Phys. **99**, 2695 (1993).
17. L. Wolniewicz and J. Hinze, J. Chem. Phys. **101**, 9817 (1994).
18. J. C. Light and R. B. Walker, J. Chem. Phys. **65**, 4272 (1976).
19. A. Alijah, J. Hinze, and L. Wolniewicz, Ber. Bunsenges. Phys. Chem. **99**, 1 (1995).
20. A. Alijah, J. Hinze, and L. Wolniewicz, J. Chem. Phys. to be published (1995).
21. W. Meyer, P. Botschwina, and P. Burton, J. Chem. Phys. **84**, 891 (1986).
22. R. Röhse, W. Kutzelnigg, R. Jaquet, and W. Klopper, J. Chem. Phys. **101**, 2231 (1994).
23. B. M. Dinelli, C. L. Sueur, J. Tennyson, and R. Amos, Chem. Phys. Lett. **232**, 295 (1995).

DYNAMICS AND SPECTROSCOPY OF HIGHLY EXCITED MOLECULES

F. BORONDO
Dep. de Química, C-IX, Universidad Autónoma de Madrid, CANTOBLANCO - 28049 Madrid (Spain).

AND

R.M. BENITO
Dep. de Física, E.T.S.I. Agrónomos, Universidad Politécnica de Madrid, 28040 MADRID (Spain).

1. Introduction.

Considerable attention has been devoted in past few years to the study of the dynamics and spectroscopy of highly vibrationally excited small polyatomic molecules. The rate at which intramolecular vibrational relaxation (IVR) can redistribute the excitation energy, in competition with unimolecular processes, such as isomerization or decomposition, is of paramount importance for example in relation with the possibility of a laser selective chemistry [1]. Modern femtosecond lasers and the possibility of frequency bandwidth tailoring ("chirping") open new interesting possibilities.

The development of new spectroscopical methods, such as stimulated emission pumping (SEP), vibrational predissociation spectroscopy, or supersonic jet FTIR, have contributed to improve the knowledge of these high vibrational states [2].

On the theoretical side there has also been a great advance in the study of these states, due to the use of methods like discrete variable representation (DVR) [3] or grid methods for solving the time dependent Schroedinger equation (wavepacket propagation) [4]. In the high energy regime the density of states is often very high, and then correspondence between quantum and classical mechanics is expected to work quite well. Classical trajectories are much easier to implement, specially as the number of atoms in the molecule increases. Recent advances in nonlinear dynamics [5] has

E. Yurtsever (ed.), Frontiers of Chemical Dynamics, 371–392.

contribute significantly to invigorate the use of classical mechanics to rationalize the study of some molecular processes. The ubiquity of chaos has configured nonlinear dynamics as a highly multidisciplinary area, and numerous publications have appeared in a variety of fields [5].

2. Classical Mechanics of Hamiltonian Systems.

The dynamics of Hamiltonian systems is defined by the Hamilton equations of motion

$$\begin{aligned} \dot{q}_i &= \frac{\partial H}{\partial P_i} \\ \dot{P}_i &= -\frac{\partial H}{\partial q_i}, \qquad i = 1, ..., N \end{aligned} \tag{1}$$

where N is the number of degrees of freedom of the system.

The $2N$ dimensional space formed by the coordinates of the system, q_i, and their conjugate momenta, P_i, is called *phase space* of the system, as opposed to that formed only by the coordinates, which is called *configuration space*.

The motion of the system is obtained by numerical integration of eqs. (1), starting from a suitable set of initial conditions, $(\vec{P}(0), \vec{q}\,(0))$. Then, the time evolution of this point, $(\vec{P}(t), \vec{q}\,(t))$ defines a trajectory in phase space, where the dynamics of the system is most clearly visualized.

3. Regular and Chaotic Motions.

For autonomous Hamiltonian systems, in which H does not depend explicitly on time, energy is a constant of motion, and then the dimensionality of phase space reduces in one. When the number of independent constants of motion is equal to N, the system is said to be *integrable* and the motion *regular*. In this case:

1. A canonical transformation from the old variables, $(\vec{P}, \vec{q}\,)$, to a new set of angle action variables, $(\vec{I}, \vec{\theta}\,)$, exists, such that H only depends on the actions. Hamilton equations in these new coordinates present an specially simple form

$$\begin{aligned} \dot{\theta}_i &= \tfrac{\partial H}{\partial I_i} = \omega_i(\vec{I}\,), \quad \theta_i = \omega_i(\vec{I}\,)t + \delta_i \\ \dot{I}_i &= -\tfrac{\partial H}{\partial \theta_i} = 0, \qquad I_i = cte \end{aligned} \tag{2}$$

where ω_i are the N frequencies characterizing the motion.

2. Trajectories in phase space are confined to the surface of a N-dimensional invariant torus. This is easily understood if one interprets I_i as the radius of a circle, and θ_i as the corresponding polar angle. The total

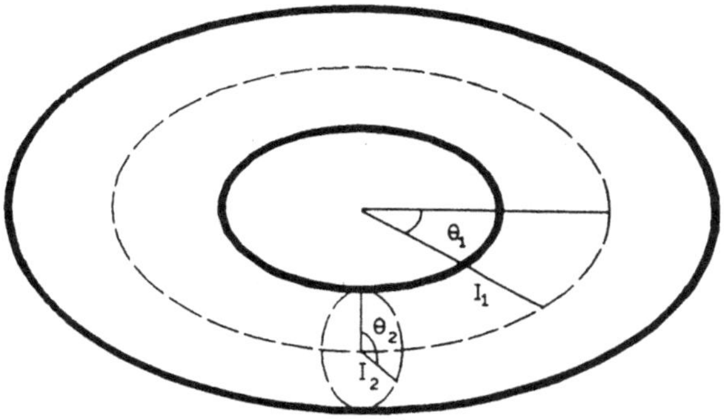

Figure 1. Invariant torus as the cartesian product of two circles.

motion will then take place in the cartesian product of all of them, which is a torus (see Fig. 1).

3. The trajectories are given by a Fourier series of the N fundamental frequencies, and their combinations

$$\vec{P}(t) = \sum_{\vec{m}} \vec{a}_{\vec{m}}(\vec{I}\,)\, e^{i\vec{m}\cdot\vec{\theta}}$$

$$\vec{q}(t) = \sum_{\vec{m}} \vec{b}_{\vec{m}}(\vec{I}\,)\, e^{i\vec{m}\cdot\vec{\theta}} \tag{3}$$

where $\vec{m}$ is a N-dimensional vector of integer numbers. The motion can be of two types. If there is a resonant relationship between the frequencies, $\vec{\omega} = \vec{M}\omega_b$, the orbit on the torus closes over itself after M_1 turns around θ_1, M_2 turns around θ_2, ... , M_N turns around θ_N, in such a way that it only fills a one dimensional region on the torus. In this case, the motion on top of being *regular* is *periodic* (see Fig. 2a). When there is no resonance, and the frequencies are not conmensurated, the orbit never closes, densely filling the surface of the torus (see Fig. 2b). In this case the motion is said to be *quasiperiodic*.

The simplest example of regular motion is the pendulum, which is trivially integrable, since it is a 1D of freedom system. The Hamiltonian of a pendulum of length l and mass m is given by

$$H = \frac{P_\theta^2}{2ml} - mgl\cos\theta \tag{4}$$

In Fig. 3 the phase space of the pendulum is shown. Two different kinds of motion are possible: *vibrations* and *rotations*. Also, the two equilibrium

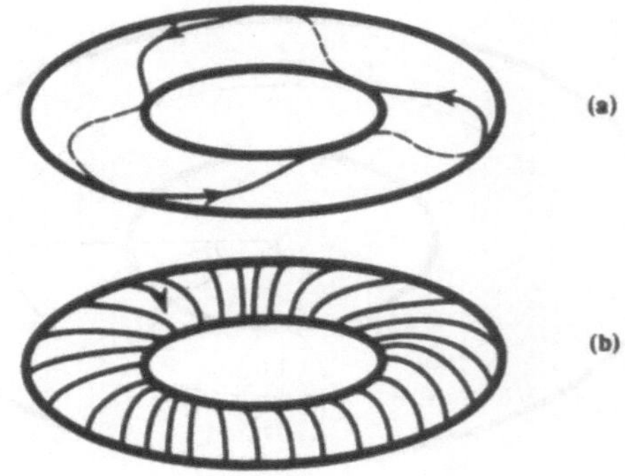

Figure 2. Trajectories on the torus: (a) periodic, (b) quasiperiodic.

points, at $\theta = 0$ (stable) and $\theta = \pi$ (unstable), are represented. In the jargon of nonlinear dynamics these points are called *fixed*; *elliptic* the first ones, since in its vicinity the trajectories describe ellipses, and *hyperbolic* the others because the neighbouring trajectories can be approximated by branches of hyperbolas. In Fig. 3 we have also included the *separatrix* (in dashed line) which separates vibrations from rotations. It corresponds to an energy of mgl, and the corresponding period would be infinite. Moreover,

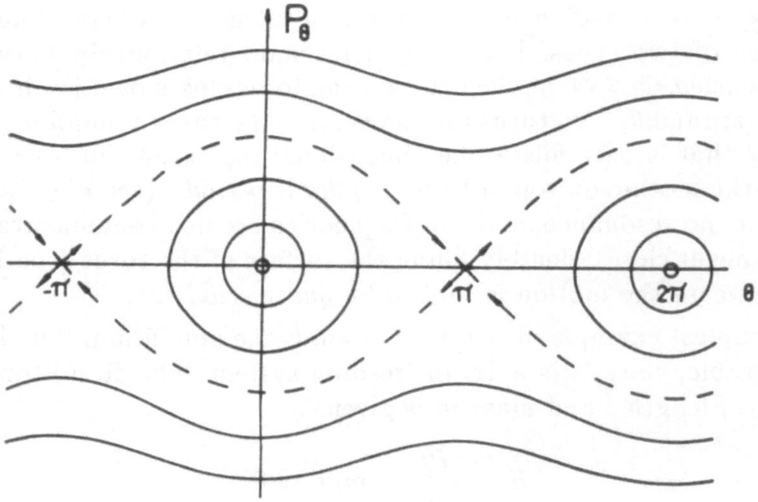

Figure 3. Phase space of the pendulum. The separatrix is shown in dashed line, the elliptic points are represented by circles and the hyperbolic points by crosses.

the two branches of the separatrix touch at the hyperbolic fixed points, originating at each one four invariants *manifolds*; two of them are *incoming*, H^+, and the other two *outgoing*, H^-. The incoming manifolds correspond to trajectories approaching the fixed point; as this happens the motion is slowed down, and the fixed point would only be reached after an infinite time. On the other hand, the outgoing manifolds correspond to trajectories in which the pendulum have been infinitesimally separated from the unstable equilibrium point, and moves exponentially away from it. Clearly, an outgoing manifold originated at a hyperbolic point becomes an incoming manifold of the next hyperbolic point.

On the other hand, when there is no constant of motion other than the energy, the motion is said to be *ergodic* or *chaotic*. Then, almost any trajectory of the system passes, sooner or later, arbitrarily close to any point of the available phase space [6]. In fact, ergodic motion is only the first step in the hierarchy of chaos [7]. This is an organization of the properties of dynamical systems, done in such a way that each step always implies the lower ones. The next upper steps in this ladder are the mixing systems, in which the evolution toward equilibrium is guaranteed, and the K-systems and Bernouilli systems, for which the dynamics, although totally deterministic, is indiscernible from the results obtained from a roulette wheel.

4. Generic Systems.

In the previous Section we have described two limiting cases for the behaviour of a dynamical system. The question of interest now is in which of these two categories do the Hamiltonians of interest in Chemical Physics fall. Unfortunately, very few interesting cases have been rigorously proven to fall in either class, and then we will resort to numerical experiments to answer this question.

A very convenient way to visualize the structure of phase space in 2D dynamical systems is the *Poincaré surface of section* (SOS). The SOS is a representation of the intersection of a given trajectory with a suitable plane in phase space, taking only those points for which the plane is crossed in a given direction (see Fig. 4). The SOS defined in this way consists of a sequence of points

$$\begin{array}{l} (P_1(0), q_1(0)) \xrightarrow{T} (P_1(1), q_1(1)) \\ (P_1(1), q_1(1)) \xrightarrow{T} (P_1(2), q_1(2)) \\ \cdots\cdots\cdots\cdots\cdots\cdots\cdots\cdots \end{array} \qquad (5)$$

and the transformation T is a *map*. For hamiltonian systems this Poincaré map is an area preserving map, due to the Liouville theorem.

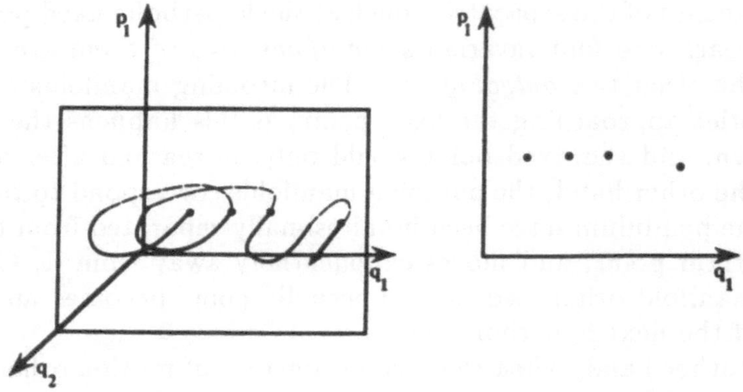

Figure 4. Construction of the Poincaré surface of section (SOS).

As we discussed above, the motion in the regular regime takes place on the surface of a torus, which renders a line in the SOS when it is cut by the sectioning plane. In the ergodic regime, the motion takes place in the 3D energy shell, and then the successive intersections of the trajectory with the SOS will fill an area (see Fig. 5). The global structure of the phase space can be visualized by the *composite SOS*, which consists in the superposition, in the same figure, of the SOS of a significant ensemble of trajectories, all of them calculated at the same energy.

Now we are in the position to answer the question that we posed at the beginning of this Section. As an example, we choose the vibrational dynamics of a triatomic molecule: the $LiNC/LiCN$ isomerizing system

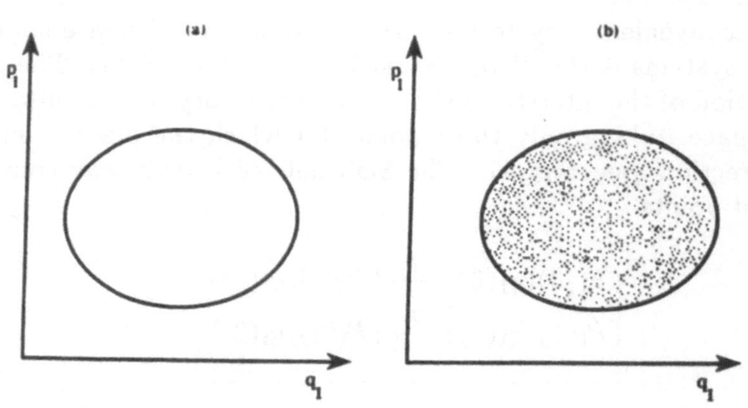

Figure 5. Poincaré surface of section: (a) quasiperiodic trajectory, (b) chaotic trajectory.

[8]. This molecule presents two stable isomers, corresponding to the linear configurations, separated by a relatively modest barrier (3454 cm^{-1}). The motion in the bending coordinate is very floppy, and then the Li atom can easily rotate around the CN fragment; chaos sets in at low values of the excitation energy. Also, the CN vibrational frequency is very high and separates from the remaining modes of the system. Accordingly, the dynamics of the system can be studied by a 2D of freedom model, where the CN distance is kept frozen at its equilibrium value ($r_e = 2.186$ a.u.). The vibrational Hamiltonian in scattering coordinates is given by

$$H = \frac{P_R^2}{2\mu_{Li-CN}} + \frac{1}{2}\left(\frac{1}{\mu_{Li-CN}R^2} + \frac{1}{\mu_{C-N}r^2}\right) P_\theta^2 + V(R,\theta) \tag{6}$$

where R is the distance between the Li atom and the centre of mass of the CN fragment, r the CN distance, and θ the angle formed by these two vectors. The potential energy surface has been taken from the literature [9].

The isomerization process in this system can be followed by considering the motion along the θ coordinate. Usually, the SOS for systems of coupled oscillators is obtained representing one coordinate and its conjugate momentum every time that the other coordinate passes through its equilibrium distance and the momentum has a predetermined sign. However, in our case, the equilibrium distance varies with θ. Then, a better way to define the SOS [10] in this case is to use a new set of coordinates

$$\begin{aligned} \rho &= R - R_e(\theta), & \psi &= \theta \\ P_\rho &= P_R, & P_\psi &= P_\theta + P_\rho \left(\frac{\partial R_e}{\partial \theta}\right)_{\theta=\psi} \end{aligned} \tag{7}$$

where $R_e(\theta)$ is the minimum energy path connecting the two isomer wells. The SOS corresponds then to the intersection with the $\rho = 0$ plane, taking only those points for which P_ρ is in a predetermined branch of the second order differential equation

$$H(\rho = 0, \psi, P_\rho, P_\psi) = E \tag{8}$$

In Fig. 6 we present the SOS for the $LiCN$ molecule for some representative values of the vibrational energy. We observe that for any energy, regions of regularity and chaos coexists, and that the proportion of chaos increases with energy. Also, the appearance of chains of islands is observed. This behaviour is general of any typical Hamiltonian system, and can be explained with the aid of the KAM and the Poincaré-Birkhoff theorems.

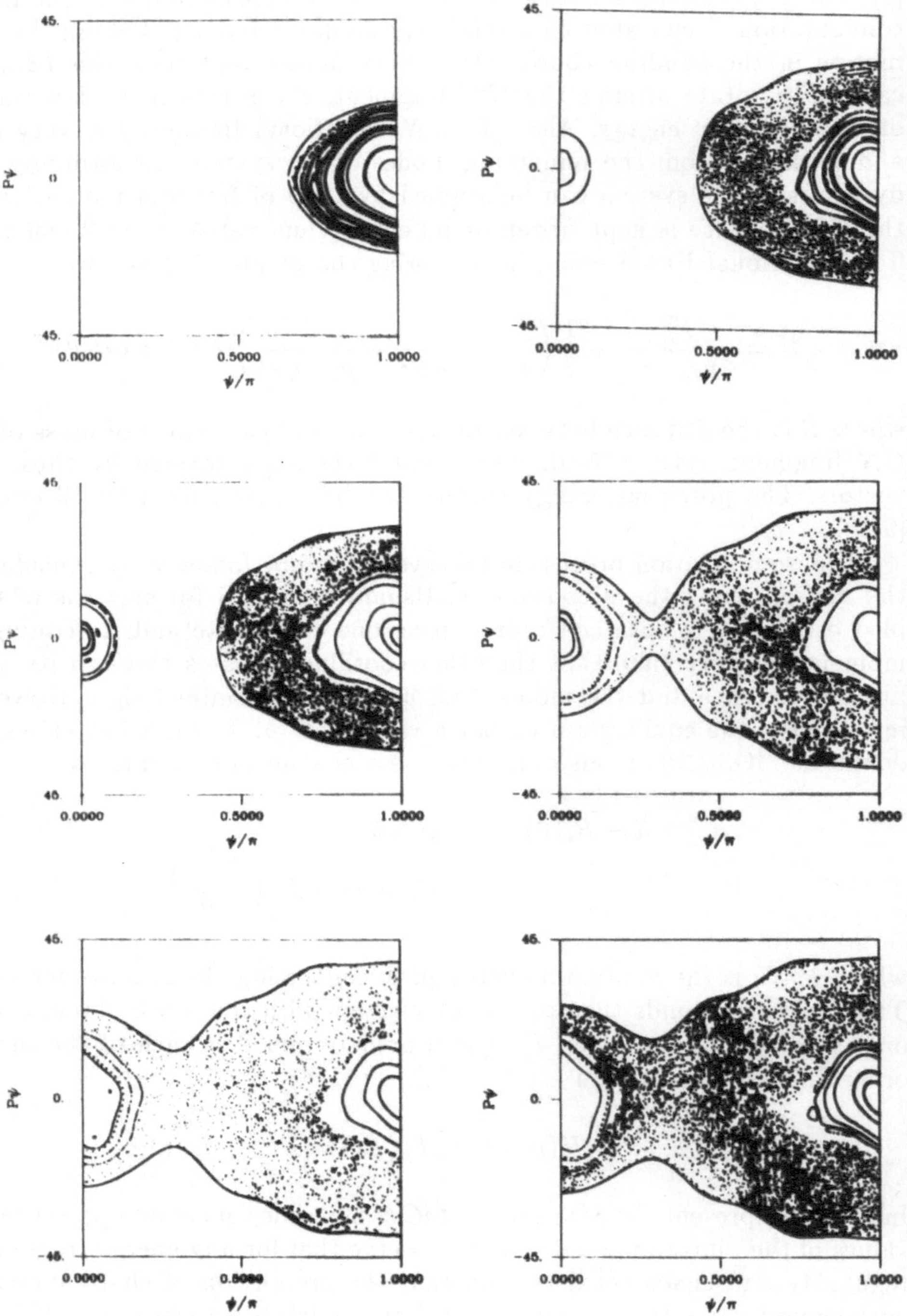

Figure 6. **Composite Poincaré surface of section for *LiNC/LiCN* at different energies: from upper left to lower right, 0.005772, 0.01162, 0.01179, 0.01589, and 0.01897 respectively. It is observed that the fraction of chaos increases with energy.**

5. KAM and Poincaré-Birkhoff Theorems.

The Hamiltonian of a generic system can be written as

$$H = H_0 + \varepsilon H_1 \tag{9}$$

where H_0 is an approximation to H corresponding to an integrable system, H_1 is a perturbation term, and ε is a parameter controlling the degree of the perturbation.

The KAM theorem [11], due to Kolmogorov, Arnold and Moser, establishes that in a slightly perturbed system a finite measure of tori, *i.e.*, those characterized by a "sufficiently irrational" frequency ratio, $\alpha = \omega_1/\omega_2$, survive. For a 2D of freedom system this irrationality condition is given by

$$\left| \frac{\omega_1}{\omega_2} - \frac{n}{m} \right| > \frac{K(\varepsilon)}{m^{5/2}} \tag{10}$$

for all integer numbers n and m. $K(\varepsilon)$ is a complicated function of the perturbation which tends to 0 as the perturbation disappears. These tori, which are only distorted by the perturbation, are called *KAM tori.*

The rest of tori are destroyed, but according to two different mechanisms depending on the value of α. If α is a rational number (resonant tori), the previous condition does not hold for any value of the perturbation. The fate of these tori is regulated by the Poincaré-Birkhoff theorem [12]. It states that under small enough perturbations, an even number of fixed points of the original torus survive; half of them are elliptic, and half of them hyperbolic. The elliptic points are surrounded by new tori (stability islands), some of them rational, and some of them irrational. Again the rationals and some of the irrationals will be destroyed by the perturbation, according to the KAM theorem. Then, the previous structure is repeated again and again at a finer and finer scale (see Fig. 7). In the neighbourhood of

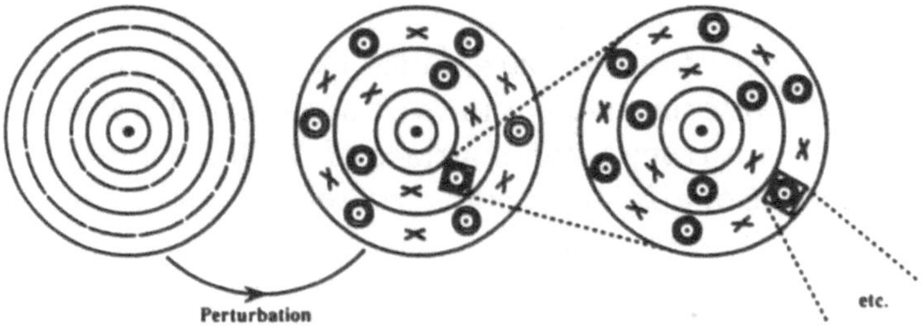

Figure 7. Mechanism for the destruction of tori. Adapted from Ref. 12.

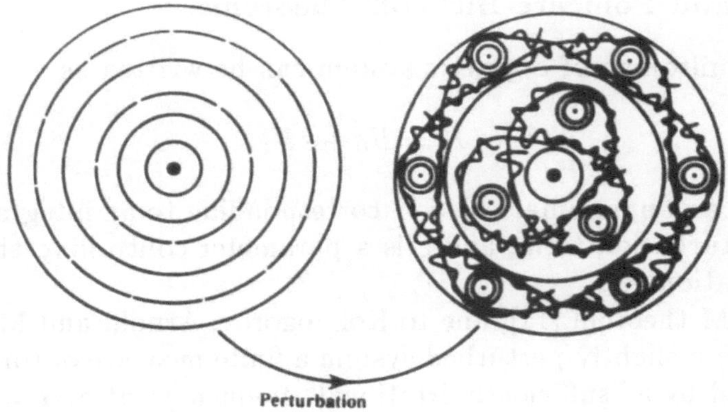

Figure 8. Homoclinic oscillations. Adapted from Ref. 12.

the hyperbolic points (related to the separatrix of H_0) the motion becomes very complex, due to the homoclinic oscillations discovered by Poincaré. The incoming and outgoing manifolds cannot cross, and then they cross to each other an infinite number of times. The points at which they cross are called homoclinic, and form an ensemble dense in a region of the SOS (see Fig. 8), which originates a band of stochasticity around the chain of stability islands.

The tori which are not irrational enough according to eq. (10), are also destroyed by the perturbation. But they turn into invariant ensembles called *cantori.* The nature of cantori is well understood in terms of their SOS. They are quasiperiodic orbits, similar to the normal invariant tori, but they only fill a fractal object, which is a Cantor set, in the SOS. Contrary, to what happens with the invariant tori, the cantori only represent partial barriers for the flux of trajectories in phase space.

6. Systems With More Than Two Degrees of Freedom.

The existence of invariant tori in 2D of freedom systems has a drastic effect in the structure of phase space. Since two trajectories cannot cross in phase space, each cantorus divide it in two unconnected regions. For systems with $N > 2$ this is not true: A hypersurface of N dimensions (corresponding to an invariant torus of a N-dimensional system) do not define an inner and an outer regions in a space of $2N - 1$ dimensions (energy shell).

Then, in despite of the existence of zones of regular motion, the chaotic areas are interconnected, constituting what is known as the Arnold web, and a single trajectory can explore the whole chaotic region, in which is known as Arnold diffusion [13].

7. Correspondence Between Classical and Quantum Mechanics.

The connection between classical and quantum mechanics has attracted a great deal of interest since the early days of quantum theory. The correspondence principle enunciated by Bohr in 1913 was a cornerstone in the development of quantum mechanics. Recently, and specially due to the advances in nonlinear dynamics that interest has been renewed, mainly in relation to the question of understanding which is the quantum analogue of the classical chaos. For example, Ford *et al.* [14] demonstrated, using algorithmic complexity theory arguments, the failure of the correspondence principle in the Arnold's cat map.

When considering the correspondence between the class`.al treatment presented in the previous Sections, and the corresponding quantum counterpart, it is very important to consider separately the cases of regular and chaotic motions.

7.1. REGULAR REGIME.

In the case of regular motion, the connection between classical and quantum mechanics is well establish in terms of the Einstein-Brillouin-Keller (EBK) quantization conditions for the corresponding invariant tori

$$\oint_{C_j} \vec{P} \cdot d\vec{q} = h\left(n_j + \frac{\alpha_j}{4}\right), \qquad n_j = 0, 1, 2, ... \tag{11}$$

In the Bohr-Sommerfeld theory the action integral were computed around closed circuits in each of the coordinates of the system. However, this is only valid for separable Hamiltonians. Einstein [15] was the first one to realize that the correct way to calculate these integrals is along the N topologically independent paths, C_j, defining the torus. α_j are the Maslov indexes [16], which account for the number of times that the trajectory touches the caustics of the system. For 1D systems the caustics are simply the turning points. Let us remark that equation (11) constitutes a typical semiclassical expression, which combines purely classical information in the left hand side with a quantum condition in the other. For each energy allowed by eq. (11) there is an associated semiclassical WKB (Wentzel-Kramers-Brillouin) wavefuntion [16]

$$\Psi_{WKB}(\vec{q}) \sim \sum_j A_j e^{iS_j(\vec{q},\vec{I})} \tag{12}$$

where S_j are the branches of Hamilton's characteristic function.

Numerous methods have been proposed in the literature to solve eq. (11), using very different strategies.

The most intuitive and straightforward method is one due to Noid and Marcus [17]. In it the quantizing invariant tori are obtained by numerically construction. Hamilton equations of motion are integrated in a suitable set of generalized coordinates, and at the same time the SOS is calculated. The actions can then be evaluated by integration of the area in the SOS. Finally, by repeating these steps, adjusting the energy and the initial conditions, iteratively the semiclassically quantized energies are obtained. The main difficulties of this method are two. In the first place, a suitable set of coordinates, adequate to the symmetry of the system, needs to be found. Also, the computational effort implied in its application to systems with more than 2D of freedom is enormous.

Other methods have been proposed based on the iterative construction of good action-angle variables [18], classical perturbation theory [19], classical variational principles [20], Fourier analysis of the trajectories [21], etc.

Most of the methods described above are difficult to apply to systems with more than 2D of freedom. One method that can, in principle, overcome this difficulty is a primitive semiclassical quantization method based on the adiabatic invariance of the action [22]. Initially proposed by Ehrenfest and Einstein, it was Solov'ev [23] who made the first numerical application to nonintegrable systems. In 1983 it was brought to the attention of the chemical physics community [24], and then Skodje, Borondo and Reinhardt [25,26] studied the method in detail and apply it to a variety of coupled oscillator systems. Since then, adiabatic switching has attracted a great deal of attention, and numerous applications have been reported in the literature [26].

The adiabatic switching method is based on an extension of the adiabatic theorem of classical mechanics [22] to multidimensional nonseparable systems. Roughly speaking the theorem asserts that if we perturb a bound 1D trajectory, the action is conserved, provided that the perturbation is slow compared to the motion of the system, and that the instantaneous frequency is never zero.

The actual application of the method is as follows. We write our Hamiltonian in the form

$$H = H_0 + H_I \tag{13}$$

where H_0 is a separable zero order approximation and H_I an interaction term (which needs not to be small). From this we construct a time dependent Hamiltonian of the form

$$H(t) = H_0 + \lambda(t) H_I \tag{14}$$

where $\lambda(t)$ is a smooth, slowly varying function of time, which goes from 0 to 1 along a switching time T. The quantization of H_0 is trivial, since we can write for each separable mode

$$I_j^0 = \oint P_j \, dq_j = h\left(n_j + \frac{\alpha_j}{4}\right); \qquad j = 1, ..., N; \; n_j = 0, 1, ... \tag{15}$$

Then, if we run a classical trajectory using $H(t)$, with starting initial conditions corresponding to a point on a quantized torus of H_0, the actions will be conserved (if T is large enough) and the final point will be on the corresponding quantized torus of H. Of course, the energy will change along the integration of the trajectory, but its final value will be the primitive semiclassical eigenvalue. From the practical point of view, instead of running a single trajectory for a long time to obtained converged results, it is more convenient to get the energy by averaging the results of an ensemble of trajectories with initial conditions chosen randomly on the surface of the zero order torus. Also, one can get an estimation of the selfconsistency of the method from the magnitude of the root mean square deviation of the final energies. A large value of this magnitude tell us that in the switching process a region of resonance or chaos has been encountered, thus building up non-adiabaticities in the process.

7.2. CHAOTIC REGIME.

For chaotic systems the semiclassical EBK quantization methods are not applicable, since they rely on the existence of invariance tori. Eighty years after the EBK theory was established we still do not completely understand the quantization of chaotic systems.

In this respect, the most valuable piece of theory is the work developed by Gutzwiller after 1967 [27]. This author proposed a completely different quantization scheme based on the use of the quantum mechanical Green function, $G(\vec{q}\,'', \vec{q}\,', E)$, which represents the probability amplitude for a particle with energy E to move from $\vec{q}\,'$ to $\vec{q}\,''$ in configuration space. This function has poles (as a function of E) at the position of the energy eigenvalues on the real axis, and the residues at the poles gives the corresponding wavefunctions. Gutzwiller constructed a semiclassical approximation to G, which results in a sum over all classical trajectories starting at $\vec{q}\,'$ and ending at $\vec{q}\,''$ at a given energy. This summation can be easily carried out in the regular regime, but it is very hard to do when the dynamics of the system is chaotic. In this case, one can resort to the trace of the Green function

$$g(E) = \int d\vec{q}\, G(\vec{q}, \vec{q}, E) = \sum_{j=0}^{\infty} \frac{1}{E - E_j} \tag{16}$$

This simpler quantity gives only information on the energy levels. Using stationary phase arguments a semiclassical version of eq. (16) can be obtained

$$g_{SC}(E) = \sum_p \frac{T_p}{i} \sum_{r=1}^{\infty} \frac{exp\left(irS_p - \frac{i\pi}{2} r\mu_p\right)}{\mid det[(\mathbf{M}_p)^r - \mathbf{I}] \mid^{1/2}} \tag{17}$$

where the index p refers to the periodic orbits (PO) of the system and r to multiple transversals of a given orbit. T_p, $\mathbf{M}_p$, S_p are respectively the period, stability matrix and action integral of a classical orbit. μ_p is the Morse index, which is the number of conjugate points along the orbits, and corresponds to twice the number of times that the stable and unstable manifolds wind around the orbit over one period. Using this equation, Gutzwiller obtained energy levels for a free particle on a surface of constant negative curvature and for the anisotropic Kepler problem. More recently, Burghardt and Gaspard [28] applied the same method to calculate the position and lifetime of the HgI_2 photodissociation resonances.

Another interesting problem in connection with the classical-quantum correspondence is the identification of signatures of chaos in quantum mechanics.

Some criteria were proposed in the literature some time ago, which were general enough to be applicable to a wide variety of systems: nodal pattern of the wavefunctions [29], dominant coefficients [30], overlapping avoided crossings [31], second differences in the energy [32], and level statistics of spectra [33].

Another way to study the quantum dynamics of a system, more in the spirit the present paper, is to look for classical-quantum correspondence in the framework of phase space. There is no unique way to define a phase space representation of quantum mechanics. Probably the most popular one is that introduced by Wigner in 1932 [34], in a paper devoted to the quantum corrections to statistical thermodynamics. The Wigner transform is defined as

$$W(\vec{P},\vec{q}) = \frac{1}{h^N} \int d\vec{x}\; e^{i\vec{P}\cdot\vec{x}}\; \psi\left(\vec{q} - \frac{\vec{x}}{2}\right) \psi^*\left(\vec{q} + \frac{\vec{x}}{2}\right) \tag{18}$$

This function has certain properties which suggest that it can be interpreted as a probability density in phase space. Namely, it gives the correct marginal probability distributions

$$\begin{aligned} \int d\vec{q}\, W(\vec{P},\vec{q}) &= \mid \langle \vec{P} \mid \psi \rangle \mid^2 \\ \int d\vec{P}\, W(\vec{P},\vec{q}) &= \mid \langle \vec{q} \mid \psi \rangle \mid^2 \end{aligned} \tag{19}$$

Also in the semiclassical limit, $\hbar \to 0$, it tends to a delta function centred on the invariant torus associated to ψ

$$\lim_{\hbar \to 0} W_\psi(\vec{P}, \vec{q}) = \frac{1}{(2\pi)^N} \delta[\vec{I}(\vec{P}, \vec{q}) - \vec{I}_\psi] \tag{20}$$

However, there are some objections to this interpretation since W can be negative. This is due to the fact that defining a function in a precise point of phase space is somehow in contradiction with the uncertainty principle. A way to overcome this difficulty was devised by Husimi [35], who proposed the use of gaussian convolutions of the Wigner function. The new function can be shown to be equal to

$$H(\vec{P}, \vec{q}) = h^{-N} \mid \langle \phi \mid \psi(\vec{q}) \rangle \mid^2 \tag{21}$$

where ϕ is a minimum uncertainty harmonic oscillator coherent state [36]. The Husimi function has been proven to be everywhere nonnegative [37], and can be interpreted in a number of ways: as a smoothed Wigner distribution [35,37], as a coarse graining of the phase space [38] or as the expectation value of the projection operator on a certain state. More properties of the Husimi function have been discussed in the literature [39].

Projection of the Wigner or Husimi functions into suitable planes of phase space gives quantum analogues to the classical Poincaré surfaces of section (QSOS).

We have obtained QSOS for the first 100 states of the $LiCN$ molecule [10], by numerical integration of eq. (21) using eigenfunctions obtained by the DVR program of Bacic and Light [3] with 416 basis functions (this gave eigenvalues converged to 0.01 cm^{-1}). In Fig. 9 we present the QSOS of the states corresponding to the energies of Fig. 6, where the classical SOS were presented. Comparing this two figures we conclude that the quantum probability densities are much more localized than the classical counterparts. In Figs. 9a and 9d the QSOS are localized on the elliptic fixed points corresponding to the $LiNC$ and $LiCN$ isomers, respectively. For states n=24 and 25 (parts b and c) the QSOS are localized, respectively, on the hyperbolic and elliptic fixed points of an eightfold chain of islands located on the boundaries of the available phase space of the $LiNC$ well (see Figs. 6b and 6c). State 65 presents a maxima of the QSOS on the neighbourhood of the hyperbolic point corresponding to the saddle of the potential energy surface. Finally, the QSOS of state 79 is highly localized on the two fixed points associated to the small island of stability which is on the left of the regular region corresponding to the $LiNC$ well. This two orbits, one stable and another unstable, appear "out of the blue" from a saddle-node bifurcations taking place at lower energies [40]. As a conclusion of this classical-quantum comparison, we have found that, for most of

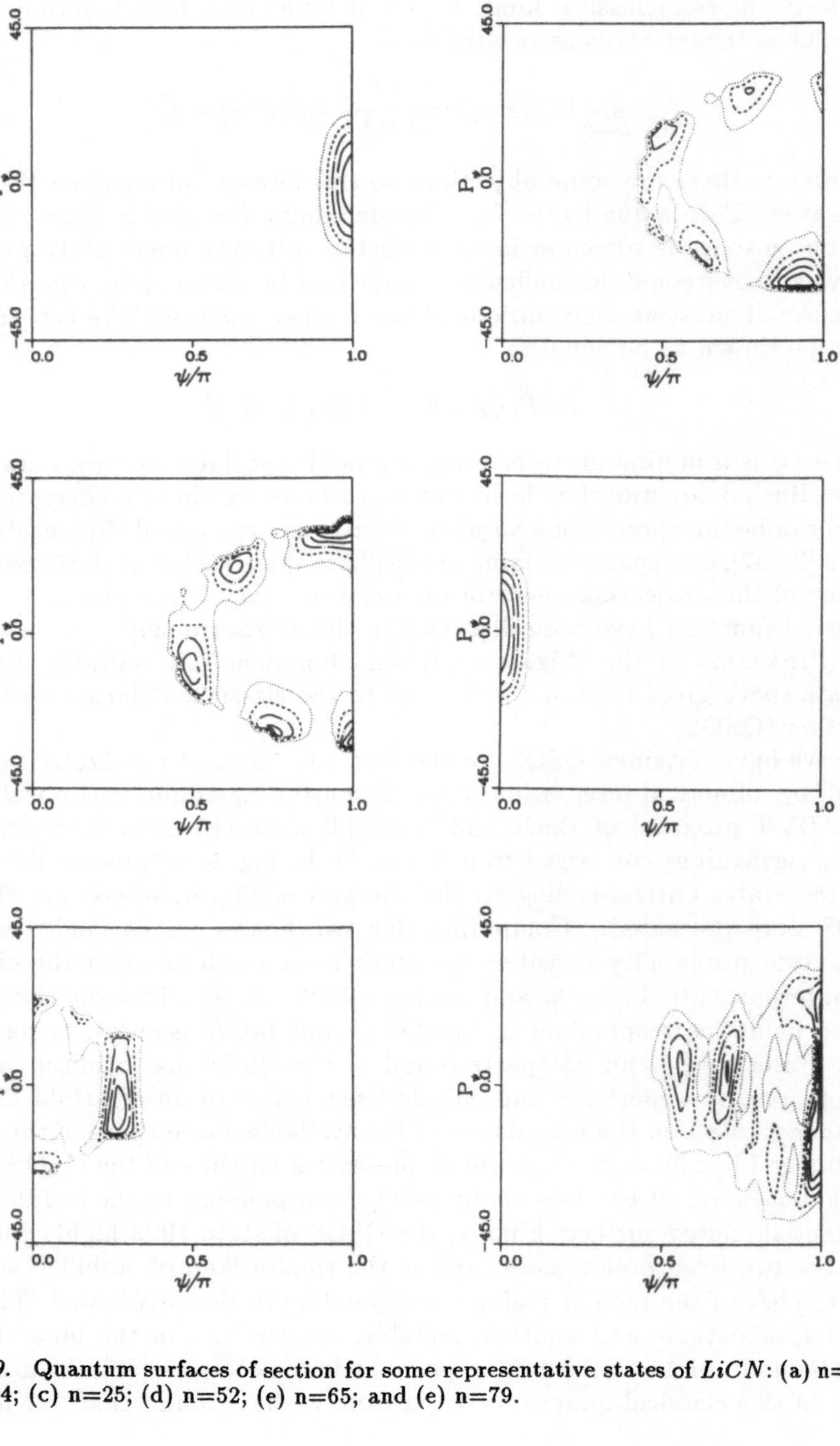

Figure 9. **Quantum surfaces of section for some representative states of *LiCN*: (a) n=5; (b) n=24; (c) n=25; (d) n=52; (e) n=65; and (e) n=79.**

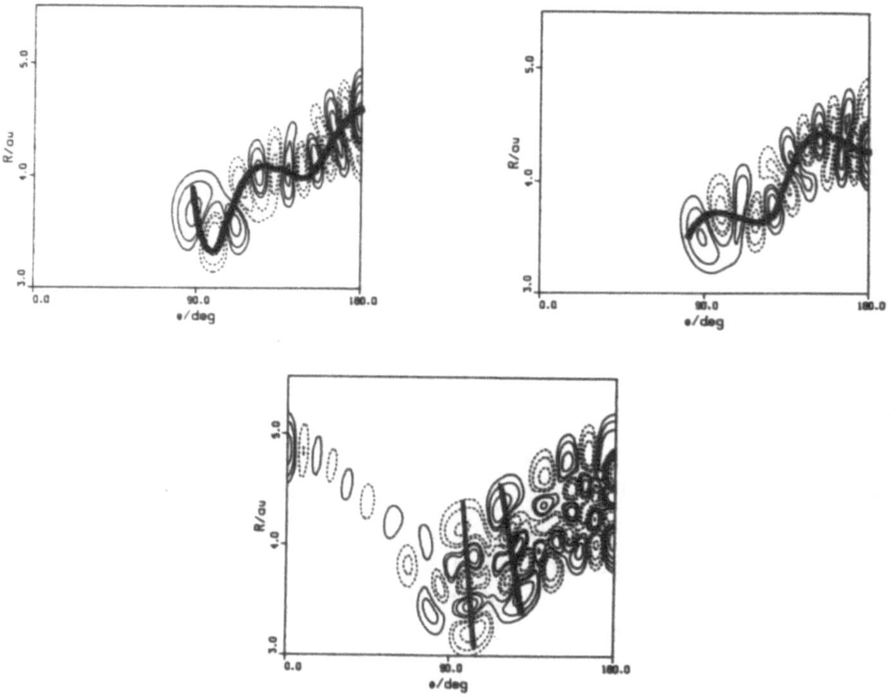

Figure 10. **Wavefunctions for the states n=24, 25 and 79 of *LiNC/LiCN*. The most relevant periodic orbits, according to the QSOS of Fig. 9, have been superimposed to show its scarring effect.**

the states, quantum probability density tends to accumulate around fixed points, corresponding to classical POs. Similar results have been described in the literature for other hamiltonian models [41].

For a better understanding of the previous result we can go one step further and compare the corresponding configuration space representations, *i.e.*, classical POs *vs.* quantum wavefunctions. This is done in Fig. 10 for states n=24, 25 and 79. It is apparent that the POs follow remarkably well the maxima of the wavefunctions. This "scarring" effect of the POs was found for the first time by McDonald and Kauffman [42] in the high lying states of the billiard stadium, and has been the subject of many interesting papers in the literature [43].

8. Wavepacket Propagation and Quantum Localization.

More information on the scarring effect of POs on eigenstates can be gained by following the dynamics of a wavepacket, $\phi(0)$, initially placed in

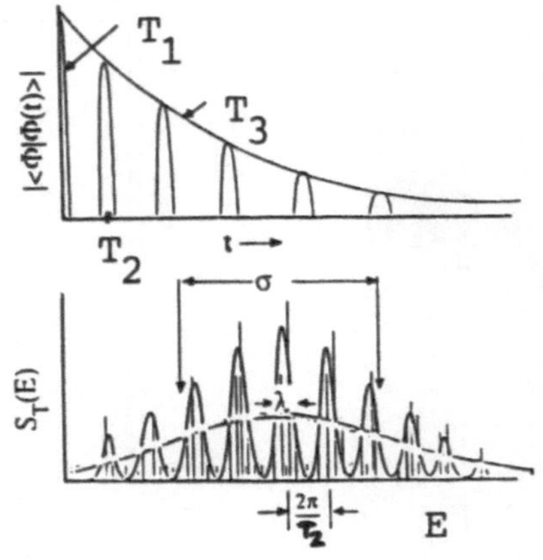

Figure 11. Schematic correlation function and corresponding spectra for a localized wavepacket. Adapted from Ref. 27.

phase space in the vicinity of a particular PO. According to the Ehrenfest theorem this packet will closely follow the classical dynamics of the system (along the PO) for a short period of time.

The usual Frank-Condon stick spectrum and wavepacket dynamics are related through the relation

$$
\begin{aligned}
I_\infty(E) &\propto \sum_f |\langle\phi|f\rangle|^2\delta(E-E_f) = \frac{1}{2\pi}\sum_f \langle\phi|f\rangle\langle f|\phi\rangle \int_{-\infty}^{+\infty} dt\ e^{i(E-E_f)t} \\
&= \frac{1}{2\pi}\int_{-\infty}^{+\infty} dt\ e^{iEt}\ \sum_f \langle\phi|e^{-iHt}|f\rangle\langle f|\phi\rangle \\
&= \frac{1}{2\pi}\int_{-\infty}^{+\infty} dt\ e^{iEt}\ \langle\phi(0)|\phi(t)\rangle \qquad (22)
\end{aligned}
$$

This equation permits to pass from the energy domain to the time domain, and constitutes the basis of the time dependent formulation of Spectroscopy. This formalism has been widely used in areas like Condensed Matter Physics, and was popularized among the Chemical Physics community by Heller [43] and others.

Returning to the dynamics of our wavepacket, it will move initially away from the starting position causing a rapid decrease in the correlation function, $\langle\phi(0)|\phi(t)\rangle$ (see Fig. 11). This fall is characterized by a time T_1. At a later time, T_2, of the order of the period of the PO, the wavepacket will come back to the neighbourhood of the initial position, causing recurrences on this timescale in the correlation function. Finally, the wavepacket will disperse, and the maxima of the correlation function will be progressively

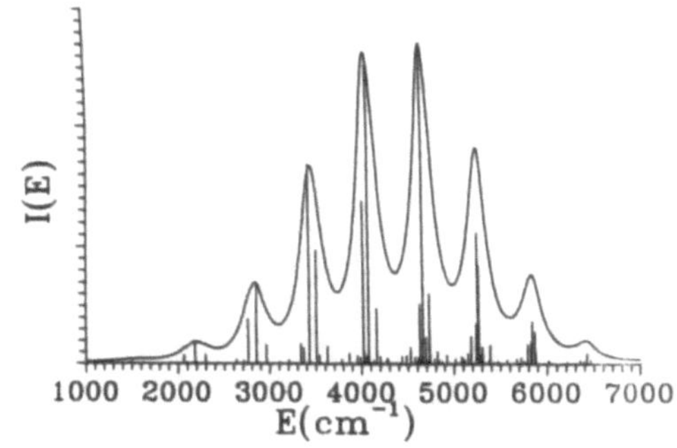

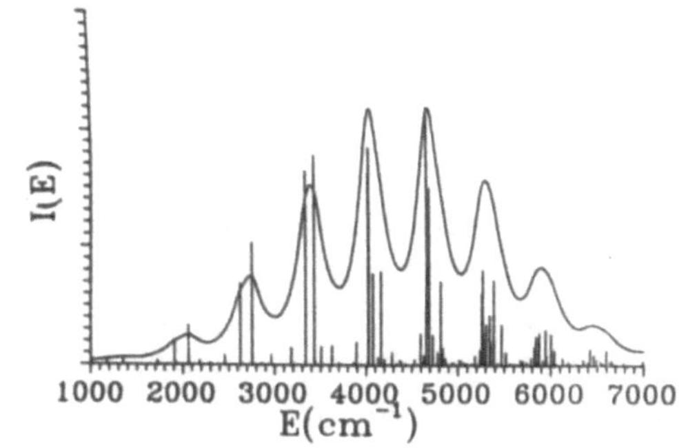

Figure 12. Simulated stick spectrum and its low resolution version for $LiCN$ corresponding to wavepackets initially located on the outer turning points of the stable (upper panel) and unstable (lower panel) scarring POs of the 79 state.

decreasing; the corresponding envelope will be characterized by a longer time, T_3, associated to the Lyapunov exponent of the PO. This dynamical characteristics can be related to the corresponding features on the low resolution spectra if we perform the Fourier integral of (22) over a finite time interval

$$I_T(E) = \frac{1}{2\pi} \int_{-T}^{+T} dt\; e^{iEt}\; \langle \psi(0) | \psi(t) \rangle \tag{23}$$

The characteristic times, $T_1 < T_2 < T_3$, will result respectively in the envelope, spacing and peak widths of the corresponding spectra (see Fig. 11), since due to the uncertainty principle shorter times correspond to bigger widths in the energy domain.

Let us apply these ideas to the study of the scars in the 79-th quantum state of $LiCN$ [41], that we described in the previous Section. In Fig. 12 we present the infinite resolution spectra corresponding to gaussian wavepackets initially located on the outer turning points of the two POs scarring the state 79 (see Fig. 10). It can be observed that both of them show all char-

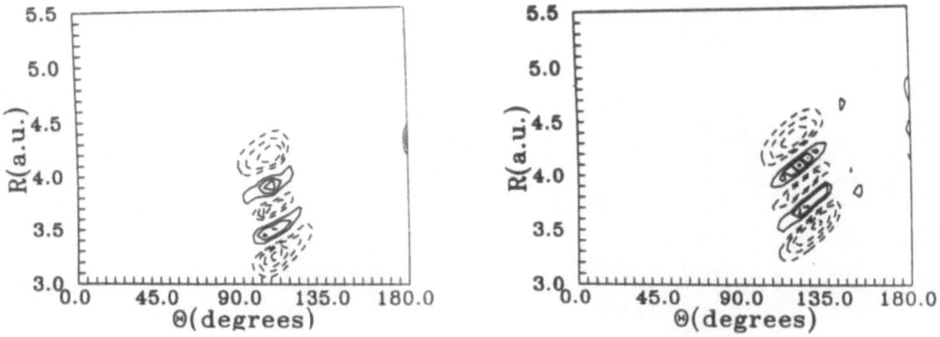

Figure 13. Localized band wavefunctions corresponding to the fourth band in the two low resolution spectra of Fig. 12.

acteristics of chaotic, unassignable spectra: level repulsion [33], distribution of intensities [44], etc. However, when a low resolution version, obtained by convolution of the stick spectrum with lorentzians, is considered a much more regular pattern appear. The results show a series of roughly equally spaced bands. The wavefunctions [45] associated to each band can be obtained using the following approximate projection operator

$$P^{BAND} = \sum_{i}^{BAND} |i\rangle\langle i| \tag{24}$$

where the sum extend to all the eigenstates falling within the energy domain of a given band. In Fig. 13 the wavefunction associated to the fourth bands in the spectra of Fig. 12, obtained as $P^{BAND}\phi(0)$, are presented. They appear as functions highly localized along the scarring POs (also shown in the figure) and with the same number of nodes as the scars in state 79.

9. Acknowledgments

This work has been supported in part by DGICYT (Spain) project PB92-181.

References

1. Kompa, K.L. and Levine, R.D. (1994) *Acc. Chem. Res.* **27**, 91.
2. Letokhov, V.S. (ed) (1989) *Laser Spectroscopy of Highly Vibrationally Excited Molecules*, Adams Hilger, New York.
3. Bacic, Z. and Light, J.N. (1989) *Ann. Rev. Phys. Chem.* **40**, 469.

4. Leforestier, C., Bisseling, R., Cerjan, C., Feit, M.D., Friesner, R., Guldberg, A., Hammerich, A., Jolicard, G., Karrlein, W., Meyer, H.D., Lipkin, N., Roncero, O., and Kossloff, R. (1991) *J. Comput. Phys.* **94**, 59.
5. Reichl, L.E., (1992) *The Transition to Chaos in Conservative Classical Systems: Quantum Manifestations*, Springer-Verlag, New York.
6. Lebowitz, J.L. and Penrose, O. (1973) *Phys. Today* **54**, 23.
7. Arnold, V.I. and Avez, A. (1968) *Ergodic Problems of Classical Mechanics*, W.A. Benjamin, New York.
8. Borondo, F., (1992), in S. Fraga (ed), *Computational Chemistry: Structure, Interactions and Reactivity*, Elsevier, Amsterdam, pp. 592-620.
9. Essers, R., Tennyson, J., and Wormer, P.E.S. (1982) *Chem. Phys. Lett.* **89**, 223.
10. Benito, R.M., Borondo, F., Kim, J.-H., Sumpter, B.G., and Ezra, G.S. (1989) *Chem. Phys. Lett.* **161**, 60.
11. Benettin, G.C., Galgani, L., and Giorgilli, A. (1985), in R. Livi and A. Politi (eds), *Advances in Nonlinear Dynamics and Stochastic Processes*, World Scientific, Singapore.
12. Berry, M.V. (1978) *AIP Conf. Proc.* **46**, 16.
13. Arnold, V.I. (1964) *Dokl. Akad. Nauk. SSSR* **156**, 9.
14. Ford, J., Mantica, G., and Ristow G.H. (1991) *Physica D* **50**, 493.
15. Einstein, A. (1917) *Verh. Phys. Ges.* **19**, 82.
16. Maslov, V. and Feodoriuk, M.V. (1981) *Semiclassical Approximations in Quantum Mechanics*, Reidel, Boston.
17. Noid, D.W. and Marcus, R.A. (1977) *J. Chem. Phys.* **67**, 559.
18. Chapman, S., Garrett, B.C., and Miller, W.H. (1976) *J. Chem. Phys.* **64**, 502.
19. Gustavson, F.G. (1966) *Astron. J.* **71**, 670; Delos, R.T. and Swimm, J.B. (1977) *Chem. Phys. Lett.* **47**, 76; Jaffe, C. and Reinhardt, W.P. (1979) *J. Chem. Phys.* **71**, 1862; Fried, L.E. and Ezra, G.S. (1988) *Comp. Phys. Comm.* **51**, 103; Farrelly, D. (1986) *J. Chem. Phys.* **85**, 2119.
20. Percival, I.C. (1974) J. Phys. A **7**, 794.
21. Noid, D.W., Koszykowski, M.L., and Marcus, R.A. (1977) *J. Chem. Phys.* **67**, 404; Eaker, C.W., Schatz, G.C., DeLeon, N., and Heller, E.J. (1984) *J. Chem. Phys.* **81**, 5913; Martens, C.C. and Ezra, G.S. (1985) *J. Chem. Phys.* **83**, 2990.
22. Goldstein, H. (1980) *Classical Mechanics*, Addison Wesley, Reading.
23. Solov'ev, E.A. (1978) *Sov. Phys. JETP* **48**, 635.
24. Johnson, B.R. (1983) unpublished Aerospace report.
25. Skodje, R.T., Borondo, F., and Reinhardt, W.P. (1985) *J. Chem. Phys.* **82**, 4611.
26. See reviews: Skodje, R.T. and Cary, J.R. (1988) *Comp. Phys. Rep.* **8**, 221; Reinhardt, W.P. (1989) *Adv. Chem. Phys.* **23**, 925; and Borondo, F. in ref. 8, and the references therein.
27. Gutzwiller, M.C. (1990) *Chaos in Classical and Quantum Mechanics*, Springer-Verlag, New York.
28. Burghard, I. and Gaspard, P. (1994) *J. Chem. Phys.* **100**, 6395.
29. Stratt, R.M., Handy, N.C., and Miller, W.H. (1979) *J. Chem. Phys.* **71**, 3311.
30. Hose, G. and Taylor, H.S. (1983) *Phys. Rev. Lett.* **51**, 947.
31. Noid, D.W., Koszykowski, M.L., and Marcus, R.A. (1983) *J. Chem. Phys.* **78**, 4018.
32. Pomphrey, N. (1974) *J. Phys. B* **7**, 1909.
33. Bohigas, O. and Giannoni, M.J. (1984) in J.S. Dehesa, J.M.G. Gomez, and A. Polls (eds) *Mathematical and Computational Methods in Nuclear Physics*, Springer-Verlag, New York.
34. Wigner, E.P. (1932) *Phys. Rev.* **40**, 749.
35. Husimi, K. (1940) *Proc. Phys. Soc. Japan* **22**, 264.
36. McDonald, S.W. (1988) *Phys. Rep.* **158**, 337.
37. Cartwright, N.D. (1976) *Physica A* **83**, 210.
38. O'Connell, R.F. (1983) *Found. Phys.* **13**, 83.
39. Harriman, J.E. (1988) *J. Chem. Phys.* **88**, 6399.

40. Radons, G. and Prange, R.E. (1988) *Phys. Rev. Lett.* **61**, 1691; Waterland, R.L., Yuan, J., Martens, C.C., Gillilan, R.E., and Reinhardt, W.P. (1988) *Phys. Rev. Lett.* **61**, 2733; Davis, M.J. (1988) *J. Phys. Chem.* **92**, 3124; Martens, C.C. (1989) *J. Chem. Phys.* **90**, 7065; Anchell, J.L. (1990) *J. Chem. Phys.* **92**, 4342; Schweizer, W., Schaich, M., Jans, W., and Ruder H. (1994) *Phys. Lett. A* **189**, 64.
41. Borondo, F., Zembekov, A.A., and Benito, R.M. (submitted).
42. McDonald, S.W. and Kaufmann, A.N. (1988) *Phys. Rev. A* **37**, 3067.
43. Heller, E.J. (1991) in M.J. Giannoni, A. Voros and J. Zinn-Justin, *Chaos and Quantum Physics*, Elsevier, Amsterdam.
44. Brody, T.A., Flores, J., French, J.B., Mello, P.A., Pandey, A., and Wong, S.S.M. (1981) *Rev. Mod. Phys.* **53**, 385.
45. Gomez-Llorente, J.M., Borondo, F., Berenguer, N., and Benito, R.M. (1992) *Chem. Phys. Lett.* **192**, 430; Polavieja, G.G., Borondo, F., and Benito, R.M. (1994) Phys. Rev. Lett. **73**, (in press).

MIXED MODE DYNAMICS WITHIN THE TDSCF APPROXIMATION

ERSİN YURTSEVER
Middle East Technical University
Chemistry Department
Ankara, Turkey

and

JÜRGEN BRICKMANN
Technische Hochschule Darmstadt
Physikalische Chemie I
Darmstadt, and
Darmstaedter Zentrum für Wissenscahftliches
Rechnen der Technischen Hochschule Darmstadt,
Germany

1. Introduction

The atomic and molecular world is ruled by a completely different set of laws than the classical mechanics of the Newtonian world. The differences exist even in their philosophy; the classical mechanics approach in its orthodox form is deterministic claiming that if the initial conditions and the forces acting on the system are known, one can predict the complete future. In contrast, the quantum mechanics which governs the atomic and molecular regime is probabilistic in nature, expressing everything in average quantities rather than exactly measurable values. There exists another important difference from a more practical point of view: the classical mechanics is computationally more accessible once the dynamical equations are defined,. There are numereous solution methods of even for systems consisting of 10,000,000 particles although the validity of the classical formalism for such small (considering the Avogadro number) ensembles can be argued. On the other hand the quantum mechanics poses much more inconquerable problems of computational aspects. In principle one obtains the correct dynamical behaviour of any molecular system by solving the appropriate form of the time-dependent Schrödinger equation. However the computation of the eigenvalue spectra of even a small three-atomic problem is a formidable task and to follow the dynamics at somewhat high energy regimes becomes computationally prohibitive. Therefore one travels between these two approaches

E. Yurtsever (ed.), Frontiers of Chemical Dynamics, 393–406.

back and forth to find the optimal methodology which is both correct in nature and computationally feasible.

One of the exciting developments in nonlinear analysis is the discovery of the chaotic behaviour in many branches of science including the physics of small systems (1-4). All formulation of the classical mechanics are deterministic and for a long time it was thought that long term predictions were possible and only restricted by computational limitations. Nowadays it is well accepted that even though the laws of our daily life are deterministic, the solutions of certain systems (not necessarly at the molecular scale) may not be stable that is one cannot predict the future after in any detail. The field of "deterministic chaos" deals with such systems and there are various mathematical tools of classifying problems and regimes. Since the famous correspondence principle ties the classical and quantum mechanics, it is well expected that one should look for the quantum analog of the classical chaos. However all the efforts have failed up to now and the common belief is that quantum chaos does not exist. The term "quantum chaology" now refers to the differences in the dynamical behaviour of the regimes whose classical dynamics is chaotic.

The classical measures of the chaotic behaviour are well established. The common definition of the chaos is the strong dependence on initial conditions,i.e. the system forgets its history very rapidly. Lyapunov exponents and their derivatives, such as Kolmogorov entropy or fractal dimension all point out to the degree of the ability of the system remembering or forgetting the past history (5). When one treats a quantum mechanical problem, none of these measures can be used since it is not possible to define them in a probabilistic philosophy. Of course one can also use correlation functions or power spectra of various observables in both methodologies for classifying dynamics. For instance, a quick decay of an autocorrelation function can be accepted as evidence to the existence of chaotic behaviour (6). The noise level of the power spectra can also be used as a measure; that is a noisy spectrum can be accepted as a fingerprint of chaos. Both of these types of measures in practice can be computed for classical observables and quantum expectation values. However the averaging process of the quantum mechanics may be destroying the extremely minute details of the dynamics which causes chaos. Then employing such functional analysis techniques may not be the optimum choice for comparing classical and quantum mechanics.

One of the commonly used approaches to a computationally difficult quantum mechanical problem is to treat the physically uninteresting parts by classical methods. One can separate vibrational modes according to their frequencies so that slow modes are treated classically and fast modes by quantum mechanics analogous to the Born-Oppenheimer approximation. This separation of the problem allows one to use standard integration techniques for a large number of modes for somewhat incorrect dynamics and apply quantum mechanical methods of various sophistication to a small number of degrees of freedom. This mixed-mode approach can also be used to study systems under partial quantization to observe the transition from classical chaos to its quantum counterpart.

In this work we study several problems under the self-consistent-formalism which allows one to solve quantum and classical mechanical equations in an approximate manner and we discuss the effects of this approximation.

TIME-DEPENDENT SELF-CONSISTENT-FIELD METHOD

We start with partitioning of a nonintegrable Hamiltonian as a sum of two terms which are each dependent on a set of disjunct coordinates and a nonlinear term which may include coupling of the kinetic energy as well as the potential energy terms. Let us use a two-dimensional system as an example since generalization to many-dimensional cases is straightforward.

$$H = H_x + H_y + V(x,y) \tag{1}$$

Within the standard time-dependent self-consistent-field (TDSCF) approximation of the quantum mechanics, the time-dependent form of the wavefunction is written as:

$$\psi(t) = \psi_x(t) \,.\, \psi_y(t) \tag{2}$$

where each function is formally the product of a function of coordinate and a phase factor (7).

$$\psi_x(t) = \varphi(x)\,.e^{iF_1(t)} \tag{3}$$

Schrödinger equation in atomic units has the form:

$$i\ \partial\psi / \partial t = H\,\psi \tag{4}$$

Replacing the exact wavefunction by the product form (2), equation (4) is reduced to two separable equations along each mode:

$$i\ \partial\psi_x / \partial t = H_x^{SCF}\ \psi_x \tag{5-a}$$
$$i\ \partial\psi_y / \partial t = H_y^{SCF}\ \psi_y \tag{5-b}$$

with the definition for the SCF Hamiltonians given as:

$$H_x^{SCF} = H_x + \langle V \rangle_y - G_1(t) \qquad (6\text{-a})$$

$$H_y^{SCF} = H_y + \langle V \rangle_x - G_2(t) \qquad (6\text{-b})$$

where $\langle\ \rangle$ denotes an average over the other mode(s). These average quantities act as perturbations to the zeroth order Hamiltonians. $\langle V \rangle$ has an implicit time-dependence, it is either an integral over a wavefunction or an average over a bundle of trajectories if classical fields are present. There is no restriction on the smoothness of time-dependence as long as the time step of the integration is small enough so that they remain constant during each integration step.

Time dependent functions $G_1(t)$ and $G_2(t)$ real quantities whose integrals give the phase factors F_1 and F_2. The actual computation of both F_i and G_i functions are quite cumbersome since each term is coupled to other modes. On the other hand the expectation values of all physical operators are independent of the phase factors, therefore one can ignore this step. But for quantities which are given as matrix elements instead of expectation values, such as absorption spectra, it has been shown that phase factors do affect the results and one has to define formalism which allows accurate computation of these terms (8).

It is well documented that the SCF approximation has serious drawbacks even in the computation of time-independent electronic properties (9). The so called the "correlation error" of the difference between the exact and SCF forms of the wavefunction could have errors in the energy expectation values usually in the order of bond energies in molecules. However difficulties in computing the correlation correction usually force us to stay with the SCF approximation with the hope that errors are of the same magnitude during the reaction so one can get reasonable answers. In the case of the mixed-mode philosophy, TDSCF becomes much more important, being the only available methodology for solving two sets of incompatible equations. Therefore it is necessary to find out the shortcomings of the approximation from a dynamical point of view.

Again keeping the same two-dimensional case, we denote one of the modes to be classical and the other one as the quantal mode. In reality this separation is based on the frequencies or for the case of a very large system, the physically more interesting mode(s) are treated quantum mechanically. The classical dynamics can be studied by any form of the mechanical equations and we usually employ the Hamilton equations since they are easy to integrate very accurately for small systems.

$$dq/dt = \partial H / \partial p \quad \text{and} \quad dp/dt = -\partial H / \partial q \tag{7}$$

where q and p stand for the position and momentum coordinates, respectively. The quantum dynamics can be studied by several methods. One can use basis set expansions or Fourier transform techniques and their derivatives (10). Recently we have suggested that quantum dynamics can also be studied by methods which are analogous to classical mechanical algorithms (11). Time-dependent coefficients can be separated into their real and imaginary components and the equations governing their time evolution can be written exactly as Hamilton-Jacobi equations. By applying this methodology or the previously suggested basis set methods (12), we were able to study dynamics of small systems within the TDSCF approximation and compare with fully classical results.

In this work we discuss findings on the mixed-mode dynamics of a two-dimensional nonlinearly coupled oscillator system and its exact and SCF quantum dynamics. Also a one-dimensional chain of anharmonic oscillators is studied by classical methods employing accurate numerical integration techniques as well as a quasi-SCF-like formalism to understand the effects of the SCF approximation.

A PARTICLE IN A TWO-DIMENSIONAL FIELD

We have employed the following Hamiltonian for comparing the classical, exact quantum, SCF quantum and mixed-mode dynamics (13-14).

$$\begin{aligned} H = 0.5\,(p_x^2 + p_y^2 + x^2 + 1.44\,y^2) - 0.05\,x^3 + 0.00140625\,x^4 \\ -0.0864\,y^3 + 0.002916\,y^4 + 0.1\,x^2y^2 \end{aligned} \tag{8}$$

This Hamiltonian is shown to be classically chaotic above a critical energy. Its quantum eigenvalue spectrum displays a much more regular statistical behaviour than what the classical dynamics suggests. Also the eigenfunctions have fairly strong nodal structures at classically chaotic regime. In order to elucidate these contradictory findings, we carried out mixed-mode calculations, that is x-mode is represented by a particle in the classical phase space (by its coordinate and momentum) and its time evolution is followed by the numerical integration of the Hamilton-Jacobi equations. On the other hand y-mode is expressed as a wavepacket of linear combination of eigenfunctions of the zeroth order Hamiltonian along the y-mode and the time-dependent Schrödinger equation is solved by basis set expansion. The explicit forms of the dynamical equations are:

$$dx / dt = \partial H_x / \partial p_x \tag{9-a}$$

$$dp_x/dt = - \ (H_x + 0.1\langle y^2\rangle\, x^2) / \partial x \tag{9-b}$$

$$i\,\partial\psi_y/\partial t = (\,H_y + 0.1\,\{\,x^2\,\}\,y^2)\,\psi_y \tag{9-c}$$

In above equations <O > represents the expectation value of the operator O over the wavepacket at time t, { S } is the instanteneous value of the classical observable S. Once the initial conditions are defined, x^2 from the classical trajectory and the $\langle y^2\rangle$ from the wavepacket are computed, then equations 9a-9c are solved and with the new trajectory and the wavepacket the process is repeated until sufficient information is stored for a later analysis. The detailed description of the solution of these mixed-mode equations by the basis set expansion and Hamilton-like integrations are given elsewhere (11-12). We have compared mixed-mode results to those of fully classical ones. Since one of the modes is always treated classically in both formalisms, one can isolate the effects of the external classical and quantum fields and study chaos under a partial quantization.

At this point we should emphasize the fact that comparison of the dynamics of a wavepacket and that of trajectories is not straightforward. For MM formalism we employ a wavepacket, for the fully classical trajectory there should be a corresponding particle with precisely defined coordinate and momentum. To the important question of how to choose this trajectory (or a bundle of trajectories), there is no single answer. The phase space representation of a wavepacket is not a uniquely resolved problem. The Wigner transformation of a wavepacket has the correct classical asymptotic behaviour however being negative in certain regions of the phase space, it cannot be treated as a rigorous probability distribution (15). A Gaussian smoothed out form of the Wigner transformation is suggested by Husimi (16).This function is positive definite hence it resembles a probability distribution but there exists no proof of its being the only (or the best) recipe for selecting classical points as counterparts of the given wavepacket (17-20). Therefore we selected initial points of our classical trajectories in the following manner: The total energy in the MM and classical calculations is chosen always equal. Then either the coordinate of the trajectory is equal to the expectation value of the coordinate <y> or the energy along y-mode is the same both for the wavepacket and classical point. Since various sets of calculations display qualitatively the same behaviour, we will not elaborate on this point any further.

It was mentioned that MM formalism allows us to study classical dynamics under a partial quantization. x-mode of our example remains to be classical particle for all calculations, then we can study its dynamics in the presence of a classical or a "similar" quantum field. Since the fully classical behaviour is chaotic even at relatively low energies, once the perturbation field is switched to a quantum field, changes in the dynamical character should give us some information on the effects of quantization. In figure 1 we present the coordinate, momentum and energy along the x direction as well as the energy of the coupling term as functions of

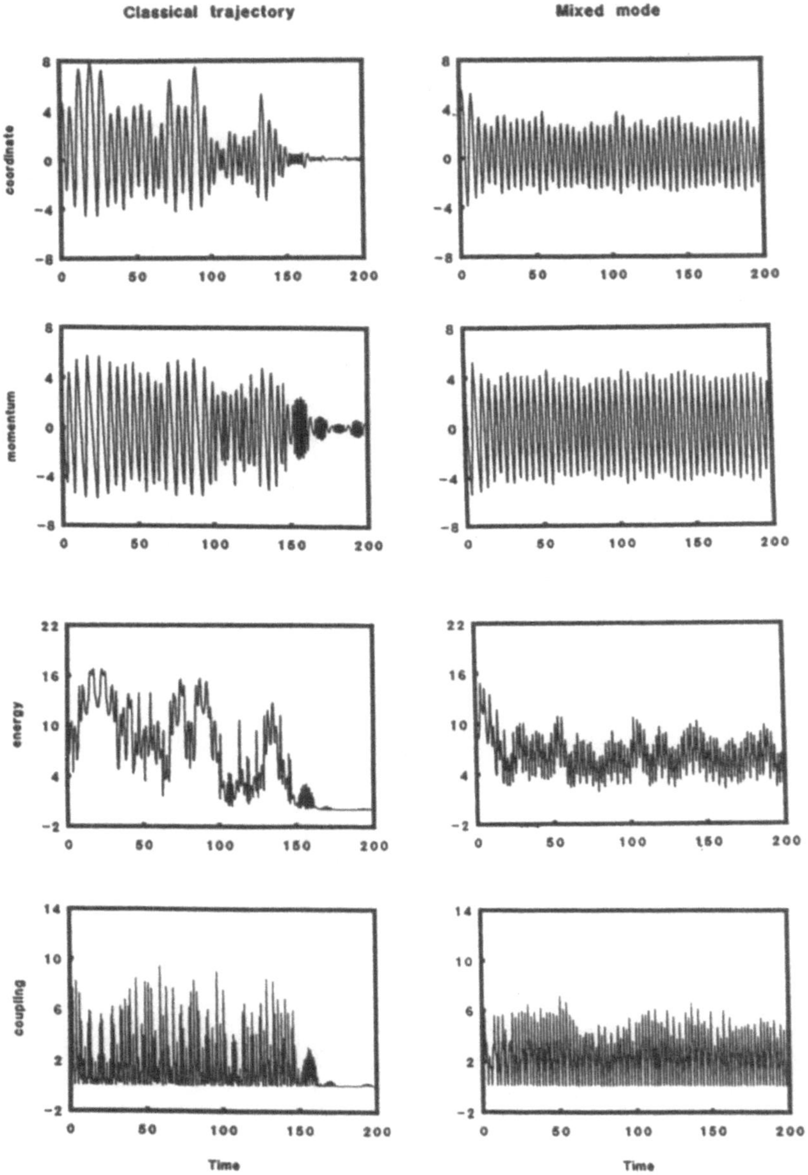

Figure 1. Coordinate,momentum,energy of the classical mode and the coupling energy as functions of time for classical and mixed-mode trajectories.

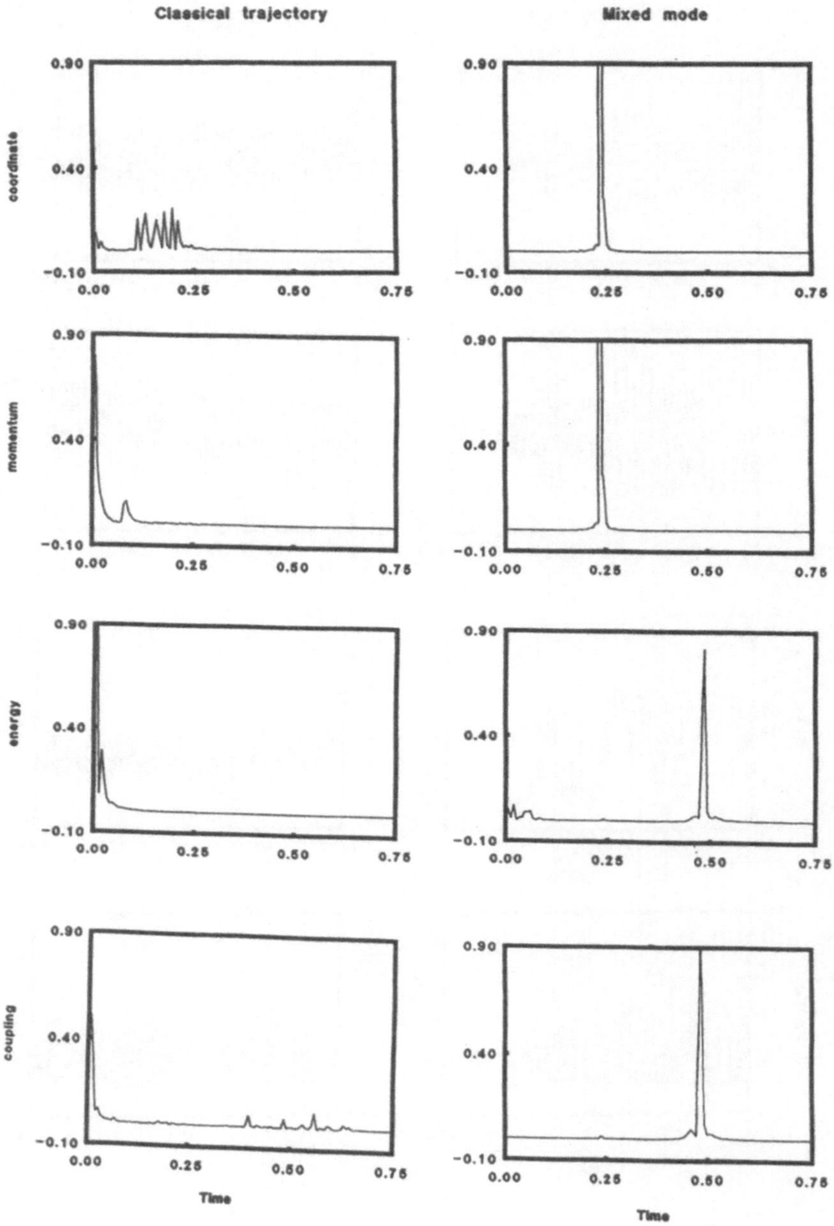

Figure 2. Power spectra of coordinate,momentum,energy of the classical mode and the coupling energy for classical and mixed-mode trajectories.

time for both sets of calculations. In figure 2, various autocorrelation functions as well as power spectra are given. Qualitative analysis of all these time series point out to a single general observation and that is the dynamical behaviour of a classical degree of freedom is much more periodic once the external field is a quantum one. In that case, all observables presented are pretty much uniform in frequency and amplitude as it is evident in their power spectra where classical time series seem to be carrying fingerprints of a well formed chaos (21).

Once these qualitatively different dynamical structures are observed, the immediate question is the magnitude of errors associated with the computational aspects, that is whether the TDSCF approximation destroys the details of dynamics which results in chaos. In order to answer to this question, one has to look at the shortcomings of the TDSCF procedure. TDSCF has two potential sources for error, it treats instanteneous potentials as average fields and it is iterative in nature. The problem of average fields is thoroughly analyzed for the electronic structure problem and it has been shown that once sufficiently large sets of configurations are included, the correlation error can be recovered. Similarly quantum dynamics can be studied correctly employing multi-configurational techniques. We have recently studied the effects of the TDSCF for the potential discussed here in case of the full quantum dynamics (22). Time evolution of mode energies and non-decay probabilities are compared for both exact and TDSCF solutions. In figure 3 we present power spectra of the non-decay probabilities which are defined through:

$$P_{\alpha\beta}(t) = \left| \langle \alpha\beta(0) \mid \alpha\beta(t) \rangle \right|^2 \qquad (10)$$

At low energy regime the power spectra of the exact and TDSCF solutions are very similar and point out to a regular periodic motion. At the high energy region, power spectra show quite different structures. This result is expected since the classical dynamics at this regime is chaotic and the trajectories should span the complete phase space; but the actual computed trajectory depends on many computational parameters such as the time step, the numerical integrator and the word length used in the programming. The general characteristics of chaotic trajectories are that they separate exponentially so that a small changes in the initial conditions would result in qualitatively same but quantitatively quite different dynamics. We believe that the quantum analog of this known characteristics is observed here, that is exact and TDSCF power spectra are different but qualitatively they describe similar periodicity.

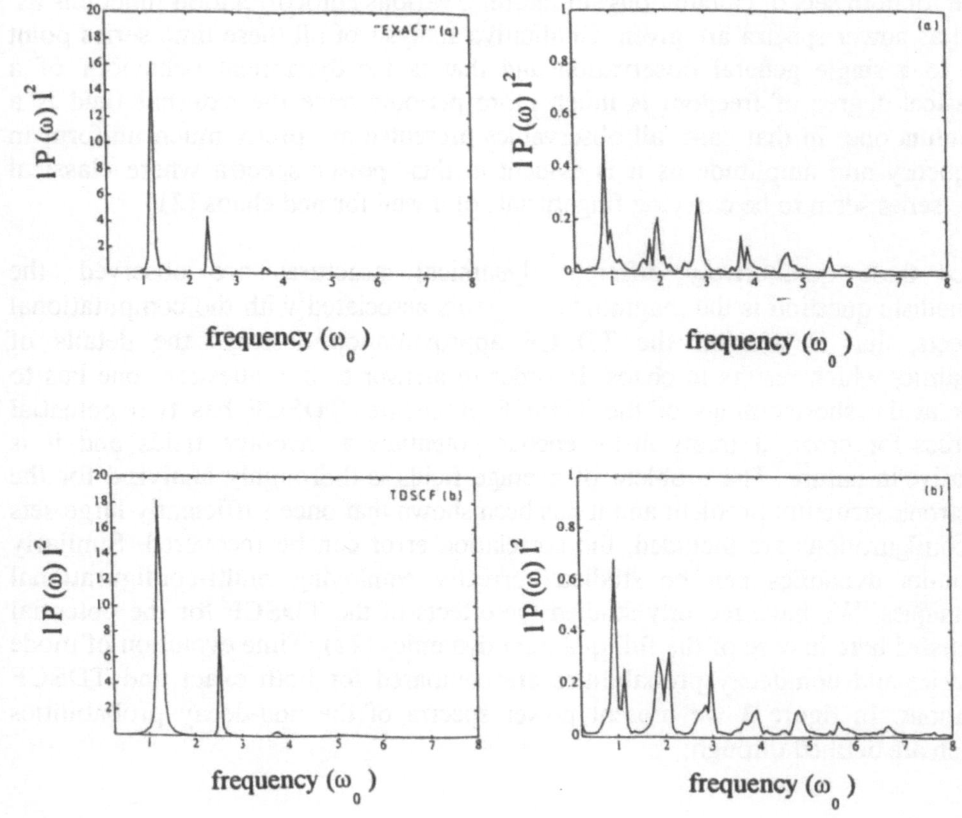

Figure 3. Power spectra of non-decay probabilities for the exact and SCF quantum methods. a) E= 3.216, b) E= 11.844

ONE-DIMENSIONAL CHAIN

We have briefly mentioned that SCF is an iterative procedure that it consists of successive updatings of the average fields as well as solutions. In order to analyze the errors introduced from such a procedure, we devised an analogous iterative method to solve the dynamical equations of classical nature and compared with the exact solutions. The model chosen is a set of nonlinearly coupled one-dimensional anharmonic oscillators.

$$H = 0.5 \sum(p^2_k + x^2_k + 1/2x^4_k) + \sum x^2_k x^2_{k+1} \qquad (11)$$

The initial conditions are chosen such that the chain is at rest with no kinetic energy. Then the oscillator at one end of the chain is stretched so that energy is stored in that bond and the transfer of energy from this bond is studied as a time series. This model could be used for the relaxation of polymer chains and with the above parameters, we have computed Lyapunov exponent spectra for various energy and coupling parameters. Once the energy reaches a critical value, then the model system is chaotic. The exact dynamics is investigated as explained before. The quasi-SCF procedure is established by solving the Hamilton's equations for each bond successively that is there are n sets of equations in which the coordinates and momenta of other bonds are treated as constants. There is an obvious advantage of studying such a methodology. In the MM formalism each class of equations are solved separately under the field of other group of modes which are frozen for a short period of time. If this separation can be accepted formally, then the only source of error is the assumption that remaining modes are stationary during the integration step. The quasi-SCF procedure should be then sufficient to find out the magnitude of this error. Again since the classical dynamics is chaotic, we should not expect to get numerically identical solutions however the qualitative nature should not be changed. We have carried out the exact and quasi-SCF calculations at two different time steps. In figure 4, we display the length of the final bond, and its power spectrum. If the time step is sufficiently low, then the qualitative nature of the energy transfer remains the same. If the time step is larger then the system becomes more regular as depicted in the third set of figures. The power spectra of the exact solution has a lot of detailed structure as well as a high frequency motion. Upon introducing the iterative approximation, at small time steps we observe a similar structure but for larger time steps, this periodic motion becomes a superposition of several frequencies. Although a definite evidence should come from extensive studies of chaotic measures such as Lyapunov exponents or Kolmogorov entropy, it seems that one may use SCF methodology for time-dependent systems with some additional care without really disturbing the characteristics of the system.

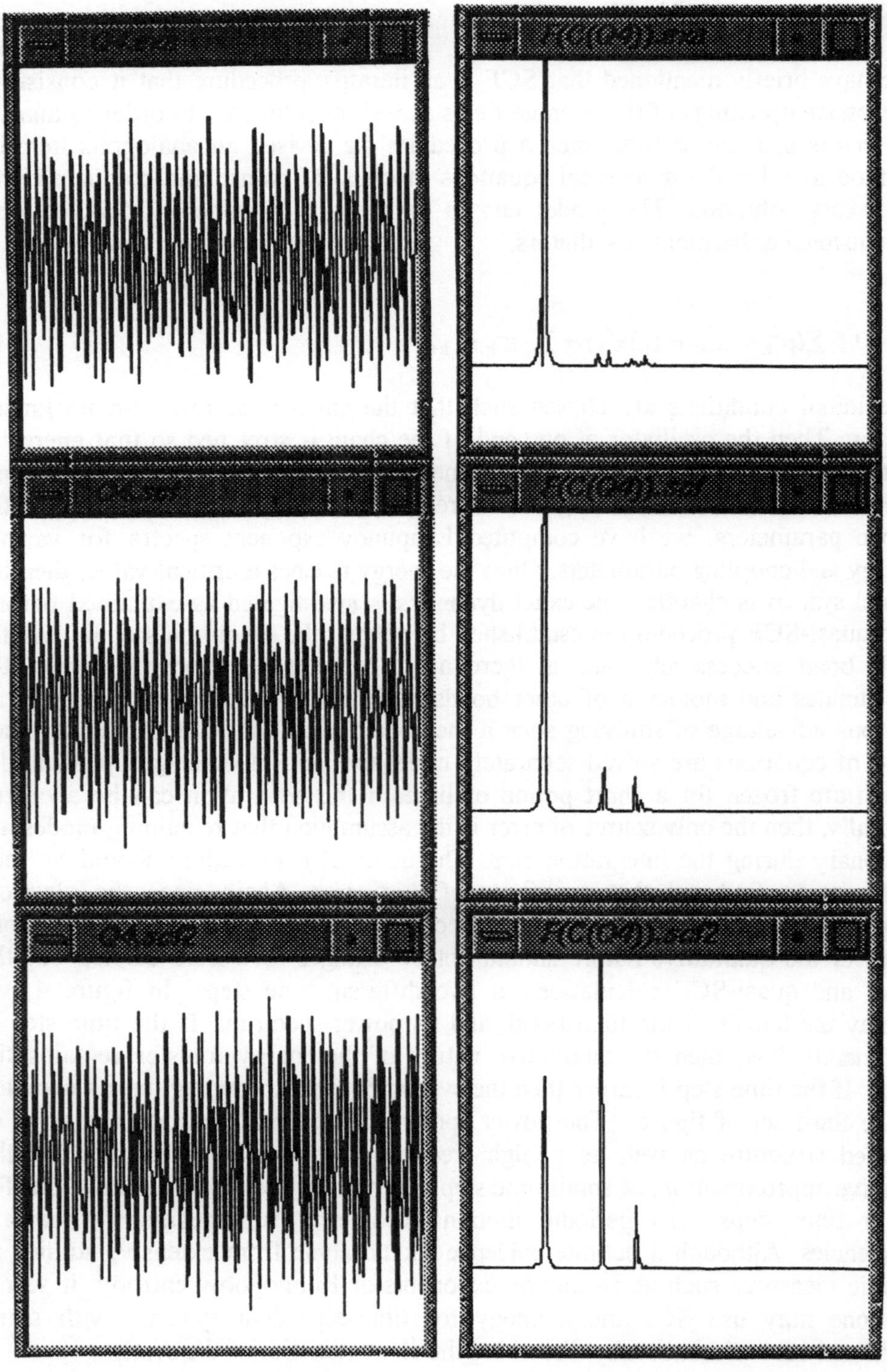

Figure 4. The length of the 4th bond and its power spectrum.

RESULTS AND DISCUSSION

The mixed-mode formalism can be used to study the dynamical behaviour of classical systems under the effect of quantum fields. In this manner one may get some additional information on the correspondence between the classical chaos and the quantum dynamics. However the combination of two different philosophies requires approximation methods which should be introducing some errors. In this work we have tried to look at these errors from a qualitative point of view as numerical comparisons do not seem to be feasible at this moment. A quantum SCF procedure and a quasi-SCF classical method are carried out and compared with exact solutions. In two different model system, we observe several properties as functions of time and study their power spectra. The SCF-like solutions seem to keep the characteristics of the exact solutions, that is they are numerically quite different but display similar qualitative behaviour. A stronger support of this conjecture can be found in the chaotic measures such as the maximum Lyapunov exponent obtained from time series of various observables. This is a rather complicated process and the work along this line is in progress. If one can definitely state that SCF procedures do not destroy the chaotic behaviour, then these mixed-mode methods would produce reasonable insight to the dynamics of large systems.

REFERENCES

1) Schuster,H.G. (1989) *Deterministic Chaos*, VCH, Weinheim
2) Haake,F. (1991) *Quantum Signatures of Chaos*, Springer-Verlag, Berlin
3) Dittrich,W. and Reuter,M. (1994) *Classical and Quantum Dynamics*, Springer-Verlag, Berlin
4) Gutzwiller, M.C. (1990) *Chaos in Classical and Quantum Mechanics*, Springer-Verlag, New York
5) Lichtenberg, A.J. and Lieberman, M.A. (1983) *Regular and Stochastic Motion*, Springer-Verlag, New York
6) Shapiro,M., Ronkin,J. and Brumer,P. (1985) *The conservation of the correlation length of quantum and classical chaotic states*, Ber.Bunsenges.Phys.Chem. 92, 212
7) Farrely, D. and Smith, A.D. (1986) *A generalized semiclassical self-consistent-field procedure for nonseparable vibrationally bound states*, J.Phys.Chem. 90, 1599
8) Haug, K. and Metiu, H. (1993) *Absorption spectrum calculations using mixed quantum-Gaussian wave packet dynamics*, J.Chem.Phys. 99, 6253
9) Hehre, W.J., Radom,L., Schleyer, P.von R., and Pople, J.A., (1986) *Ab initio molecular orbital theory*, John Wiley, New York
10) Kosloff,R. (1988) *Time-dependent quantum-mechanical methods for molecular dynamics*, J. Phys. Chem. 92, 2087
11) Yurtsever, E. and Brickmann,J. (1994), *Hamilton-Jacobi dynamics for the solution of time dependent quantum problems I. formalism and wave packet propagation in one dimension*, Ber.Bunsenges.Phys.Chem. 98,554

12) Yurtsever,E. and Brickmann,J. (1992) *Quantal-classical mixed mode analysis of nonlinearly coupled oscillators: A time-dependent self-consistent-field approach*, Ber.Bunsenges.Phys.Chem., 96, 142

13) Yurtsever,E. and Brickmann,J. (1990) *Regularity in nonlinearly coupled quantum oscillators far from the semiclassical limit*, Phys.Rev.A, 41, 6688

14) Yurtsever,E. and Brickmann,J. (1990) *Does quantum mechanics select out regularity and local mode behaviour in nonlinearly coupled vibrational systems?*, Ber.Bunsenges.Phys.Chem., 94, 804

15) Wigner,E. (1932) *On the quantum correction for thermodynamic equilibrium*, Phys.Rev. 40, 749

16) Husimi, K. (1940) *Some formal properties of the density matrices*, Proc.Phys.Math.Soc.Jpn. 22, 264

17) Zyczkowski, K. (1987) *Quantum chaotic systems in the generalized Husimi representation*, Phys.Rev.A 35, 3546

18) Takahashi, K. and Saito N., (1985) *Chaos and Husimi distribution function in quantum mechanics*, Phys.Rev.Lett. 55, 645

19) Harriman, J.E., (1988) *Some properties of Husimi function*, J.Chem.Phys. 88, 6399

20) Jalil, Al P. and Yurtsever, E. (1994) *Phase space representation of wavepackets*, Tr.J.Phys. 18, 1095

21) Yurtsever, E., (1994) *Quantal-classical mixed-mode dynamics and chaotic behaviour*, Phys.Rev.E 50, 3422

22) Gunkel,T., Baer,H.-J., Engel, M., Yurtsever, E. and Brickmann,J.,(1994), *Hamilton-Jacobi dynamics for the solution of time dependent quantum problems II.Wave packet propagation in two dimensional, nonlinearly coupled oscillators - exact and time-dependent SCF solutions-* Ber.Bunsenges.Phys.Chem., 98, 1552